WITHDRAWN
WRIGHT STATE UNIVERSITY LIBRARIES

VENOUS ULCERS

St. Peregrine, O.S.M.

Saint Peregrine dedicated his life to labors among the poor, the sick, and the sinful. He accepted a commission to re-establish the Servite Monastery in his home town of Forli. His tender care of the sick during a calamitous plague in Forli in 1323 secured for him the everlasting gratitude and respect of the citizens.

One of the special penances he imposed on himself for 30 years was that he never sat down and when he was forced to sleep he did so by leaning against a wall. Peregrine spent 62 years of his life in his religious order. In the course of this period Peregrine developed varicose veins and a huge ulcerative growth appeared on his leg, exposing the bone and emanating the odor of gangrene. All medical treatment failed. He was examined by the best physicians of the time who pronounced the lesion to be cancerous and advised amputation. He is in the background carrying his saw. The night before surgery was scheduled Peregrine spent much time in prayer before the crucified Jesus asking God to heal him if it was God's will to do so.

In this painting from a Catholic Church in Vienna, we see an angelic nurse changing the dressing on what appears to be a venous leg ulcer. Other paintings of Saint Peregrine also show that the lesion was on the medial lower leg in what Cockett described was "the gaiter area". Today we would attribute healing to effective compression dressings applied by a more modern but still angelic nurse.

To continue the Peregrine legend, falling asleep at one point, Peregrine had a vision of the crucified Jesus leaving the cross and touching his leg. When Peregrine awoke, the wound was healed and his foot and leg, seemingly miraculously cured, were saved. He lived another 20 years, continued his work among the sick until he died on May 1, 1345, his 80th birthday.

St. Peregrine has been declared the universal patron Saint of Spain, and in Vienna is cherished as the most popular saint. In 1967, George Pack, MD, who developed the Gastric and Mixed Tumor services at the Memorial Hospital, New York eulogized Saint Peregrine as the patron Saint of Cancer Patients in an issue of Ca A Cancer Journal for Clinicians (1967;17;183–184) pointing out that this was the first reported case of spontaneous cure of cancer.

This figure was furnished by Dr Hugo Partsch of Vienna.

John Bergan, MD   La Jolla, 2007

Venous Ulcers

Editors

John J. Bergan
VEIN INSTITUTE OF LA JOLLA, CALIFORNIA
UNIVERSITY OF CALIFORNIA, SAN DIEGO,
SCHOOL OF MEDICINE

Cynthia K. Shortell
DUKE UNIVERSITY SCHOOL OF MEDICINE

AMSTERDAM · BOSTON · HEIDELBERG · LONDON
NEW YORK · OXFORD · PARIS · SAN DIEGO
SAN FRANCISCO · SINGAPORE · SYDNEY · TOKYO
Academic Press is an imprint of Elsevier

WG
620
V464
2007

Elsevier Academic Press
30 Corporate Drive, Suite 400, Burlington, MA 01803, USA
525 B Street, Suite 1900, San Diego, California 92101-4495, USA
84 Theobald's Road, London WC1X 8RR, UK

This book is printed on acid-free paper. ∞

Copyright © 2007, Elsevier Inc. All rights reserved.

No part of this publication may be reproduced or transmitted in any form or by any means, electronic or mechanical, including photocopy, recording, or any information storage and retrieval system, without permission in writing from the publisher.

Permissions may be sought directly from Elsevier's Science & Technology Rights Department in Oxford, UK: phone: (+44) 1865 843830, fax: (+44) 1865 853333, E-mail: permissions@elsevier.co.uk. You may also complete your request on-line via the Elsevier homepage (http://elsevier.com), by selecting "Customer Support" and then "Obtaining Permissions."

Library of Congress Cataloging-in-Publication Data

Venous ulcers / editors, John Bergan, Cynthia Shortell.
   p. ; cm.
Includes bibliographical references and index.
ISBN-13: 978-0-12-373565-2 (hardcover : alk. paper)
ISBN-10: 0-12-373565-3 (hardcover : alk. paper)
1. Venous insufficiency. 2. Veins–Ulcers. 3. Leg–Ulcers. I. Bergan, John J., 1927– II. Shortell, Cynthia, K.
[DNLM: 1. Varicose Ulcer–diagnosis. 2. Varicose Ulcer–therapy.
WG 620 V464 2008]
RC700.V45V46 2008
616.1′43–dc22

                                                  2007002963

British Library Cataloguing in Publication Data
A catalogue record for this book is available from the British Library

ISBN 13: 978-0-12-373565-2
ISBN 10: 0-12-373565-3

For all information on all Elsevier Academic Press publications
visit our Web site at www.books.elsevier.com

Printed in China
07 08 09 10 9 8 7 6 5 4 3 2 1

Working together to grow
libraries in developing countries

www.elsevier.com | www.bookaid.org | www.sabre.org

ELSEVIER   BOOK AID International   Sabre Foundation

TABLE OF CONTENTS

EMBLEM OF HUNCZOVKSY   XI

CONTRIBUTORS   XIII

FOREWORD   XVII
MARK A. TALAMINI

PREFACE   XIX
JOHN J. BERGAN AND CYNTHIA K. SHORTELL

SECTION I

ETIOLOGY AND INITIAL EVALUATION OF THE PATIENT WITH CHRONIC VENOUS INSUFFICIENCY

1

INFLAMMATION AND THE PATHOPHYSIOLOGY OF CHRONIC VENOUS DISEASE: MOLECULAR MECHANISMS   3

GEERT W. SCHMID-SCHÖNBEIN AND LUIGI PASCARELLA

2

MICROCIRCULATION IN CHRONIC VENOUS INSUFFICIENCY   15

KEVIN BURNAND AND SAID ABISI

3
Epidemiology of Venous Ulcers 27
Olle Nelzén

4
Lower Extremity Ultrasound Evaluation and Mapping for Evaluation of Chronic Venous Disease 43
Stylianos K. Tsintzilonis and Nicos Labropoulos

5
Hypercoagulable States Associated with Chronic Venous Insufficiency 55
Jose L. Trani Jr. and Jeffrey H. Lawson

6
The Chronically Swollen Leg: Finding the Cause: Theory and Practice 67
Warner P. Bundens

Section II
Nonoperative Management of Chronic Venous Insufficiency

7
Compression Therapy in Venous Leg Ulcers 77
Hugo Partsch

8
WOUND HEALING: ADJUVANT THERAPY AND TREATMENT ADHERENCE 91
DALE BUCHBINDER AND SHARON B. BUCHBINDER

9
NEGATIVE PRESSURE DRESSINGS IN VENOUS ULCERS 105
BRIAN D. LEWIS AND JONATHAN B. TOWNE

10
DEEP VEIN THROMBOSIS AND PREVENTION OF POST-THROMBOTIC SYNDROME 113
JOSEPH A. CAPRINI

SECTION III
THERAPUTIC PROCEDURES FOR CHRONIC VENOUS INSUFFICIENCY

11
RESULTS COMPARING COMPRESSION ALONE VERSUS COMPRESSION AND SURGERY IN TREATING VENOUS ULCERATION 135
MANJIT S. GOHEL, JAMIE R. BARWELL, MARK R. WHYMAN, AND KEITH R. POSKITT

12
SUPERFICIAL SURGERY AND PERFORATOR INTERRUPTION IN THE TREATMENT OF VENOUS LEG ULCERS 149
WILLIAM MARSTON

13
Endovascular Techniques for Superficial Vein Ablation in Treatment of Venous Ulcers 173
Michael J. Stirling and Cynthia K. Shortell

14
Treatment of Chronic Venous Insufficiency with Foam Sclerotherapy 185
Alessandro Frullini

15
Foam Treatment of Venous Leg Ulcers: The Initial Experience 199
Juan Cabrera and Pedro Redondo

16
Foam Treatment of Venous Leg Ulcers: A Continuing Experience 215
Van L. Cheng, Cynthia K. Shortell, and John J. Bergan

17
Ultrasound Guidance for Endovenous Treatment 227
Luigi Pascarella, Lisa Mekenas, and John J. Bergan

18
Repair of Venous Valves in Severe Chronic Venous Insufficiency 243
Michel R. Perrin

SECTION IV

SPECIAL TOPICS

19

Treatment of Recalcitrant Venous Ulcers with Free Tissue Transfer for Limb Salvage 261

Stephen J. Kovach and L. Scott Levin

20

The Diagnosis and Treatment of Major Venous Obstruction in Chronic Venous Insufficiency 275

Francisco José Osse and Patricia Ellen Thorpe

21

The Venous Ulcer and Arterial Insufficiency 299

Robert B. McLafferty and Colleen M. Johnson

22

Topical Negative Pressure Techniques in Chronic Leg Ulcers 307

Jeroen D. D. Vuerstaek, Tryfon Vainas, and Martino H. A. Neumann

Epilogue 327
John J. Bergan

Index 329

Emblem of Hunczovksy*

This graphic symbol with its caption, "In Union Health," depicts the age-old emblems of medicine: the Aesculapian rod with a serpent and another rod showing a hand with an eye in its palm. Bringing these two emblems together in this way represents a unification of the disciplines of medicine and surgery to form a combined medical practice. In 1787 such a unification was hoped to be successful in improving total patient care and offering an advantage to all who were sick.

The occasion for the development of this symbol was the second anniversary of the opening of the 2000-bed Vienna General Hospital and its merger with the 1200-bed Academia Medico-Chirurgica "Josephina." These two institutions were combined to provide the best possible care with the best equipment available. The new institution was founded by Emperor Joseph II, who had traveled throughout Europe and learned of the improvements in education of military surgeons in France and the elevation of surgeons to academic rank by King Louis XV.

The great Vienna General Hospital eventually became the site of a massive collection of precious medical books, surgical equipment, and the historical anatomic wax museum. It was also the location of the military medical school known as the Josephinum. In the twentieth century the symbol of the Vienna General Hospital became the signet of the Austrian Society of Surgeons as a reminder of the original attempt to create modern surgery approximately 200 years earlier.

*Modified from Holubar K. In unione salus: A short historical perspective on the development of surgical education in Austria in the 18th century. *Vasa* 1991; 20:10–11.

In its original form, the emblem was chosen by the Working Group for Phlebology of the Austrian Society of Dermatology. This choice was appropriate because the rod of Aesculapius represents academic theoretical traditions of medicine and conservative nonsurgical care. The seeing hand, which "sees" what the hand does, is the symbol for both experience in medicine and the manual dexterity associated with surgical skill.

To physicians dedicated to the care of the patients with varicose veins and telangiectasias, the symbol is a reminder of the union of surgery and sclerotherapy, and alliance that is necessary for optimum care of the venous insufficiency produced by microphlebectasias and macrophlebectasias. Chosen to be the logo for the Fifth European-American Venous Symposium in Vienna, this symbol is an appropriate cover for this book.

Contributors

Said Abisi, Department of Academic Surgery, St. Thomas Hospital, SE1 7EH London, UK

Jamie R. Barwell, Department of Vascular Surgery, Cheltenham General Hospital, GL53 7AN Gloucestershire, UK

John J. Bergan, Vein Institute of La Jolla, La Jolla, CA 92037

Dale Buchbinder, Vascular Surgery, The Greater Baltimore Medical Center, Towson, MD. 21204

Sharon B. Buchbinder, Department of Health Sciences, Towson University, Towson, MD 21204

Warner P. Bundens, Departments of Surgery, Family and Preventative Medicine, University of California-San Diego, La Jolla, CA 92037-0974

Kevin Burnand, Department of Academic Surgery, St. Thomas Hospital, SE1 7EH London, UK

Juan Cabrera, Phlebology Unit, University Clinic of Navarra, 31008 Pamplona, Spain

Joseph A. Caprini, Northwestern University Feinberg School of Medicine, Evanston Northwestern Healthcare, Skokie, IL 60077

Van L. Cheng, Vein Institute of La Jolla, La Jolla, CA 92037

Alessandro Frullini, Studio Flebologico, Incisa Valdarno, Florence, Italy

Manjit S. Gohel, Department of Vascular Surgery, Cheltenham General Hospital, GL53 7AN Gloucestershire, UK

Colleen M. Johnson, Southern Illinois University School of Medicine, Springfield, IL 62794

Stephen J. Kovach, Division of Plastic, Reconstructive, and Maxillofacial Surgery, Duke University Medical Center, Durham, NC 27710

Nicos Labropoulos, Division of Vascular Surgery, University of Medicine and Dentistry of New Jersey, Newark, NJ 07103

Jeffrey H. Lawson, Department of Pathology, Duke University Medical Center, Durham, NC 27710

L. Scott Levin, Division of Plastic, Reconstructive, and Maxillofacial Surgery, Duke University Medical Center, Durham, NC 27710

Brian Lewis, Medical College of Wisconsin, Milwaukee, WI 53226

William A. Marston, Division of Vascular Surgery, University of North Carolina at Chapel Hill, Chapel Hill, NC 27599

Lisa Mekenas, Vein Institute of La Jolla, La Jolla, CA 92037

Robert B. McLafferty, Southern Illinois University School of Medicine, Springfield, IL 62794

Olle Nelzén, Skaraborg Leg Ulcer Center and Vascular Surgery Unit, Skaraborg Hospital/KSS, 541 85 Skövde, Sweden

Martino HA Neumann, Department of Dermatology, Erasmus Medical Centre Rotterdam, Rotterdam, The Netherlands

Francisco J. Osse, Department of Surgery, University of Iowa Health Care, Iowa City, IA 52242

Hugo Partsch, Medical University of Vienna, A-1160 Vienna, Austria

Luigi Pascarella, The Whitaker Institute of Biomedical Engineering, University of California-San Diego, La Jolla, CA 92093-0412

Michel R. Perrin, Clinique du Grand Large, F-69680 Chassieu, France

Keith R. Poskitt, Department of Vascular Surgery, Cheltenham General Hospital, GL53 7AN Gloucestershire, UK

Pedro Redondo, Department of Dermatology, University Clinic of Navarra, 31008 Pamplona, Spain

Geert W. Schmid-Schönbein, The Whitaker Institute of Biomedical Engineering, University of California-San Diego, La Jolla, CA 92093-0412

Cynthia K. Shortell, Department of Surgery, Duke University Medical Center, Durham, NC 27710

Michael J. Stirling, Duke University Medical Center, Durham, NC 27710

Patricia Thorpe, Endovenous Reconstruction, Arizona Heart Hospital, Phoenix, AZ 85016

Jonathan B. Towne, Medical College of Wisconsin, Milwaukee, WI 53226

Jose L. Trani, Jr., Department of Pathology, Duke University Medical Center, Durham, NC 27710

Stylianos K. Tsintzilonis, University of Medicine and Dentistry of New Jersey, Newark, NJ 07103

Tryfon Vainas, Department of Surgery, Maastricht University Hospital, Cardiovascular Research Institute of Maastricht, 6202 AZ Maastricht, The Netherlands

Jeroen DD Vuerstaek, Department of Dermatology and Phlebology, DermaClinic, 3600 Genk, Belgium

Mark R. Whyman, Department of Vascular Surgery, Cheltenham General Hospital, GL53 7AN Gloucestershire, UK

FOREWORD

Development of endovascular surgery during the past 25 years has been at a fast and furious pace. It is a tribute to those who have worked in this exciting field that the new techniques have been so well accepted and universally employed.

In contrast, until recent years, the venous system has been relatively neglected clinically. Dogma, rooted in the pronouncements of influential surgeons of 100 years ago, still permeates clinical decision making. However, developments in understanding the pathophysiology of the venous system have kept pace with the developments in endovascular surgery. This has been due to the universal acceptance of ultrasound as a clinical imaging tool as well as its utility in documenting pathologic anatomy.

This book, utilizing the new information uncovered by ultrasound, takes a long step forward in the right direction to bring new developments in diagnosis and treatment of venous leg ulcers into one volume. Here, in one place, an outstanding group of contributors has set down their best current concepts. Nearly all of the acknowledged experts in the field of venous investigations have made their contributions.

Even a cursory look at the table of contents shows the names of many who have made basic observations and also those who have vast clinical knowledge and experience in treating venous leg ulcers. Innovations in patient care are included. The coverage is comprehensive, succinct, and complete. Attention has been given to the several disorders that lead to venous leg ulcers, and the problems of patient care are covered thoroughly. The latest diagnostic tests and indications for their use are well described.

So often, medical textbooks suffer from the presentation of familiar findings and repetition of data that are universally available. It is to the credit of the contributors of the chapters in this book that their texts present current concepts and new data to challenge the reader. Contributions from international authors from many countries enhance the value of this volume as a source of reference material. This is a timely publication that should be of interest to many individuals, as well as those who have a specific interest in venous disorders.

Mark Talamini, MD
Professor and Chairman
Department of Surgery
UCSD Medical School, San Diego

Preface

Venous ulceration is the most common and serious consequence of severe chronic venous insufficiency. Unfortunately, even today it is one of the most poorly managed medical conditions. Interestingly, the management of venous ulcers has not progressed very far beyond that advocated by John Hunter over 200 years ago. Hunter said, "the sores of poor people ... are often mended by rest, a horizontal position, fresh provisions, and warmth."

Also, 100 years ago, John Gay, a London surgeon, described the relationship of venous ulceration to ankle perforating veins. The concept was supported by Homans at Harvard, Turner Warwick at the Middlesex Hospital in London, Robert Linton at the Massachusetts General Hospital, and, of course, Frank Cockett at St. Thomas' Hospital who generously lent his name to the three main ankle perforators. These four names remain famous in history of the treatment of venous ulcers, and each surgeon had his own concept of care. Linton's, for example was subfascial exploration; Cockett's was suprafascial.

In countries other than the United States, a heavy workload is imposed on nurses in dressing ulcers. What is less appreciated or well known is the long-term ineffectiveness of such ulcer treatment. Negus found in his clinic that 77 patients with 109 ulcerated legs had continuous or intermittent ulceration for prolonged periods of time ranging from a few months to more than 50 years. In Negus's clinic in the center of a well-doctored area, all of his patients were treated by conventional methods before arriving at his clinic.

Since the last century a great deal has been learned about the pathogenesis and, therefore, the treatment of venous ulceration, but it still remains a problem where wound care clinics treat the wound and not the cause of the problem. In this volume, the fundamental molecular mechanisms, which lead to chronic venous insufficiency on one hand and venous ulceration on the other, are well described by the essays of Geert Schmid-Schönbein and Luigi Pascarella. Taking a cue from those studies, Kevin Burnand has described the microenvironment of venous ulcerations and Olle Nelzén has collected information on the epidemiology of such ulcerations.

Dominating the new knowledge of cause and development of venous ulceration have been the studies by ultrasound of the lower extremities. Among the leading investigators in this field is Nicos Labropoulos, who has contributed a chapter here on the evaluation of limbs with venous ulceration.

In many instances, previous deep venous thrombosis has led to chronic venous insufficiency, which eventuates in venous ulceration. Jeffrey Lawson of Duke University has described the hypercoagulable states that lead to venous ulcer, and to round out the evaluation of such problems, Warner Bundens of UCSD has provided a chapter on the differential diagnosis of the swollen leg.

The fundamental treatment of venous ulceration begins with compression therapy, and the best-informed individual in that area of treatment is Hugo Partsch of Vienna, who has provided a succinct chapter on the subject of compression in treating chronic venous insufficiency. Such compression is most often combined with topical agents in wound healing, and Dale Buchbinder of Baltimore has summarized important information on this subject in his contribution.

To prevent the need for treatment of venous ulcers, it is necessary to prevent recurrent deep venous thrombosis, and this is the subject of Joseph Caprini's essay from the Northwestern Healthcare System where his influence is pervasive.

With the new knowledge of venous ulcers and their cause have come new methods of treatment. Negative pressure dressings and wound vacuum systems have come into play, and Jonathan Towne has summarized his experience with these in his chapter. But, of course, compression combined with surgery has been the mainstay of treatment of venous ulcerations during the last century. Results comparing compression to compression and surgery have been an interest of Keith Poskitt of Gloucester in the UK. He has described his studies on this subject in an erudite chapter seen here. But, of course, harking back to experience of the last century in treatment of venous ulcers with perforator interruption, William Marston of Chapel Hill has described his experience with this technique. In addition to treatment of perforating veins and superficial incompetence are the endovenous techniques, which are used to treat axial incompetence in the care of venous ulcers. Cynthia Shortell, also of North Carolina, has provided a chapter on this subject.

Among the new methods of leg ulcer treatment, and perhaps the most important, is the use of foam sclerotherapy. This technique has now crossed the Atlantic, but much of the new knowledge still resides overseas. Alessandro Frullini, of Florence, Italy, has explained his experience with the development of foam therapy for venous insufficiency. The modern era of foam treatment of venous insufficiency began with Juan Cabrera in Granada, Spain, and here he has described the initial experience in treating venous ulcerations with sclerosant foam. This experience has been carried further at the Vein Institute in La Jolla, where we described the foam treatment of venous leg ulcers not knowing of the experience of Cabrera.

Both the endovenous techniques and foam sclerotherapy depend on ultrasound evaluation and management. Luigi Pascarella has long experience with these techniques and has described them in his chapter on this subject.

There has been some experience with the repair of venous valves in severe chronic venous insufficiency, and Michel Perrin, a truly distinguished French surgeon, has described his experience and that of others in developing these techniques.

Excision of the ulcer bed and then finding coverage has drifted from general surgery into plastic surgery, where Scott Levin of Duke University has a provided an interesting chapter on this subject.

A few cases of venous ulcer have as the fundamental cause venous obstructive disease. Patricia Thorpe has a long experience with this subject and has described that in her contribution on treatment of major venous obstruction in chronic venous insufficiency.

Lastly, venous ulcers are not isolated entities. They can be present in limbs with arterial insufficiency. It is the arterial insufficiency and its treatment that dominate. An experience with these conditions is well described by Robert McLafferty of Springfield.

While the final words on diagnosis and treatment of venous ulceration will not be provided in this volume, in fact much light is thrown on various entities by the information supplied in these chapters. We, as editors, hope that the contents of this book will stimulate investigations in the solution of the problems caused by venous ulcerations so that diagnosis can be more precise, treatment more standardized, and results to the patient more uniformly beneficial.

John Bergan, La Jolla
Cynthia Shortell, Durham

SECTION I

Etiology and Initial Evaluation of the Patient with Chronic Venous Insufficiency

1

Inflammation and the Pathophysiology of Chronic Venous Disease: Molecular Mechanisms

Geert W. Schmid-Schönbein and
Luigi Pascarella

INTRODUCTION

Increasing evidence supports the hypothesis that chronic venous disease is associated with an inflammatory cascade. Classical signs of inflammation have been observed in human biopsies and plasma samples and in acute and chronic experimental models of venous disease. The cascade starts with a relatively innocent elevation of permeability, mast cell degranulation, and progresses to adhesion of leukocytes and platelets to the endothelium. There is early release of inflammatory mediators, such as cytokines, oxygen radicals, nitric oxide products, and lipid mediators. Eventually, the disease may progress to overt cell apoptosis, enzymatic degradation of basement membrane and extracellular matrix proteins, accumulation of extracellular tissue fragments, a lack of cell responsiveness in the skin, and a prominent vascular restructuring in the form of telangiectases and venous varicosities, which serve as a key index for classification of different stages of the disease. Lymphatic transport to remove the extracellular fragments may be compromised, and in the most severe manifestations of the disease, complete breakdown of the skin and subdermal tissues leads to ulceration.

The progression of these steps in the inflammatory cascade is different among individual patients, both in terms of the time course and their severity. The individual steps progress spatially nonuniform so that early and late stages may occur simultaneously in the same leg. Current macroscopic clinical manifestations of the disease are relatively late-stage manifestations of the inflammatory reactions; observations at the microvascular level may serve to detect the inflammatory events at earlier stages. In the following we will focus on selected inflammatory steps in which molecular mechanisms have been identified. The analysis is still at an early stage, but opportunities arise with new ideas for early intervention.

LEUKOCYTE ACTIVATION AND MICROVASCULAR INFILTRATION

Chronic venous disease is accompanied by enhanced infiltration of circulating leukocytes into the affected leg.[1] Circulating leukocytes infiltrate the microcirculation predominantly by entrapment in true capillaries and by attachment to the endothelium of venules. The two processes are different in several respects. Entrapment in capillaries occurs as a result of either a reduction of the pressure drop across the length of capillaries and an inadequate pressure to deform the relatively stiff viscoelastic leukocytes in the circulation and/or by activation of endothelial cells or leukocytes. The activation may be accompanied by polymerization of the cytoplasmic actin with formation of rigid pseudopods, a process that is under control of soluble agonists (e.g., complement fragments, leukotrienes, platelet activating factor, interleukin-8) by means of G-protein coupled receptors.[2] Recent evidence indicates that pseudopod formation in circulating leukocytes is also under the control of purely mechanical factors, such as the fluid *shear stress* (force parallel to the area/area, dyn/cm^2) generated by the flow of plasma.[3] Reduced fluid shear stresses, the presence of inflammatory agonists, or their combination serves to enhance entrapment of leukocytes in capillaries. The capillaries become permanently obstructed since the leukocyte completely fills the vessel lumen, and the process is frequently followed by red cell aggregation and thrombus formation.

In contrast, attachment of leukocytes to endothelium in postcapillary venules or larger veins may occur without vessel obstruction. In this case membrane attachment between leukocyte and endothelium is required and facilitated by selectins (e.g., P-, L-selectin, P-selectin glycoprotein ligand-1, sialyl Lewis X), integrins (CD11/CD18), and members of the immunoglobulin family (ICAM-1).[4] Their expression in turn is constitutive either on the membrane of postcapillary venules, in the endothelial Weibel-Palade bodies of endothelial cells, or in secondary granules in the case of granulocytes.

Entrapment of neutrophilic leukocytes in the microcirculation reduces local capillary perfusion and at the same time serves to enhance oxygen free radical formation and delivery of proteolytic enzymes, most of which enhance tissue degradation. While acute interventions against the accumulation of the leukocytes in the microcirculation have been shown repeatedly to reduce tissue injury, the utility of chronic intervention against leukocyte infiltration remains to be explored. In general, methods that serve to reduce progression of venous disease (e.g., compression banding, anti-inflammatory medication) are accompanied by a reduction of leukocyte activation and infiltration.[5]

Besides the acute infiltration of the venous and venular wall with granulocytes, inflammation in chronic venous disease is accompanied by infiltration with B- and T-lymphocytes, monocytes, and by mast cells. There is also infiltration of monocyte/macrophages into the venous valves,[6] and valve destruction may in fact be mediated by these cells. The particular membrane adhesion molecules involved are less well defined, although there is evidence for VCAM-1, ICAM-1, and E-selectin on the endothelial surface binding specific leucocytes receptors (e.g., L-selectins, VLA 4, and LFA 1).[7,8] The adhesion molecules may be vessel specific, and effective interventions against enhancement of the inflammatory reaction by leukocyte infiltration may require a more detailed understanding of the actual sequence of adhesion molecules expressed in specific tissues.[9]

INFLAMMATORY MEDIATORS

Patient plasma contains an activating factor for granulocytes and also spontaneously produces a higher level of hydrogen peroxide.[5,10] The identification of the particular factor(s) involved is an important unresolved issue.

Leukocyte activation is characterized by synthesis and release of many inflammatory molecules such as leukotrienes, prostaglandin, bradykinin, free oxygen radicals, and cytokines. Cytokines act to regulate and perpetuate the inflammatory reaction by paracrine and autocrine mechanisms. TNF-α, whose expression is enhanced in many inflammatory reactions, stimulates the expression of membrane adhesion molecules, the synthesis and release of other cytokines, and the chemotactic migration of neutrophils and macrophages. The expression of TNF-α appears to be upregulated in patients with venous ulcers, and healing of the ulcer is accompanied by reduced levels of TNF-α.

Vascular endothelial growth factor (VEGF), by definition, is a potent angiogenic factor and also enhances endothelial permeability. Both VEGF expression and its receptor expression (Flk-1/KDR) are upregulated during the inflammatory reaction. Patients with chronic venous insufficiency and skin changes (CEAP 4) exhibit higher levels of VEGF than those in CEAP Classes 2 and 3 and normal control individuals. The expression of VEGF and other growth factors does not correlate with capillary density of liposclerotic skin or with the slow or poor healing of venous ulcers. Venous ulcer exudates have been found to inhibit the growth of human endothelial cells. Possible degradation of VEGF and other growth factors may be related to these observations.

Transforming growth factor beta (TGF-β) is another cytokine whose expression has been found upregulated in patients with venous ulcers. TGF-β is related to tissue remodeling by stimulating the formation of the granulation tissue, proliferation of fibroblasts, and synthesis of collagen fibers.[11]

TISSUE RESTRUCTURING

Varicose veins undergo a major collagen and elastin fiber restructuring with synthesis as well as degradation.[12] There is evidence for increased levels of hydroxyproline, suggesting an increased content of collagen type I but a decrease of collagen type III.[13] Restructuring also involves the integrin binding of tissue cells to extracellular matrix proteins. Besides the control of the acute reactions of smooth muscle cells in the vascular wall, such integrin binding sites have the ability to slide under mechanical stresses.[14]

PROTEOLYTIC DEGRADATION

An important link between inflammation and skin changes may be by way of Ca/Zn-dependent endoproteinases (matrix metalloproteinases, MMPs) and serine proteinases. Chronic dermal ulcers are characterized by excessive proteolytic activity, which degrades extracellular matrix and growth factors and their receptors. The MMP family of proteases (there are at least 26 MMPs, in addition to six membrane-bound membrane-type metalloproteinases, MT-MMPs, and in addition to four tissue inhibitors of metalloproteinases, TIMPs) are positioned on extracellular matrix proteins in an inactive proform. MMPs play an important role in cell differentiation and development. MMPs may be released from preexisting pools (cytoplasmic granules) upon stimulation and endocytosis or may be newly synthesized by several types of cardiovascular cells. The inactive proenzymes can be activated by other proteinases, including those produced by mast cells. MMPs have multiple binding sites but cleave collagen at unique sites. The MMP levels in wound fluids from chronic wounds tend to be significantly higher than from acute wounds, and healing is associated with reduced MMP activity.

The expression levels of MMPs can be controlled by mechanical stretch (strain), but in a way that depends on the time course of the strain. This has been studied on smooth muscle cells in an *in vitro* chamber subject to oscillatory and constant strain.[15] Stationary strain significantly increases MMP-2 mRNA levels at all time points, whereas cyclic strain decreases it after 48 hours. Both secreted and cell-associated pro-MMP-2 levels are increased by stationary strain at all times, whereas cyclic strain decreases secreted levels after 48 hours. MMP-9 mRNA levels and pro-MMP-9 protein are increased after 48 hours of stationary strain compared with both no strain and cyclic strain.

There is an increase of neutrophil elastase and lactoferrin from activated neutrophils in patients under transient conditions of venous hypertension and with chronic venous insufficiency. Neutrophils and macrophages also release several MMPs. There is gelatinolytic activity discharged by MMP-9; MMP-8, a neutrophil collagenase and leukolysin, a membrane-type MT-MMP. Evidence from knockout experiments suggests that MMP-9 acts upstream of neutrophil elastase by proteolytically inactivating neutrophil elastase inhibitor a1-PI[16] and that it can activate other MMPs. Extracellular MMP inducer (EMMPRIN; CD147) has been observed to increase MMP expression, and membrane type 1 MMP (MT1-MMP) has been implicated in the activation of MMPs. Venous leg ulcers have elevated expression of EMMPRIN, MMP-2, MT1-MMP, and MT2-MMP.[17]

Clearly, these proteases will remain of major interest because of their involvement in the remodeling of cutaneous tissue.[18] Examples of interest are MMP-1, MMP-2, MMP-9, MMP-12, MMP-13, and the tissue inhibitors of metalloproteinases (e.g., TIMP-1 and TIMP-2). Overexpression of MMP-3 (stromelysin-1) and MMP-13 (collagenase-3) is associated with nonhealing wounds. Venous leg ulcers have an increased expression of MMP-2 and TIMP-1 in liposclerotic skin and in wound fluid from nonhealing venous ulcers.[17,19] Expression of MMP-9 has been observed to be upregulated on the edges of intractable venous ulcers, and the rate of MMP-9 activation in plasma of patients with severe CVD is elevated. Levels of TIMP-2 are lower in lipodermatosclerotic skin and ulcers. Unrestrained MMP activity may contribute to impaired healing. The presence of gelatinase (MMP-2 and MMP-9) activity may lead to breakdown of interstitial collagen (e.g., III, IV, V, and IX). MMP-2 and MMP-9 are also expressed in venous valves preceding the time period of valve remodeling,[20] supporting the hypothesis that enzymatic degradation of the valve leaflets and venous wall by MMPs plays a key role in valve degradation, the inability of the valve to close, and therefore the venous pressure elevation.

A critical issue is not only the mechanisms that cause *expression* of these proteases, but also the mechanisms that cause their *activation* from the inactive proform and the potential lack of antiprotease activity. Among the mechanisms that have been proposed to activate MMPs are plasmin, serine proteases (e.g., trypsin), MMP-3, and MMP-13. These mechanisms may involve indirect pathways, indicating the possible complexity of the activation process involved. Plasmin converts pro-MMP enzyme to the active form. Plasmin hyperactivity due to decreased plasminogen activator inhibitor-1 (PAI-1) may thus cause MMP overactivity. These important regulatory pathways need to be clarified to understand the activation of the proteolytic process in venous disease since they may constitute a therapeutic target.

Inflammatory cells such as macrophages produce an array of proteolytic enzymes including the MMPs. *In situ* zymography, a technique that permits effective detection of proteolytic enzymatic activity in living tissues, indicates an early increase in MMP-2 and MMP-9 activity in an acute model of venous hypertension with femoral arterio-venous fistula.[20] The enhanced activity is detected in the greater saphenous vein, in the fistula per se, in the femoral vein, and in the sapheno-femoral junction, but not in the surrounding interstitial tissue, indicating progressive inflammation in the wall of these stretched veins.

Another factor that may contribute to the lack of proper restructuring in venous ulcer fibroblasts demonstrates decreased proliferative responses to growth factor stimulation.[21]

INFLAMMATION AS A REPAIR MECHANISM AND ITS RESOLUTION

The inflammatory cascade serves a lifetime as a repair mechanism that leads after removal of injured tissue to formation of new tissue. This cascade leads to return of basic organ functions or may just end up in the formation of a connective tissue (i.e., a scar). After removal of cell debris during resolution of inflammation, macrophages generate new growth factors, fibroblast and other connective tissue cells migrate and differentiate, new blood vessels grow, and new tissue is generated. Likely, there is an important role for stem cells that determine what particular functions the new tissue will be able to carry out. While in many cases venous disease leads eventually to resolution of the inflammation, we are concerned with cases that have recurrent inflammation without definitive resolution.

The lack of a resolution of inflammation indicates that the stimulus that activates the inflammatory cascade remains active and has not been brought under control. Thus, as telangiectases develop into varicose veins, skin edema, pigmentation, venous eczema, and eventually into active venous ulcers, no permanent—although temporary—resolution is achieved. This may have multiple reasons but is likely linked to the fact that the main stimulus for formation of the original inflammation remains active. Treatments need to interfere with the inflammatory cascade that causes tissue damage but cannot be targeted against the tissue repair mechanism. Strong interventions against the inflammation, especially interventions that interfere with the growth of new tissue, may need to be kept at a minimum. This is a challenge and, while currently unknown, the number of options may be limited. An effective way to circumvent the dilemma between a blockade of tissue damage, but not the repair mechanisms themselves, is to identify the trigger mechanisms for the inflammation cascade and interfere at the earliest possible state, ideally before the disease becomes a clinical manifestation. In the following, we will focus on this important issue.

TRIGGER MECHANISMS FOR INFLAMMATION IN VENOUS DISEASE

There are numerous ways to induce cell activation in the circulation and trigger an inflammatory cascade. Besides the well-recognized activation triggered by inflammatory mediators (e.g., complement fragments, a broad variety of polypeptides, fibrinolytic fragments, lipid fragments, viral and bacterial products, cytokines, lymphokines, histamine, and oxidized products), depletion of anti-inflammatory mediators (e.g., nitric oxide, adrenal glucocorticoids, adenosine, selected cytokines, and albumin) can trigger activation. Short-term transients in partial pressure of gases or temperature stimulate cell activation; so do hyper- or hypoglycemia, or insulinemia.

In the context of venous disease, it is important to note that mechanical stresses control cell activation in the circulation. Both mechanical stress of the wall of venous blood vessels (due to a pressure mediated hoop strain) above physiological values as well as the small fluid shear stress (due to fluid flow) control inflammation. The impact of a change in blood pressure and fluid shear stress on microvascular inflammatory events are observed within less than an hour on both the endothelium and on circulating cells, and they depend on the magnitude of the pressures involved.[22] Thus, an acutely or chronically elevated venous blood pressure is likely to cause on its own production of oxygen free radicals and thus serves to induce an inflammatory cascade with expression of proinflammatory and prothrombotic genes and synthesis of protein products. For example, an elevated blood pressure can itself induce synthesis of MMPs.

Another mechanical stimulus that controls the inflammatory and thrombotic phenotype is fluid shear stress. Its effect on the endothelium and on other cells in the circulation seem to be independent of the effects of pressure, even though in the circulation pressure is always coupled to shear stress (due to a basic equilibrium). The response depends on the time course and the amplitude of the shear stress applied to the endothelium.

A constant shear stress (e.g., at 10 to 12 dyn/cm^2), which is typical for a normal circulation) generates a stable endothelial cell that expresses predominantly anti-inflammatory genes. This shear stress produces in many respects the ideally stable phenotype of the endothelium associated with the least degree of restructuring, optimally cross-linked cytoplasmic proteins, minimal permeability, and a least thrombotic state. In contrast, unsteady shear stresses—for example, shear stresses with cyclic reversal of the shear direction or even turbulent shear stress with a random orientation and magnitude—generate an endothelium with elevated permeability and expression of proinflammatory and -thrombotic genes. This is currently a topic of extensive research involving intracellular signaling cascades with control of kinase and phosphatase activity under the control of fluid mechanical parameters.[23-25] The issues are especially relevant with regard to a situation in which the distension of venous vessels reaches a point at which the closure of the valve leaflets becomes incomplete and there are not only instances of time in which the venous vessel and the valves are distended but also instances when reflux develops. The endothelium on the valve leaflets is then subject to a reversed shear stress, a process that may accelerate inflammation and lead to eventual breakdown of its collagen fiber matrix.

Could it be possible that fluid shear stress and blood pressure per se trigger venous disease without any other cofactor? The answer to this question depends on the details of the pressure– and shear stress–time course involved, including events that in many cases precede the clinical manifestations of the disease and may not be readily accessible. In situations in which pressure and shear stress stay within normal limits during the early stages of the disease (e.g., an early phase with elevated permeability or restructuring with initial enlargement of capillaries to postcapillary venules), we can hardly blame shear stress or pressure as trigger mechanisms. Shear stresses and pressure serve to maintain and may enhance the progression of the disease after development of enlarged venules and veins, since at that time their local values in individual veins may shift during vascular restructuring even though their average values stay unchanged.

Development of abdominal obesity—a well-established risk factor for venous disease—may serve as a mechanism that could lead to transient or complete obstruction of venous pathways from the legs and central veins. Such a situation leads to enhancement of blood pressure in the legs together with a shift in fluid shear stresses (with reduced fluid shear stresses in some parts of the venous system and enhanced values in other parts) and may promote the disease. Clearly, such cases are candidates for recanalization and stenting of obstructed venous segments,[26] interventions that should lead to stabilization of the venous system including healing of open ulcers.

CONCLUSION AND SYNOPSIS

If there is no significant shift in fluid mechanical stresses in the early stages of the disease, we need to examine other factors that influence the biology of the venous wall and the leaflets. There is a variety of possibilities; compatibility with existing risk factors for venous disease may serve as a lead. A case in point is that abdominal obesity may also be associated with production of proinflammatory mediators and macrophage activation. Along similar lines, it is interesting to note that one of the sources of powerful inflammatory mediators may be derived from digestive enzymes in the intestine,[27] and therefore an elevated gut permeability may cause production of inflammatory mediators[28] as well as permit entry of serine proteases into the circulation, which in turn are effective activators of pro-MMP.[29]

Development of telangiectases early in the first trimester suggests an influence of pregnancy hormones, well known to affect venous compliance.[30] The sex hormone receptor distribution in the veins of the lower extremities in patients susceptible to first trimester telangiectases is an important unresolved issue in this regard and may lead to identification of an at-risk patient population. Another interesting factor that may be involved in trigger mechanisms is the ambient temperature. Elevated temperatures are effective in raising venous distension due to reduced constriction of vascular smooth muscle. Finally, there is a large contingent of potential genetic risks. Their mechanisms of action require analysis at the molecular level. Gene therapy may in the future be a potential option.

ACKNOWLEDGMENT

The work summarized here was supported in part by the Vein Institute of La Jolla and by a grant from the NHLBI HL 43026.

REFERENCES

1. Coleridge Smith PD, et al. Causes of venous ulceration: A new hypothesis. *Br Med J.* 1988;296:1726–1727.
2. Niggli V. Signaling to migration in neutrophils: Importance of localized pathways. *Int J Biochem Cell Biol.* 2003;35(12):1619–1638.
3. Makino A, et al. G protein-coupled receptors serve as mechanosensors for fluid shear stress in neutrophils. *Am J Physiol Cell Physiol.* 2006;290(6):C1633–1639.
4. Takase S, Bergan JJ, Schmid-Schonbein G. Expression of adhesion molecules and cytokines on saphenous veins in chronic venous insufficiency. *Ann Vasc Surg.* 2000;14(5): 427–435.

5. Smith PD. Neutrophil activation and mediators of inflammation in chronic venous insufficiency. *J Vasc Res.* 1999;36(Suppl 1):24–36.
6. Ono T, et al. Monocyte infiltration into venous valves. *J Vasc Surg.* 1998;27(1):158–166.
7. Johnston B, et al. Alpha 4 integrin-dependent leukocyte recruitment does not require VCAM-1 in a chronic model of inflammation. *J Immunol.* 2000;164(6):3337–3344.
8. Rivera-Nieves J, et al. L-selectin, alpha 4 beta 1, and alpha 4 beta 7 integrins participate in CD4+ T cell recruitment to chronically inflamed small intestine. *J Immunol.* 2005;174(4):2343–2352.
9. Norman MU, Kubes P. Therapeutic intervention in inflammatory diseases: A time and place for anti-adhesion therapy. *Microcirculation.* 2005;12(1):91–98.
10. Takase S, Schmid-Schonbein G, Bergan JJ. Leukocyte activation in patients with venous insufficiency. *J Vasc Surg.* 1999;30(1):148–156.
11. Pappas PJ, et al. Causes of severe chronic venous insufficiency. *Semin Vasc Surg.* 2005;18(1):30–35.
12. Wali MA, Eid RA. Changes of elastic and collagen fibers in varicose veins. *Int Angiol.* 2002;21(4):337–343.
13. Sansilvestri-Morel P, et al. Chronic venous insufficiency: Dysregulation of collagen synthesis. *Angiology.* 2003;54(Suppl 1):S13–18.
14. Martinez-Lemus LA, et al. Integrins and regulation of the microcirculation: From arterioles to molecular studies using atomic force microscopy. *Microcirculation.* 2005;12(1):99–112.
15. Asanuma K, et al. Uniaxial strain upregulates matrix-degrading enzymes produced by human vascular smooth muscle cells. *Am J Physiol Heart Circ Physiol.* 2003;284(5):H1778–1784.
16. Liu Z, et al. The serpin alpha1-proteinase inhibitor is a critical substrate for gelatinase B/MMP-9 in vivo. *Cell.* 2000;102(5):647–655.
17. Norgauer J, et al. Elevated expression of extracellular matrix metalloproteinase inducer (CD147) and membrane-type matrix metalloproteinases in venous leg ulcers. *Br J Dermatol.* 2002;147(6):1180–1186.
18. Herouy Y, et al. Lipodermatosclerosis is characterized by elevated expression and activation of matrix metalloproteinases: Implications for venous ulcer formation. *J Invest Dermatol.* 1998;111(5):822–827.
19. Mwaura B, et al. The impact of differential expression of extracellular matrix metalloproteinase inducer, matrix metalloproteinase-2, tissue inhibitor of matrix metalloproteinase-2 and PDGF-AA on the chronicity of venous leg ulcers. *Eur J Vasc Endovasc Surg.* 2005;31(3):306–310.
20. Pascarella L, Schmid-Schonbein GW, Bergan J. An animal model of venous hypertension: The role of inflammation in venous valve failure. *J Vasc Surg.* 2005;41(2):303–311.
21. Lal BK, et al. Altered proliferative responses of dermal fibroblasts to TGF-beta1 may contribute to chronic venous stasis ulcer. *J Vasc Surg.* 2003;37(6):1285–1293.
22. Takase S, et al. Enhancement of reperfusion injury by elevation of microvascular pressures. *Am J Physiol Heart Circ Physiol.* 2002;282(4):H1387–1394.
23. Busse R, Fleming I. Pulsatile stretch and shear stress: Physical stimuli determining the production of endothelium-derived relaxing factors. *J Vasc Res.* 1998;35(2):73–84.
24. Davies PF, Spaan JA, Krams R. Shear stress biology of the endothelium. *Ann Biomed Eng.* 2005;33(12):1714–1718.
25. Resnick N, et al. Fluid shear stress and the vascular endothelium: For better and for worse. *Prog Biophys Mol Biol.* 2003;81(3):177–199.
26. Neglen P, Hollis KC, Raju S. Combined saphenous ablation and iliac stent placement for complex severe chronic venous disease. *J Vasc Surg.* 2006;44(4):828–833.

27. Schmid-Schonbein GW, Hugli TE. A new hypothesis for microvascular inflammation in shock and multiorgan failure: Self-digestion by pancreatic enzymes. *Microcirculation.* 2005;12(1):71–82.
28. Cordts PR, et al. Could gut-liver function derangements cause chronic venous insufficiency? *Vasc Surg.* 2001;35(2):107–114.
29. Rosario HS, et al. Pancreatic trypsin increases matrix metalloproteinase-9 accumulation and activation during acute intestinal ischemia-reperfusion in the rat. *Am J Pathol.* 2004;164(5):1707–1716.
30. Meendering JR, et al. Effects of menstrual cycle and oral contraceptive use on calf venous compliance. *Am J Physiol Heart Circ Physiol.* 2005;288(1):H103–110.

2

MICROCIRCULATION IN CHRONIC VENOUS INSUFFICIENCY

KEVIN BURNAND AND SAID ABISI

The relationship between chronic venous insufficiency, ambulatory venous hypertension, and ulceration is still not understood. The sequence of events leading to venous ulceration is equally contentious.

Venous ulcers develop in patients who have only incompetence of the superficial or perforator veins but also develop in patients with deep venous reflux or post-thrombotic damage. The histological findings associated with venous hypertension include atrophy and scarring of the dermis associated with a loss of papillary structures at the dermal-epidermal junction. There is enlargement of the dermal capillary bed and associated fibrosis of the veins and dermis. Around the capillaries of the ulcer-bearing skin, there is a thick amorphous perivascular cuff composed of fibrin, fibronectin, laminin, tenascin, and collagen.

Various hypotheses have been proposed to explain the mechanisms that cause venous ulceration. Some theories are now refuted and regarded as historical, while other theories are still debated. This chapter provides an overview of the various microcirculatory changes found in association with venous ulcers.

HISTORICAL THEORIES

Appreciation of the relationship between calf pump dysfunction and venous ulceration is attributed to the Hippocratic school in Adams's translation of the ancient Greek book *De Ulceribus*.

Almost 20 centuries later, John Gay recognized that ulceration was a direct consequence of varicosity and described the presence of "coagula" (presumably a thrombus) in the deep veins of patients with ulceration of the legs. He also noted the presence of communicating veins that were probably first described by Verneuil in 1855.

In 1918 John Homans proposed a specific role for the valves of deep veins of the calf in the genesis of ulceration. Homans suggested that blood "stagnation or stasis" in the superficial veins leads to lack of nutrients and tissue hypoxia, which eventually causes ulceration. His work on incompetence of perforating veins was later amplified by Linton and Cockett. Homans, like Gay, also recognized the presence of "matter" in the deep veins and went on to suggest that this was the result of organization of a previous thrombosis, coining the term "post-phlebitic" or the more semantically correct "post-thrombotic leg."

Although Homans's "stasis" hypothesis is now discredited, it provided the stimulus for further studies and theories into the causes of venous ulceration.

ARTERIOVENOUS SHUNTING

Holling et al., in 1938, were probably the first to suggest that ulcers may develop as a consequence of the direct shunting of the blood from arterioles to venules in ulcerated limbs. Pratt also reported that arteriovenous communications were present in the calf skin of patients with venous ulcers. This led to the suggestion that nutrients and oxygen were shunted away from the dermal skin causing ulceration.[1] This concept was then taken up by Gius, Piulachs, and Barraquer and Fontaine, whose studies were mainly based on clinical observations. Some advocates of this theory have even suggested that all varicose veins are caused by arteriovenous fistulae.[2] Ryan and Copeman suggested that normal physiological shunts, involved in temperature control, are opened up by raised venous pressure. All these observations have been challenged and refuted by the work of Partsch and his colleagues and the work of Hehne et al. Increased trapping of isotopically labeled albumin aggregates in the ulcer skin was found using a gamma camera while there was no evidence of increased uptake in the lungs.[3-5]

DIFFUSION BARRIER

Studies at St. Thomas' Hospital showed that increasing venous pressure in an animal model increased capillary permeability for macromolecules, allowing them to accumulate in the subcutaneous tissue. This led to the "fibrin cuff hypothesis."

The initial experimental work in which an arteriovenous fistula was made between the femoral artery and vein in the groin of a dog produced elongation and dilatation of dermal capillaries in the animal's calf skin. There was increased leakage of labeled fibrinogen into the subcutaneous fluid of the animal's limbs in which venous hypertension had been induced.[6] It was hypothesized that increased venous pressure led to a dilatation of the pores between the endothelial cells of the capillaries, increasing the permeability of large molecules.[7] As a consequence, fibrin and other macromolecules accumulated in the interstitial pericapillary space. It was suggested that the perivascular infiltrate acted as a diffusion barrier to oxygen and other nutrients.

Histological examination of skin biopsies from patients with lipodermatosclerosis confirmed the deposition of fibrin in the pericapillary bed (Figure 2-1) and was associated with reduced local and systemic fibrinolytic activity.[8]

Others have confirmed the presence of pericapillary fibrin but have emphasized the fact that many other proteins are present within the cuff, including fibronectin, laminin, and deposits of collagen.[9] Stacey et al. showed that the fibrin cuffs were present before any skin changes were visible. The extent and number of fibrin cuffs seen on biopsies correlated with local tissue hypoxia.[10] Some aspects of this hypothesis were subsequently challenged including the thickness of the fibrin layer that was required to block diffusion and also the presence or absence of hypoxia in the ankle skin.[10,11]

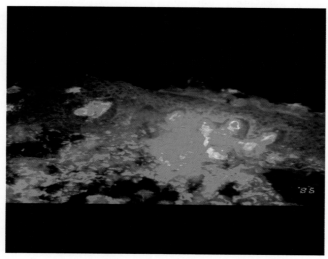

FIGURE 2-1 Fibrin deposition in the pericapillary bed as demonstrated using immune fluorescence staining of a section taken from lipodermatosclerotic skin.

The fibrin cuff hypothesis was modified by Falanga and Eaglstein,[12] who confirmed that fibrinogen, α_2-macroglobulin, and other large molecules were present in the pericapillary cuff. They suggested that this cuff might interfere with and "trap" growth factors and other stimulatory factors. The trapped molecules were unavailable for the maintenance of tissue strength and the repair process. α_2-macroglobulin is known to be a scavenger of growth factors such as transforming growth factor-β (TGF-β). An absence of TGF-β in the venous ulcer bed compared with normally healing graft donor sites has been found. Markedly increased levels of TGF-β within the pericapillary fibrin cuffs have provided further evidence for a possible "trapping" of growth factors in venous ulcers.[13]

WHITE CELL TRAPPING

Moyses, in 1987, demonstrated that blood returning from the lower limb following a period of dependency contained fewer leukocytes than the normal control limbs. Thomas et al. found that these changes were more marked in patients with chronic venous disease. Leukocytes are larger and less deformable than red blood cells and have the potential to block capillaries (Figure 2-2). The reduced pressure gradient between the arterial and venous systems in venous hypertension might encourage leukocytes to plug in capillaries and block the passage of erythrocytes.[14] This phenomenon could cause local ischemia.[15] A hypothesis known as the "leukocyte trapping hypothesis" was proposed in which leukocytes were thought to occlude venular capillaries and predispose to venous (ischemic) ulceration.[16] Attempts to directly demonstrate capillary occlusion by leukocytes were

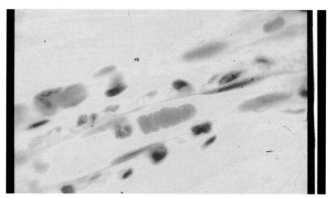

FIGURE 2-2 Illustration of one leukocyte and erythrocytes aggregating behind it in normal venular capillary. This has never been directly observed in capillaries in the vicinity of venous ulcers.

unsuccessful, casting doubts on the validity of this theory.[17] This theory therefore was adjusted, and it was proposed that trapped leukocytes migrated into the pericapillary and dermal tissue, activating the inflammatory response.

LEUKOCYTES AND ENDOTHELIAL CELLS' ACTIVATION

The theory that leukocytes had a role in the development of venous ulceration has led to further studies to identify the types and functions of leukocytes that might be responsible.

Studies reporting the association of leukocytes and venous ulcers vary. Skin biopsies taken from patients with venous ulcers have demonstrated abundant T-lymphocytes and macrophages with few neutrophils and B-lymphocytes.[17] Macrophages and mast cells were the main leukocytes found in the environs of venous ulcers in a quantitative morphometric study using electron microscopy.[18] Mast cells were increased around arterioles and postcapillary venules but not increased in the capillaries of ulcer-bearing skin of patients with lipodermatosclerosis or healed ulcers. Patients with active ulceration had a similar number of mast cells compared with controls. Macrophages were increased around the arterioles and postcapillary venules in patients with active and healed ulcerations.

Pappas et al. also assessed the capillary endothelial cells structure in the vicinity of venous ulcers. The endothelial cells had euchromatic nuclei indicative of active transcription and protein production. There were no other structural changes found in the endothelial cells.[18] The tight gaps between the endothelial cells junction were well maintained, which seemed to refute the stretch pore phenomenon.

Endothelial cell binding and interaction with leukocytes is dependent on adhesion molecules. Changes in adhesion molecules can therefore modulate leukocytes' migration and activation. Expression of intracellular adhesion molecules (ICAM-1) and vascular adhesion molecules (VCAM-1) on the endothelial cell increases as lipodermatosclerosis deteriorates.[19,20]

Pappas et al. measured the activity of circulating leukocytes in patients with venous ulceration using flow cytometric analysis of activation markers on the monocytes and T-lymphocytes. A decrease in activated T-lymphocytes and an increase in activated monocytes were found in patients with venous ulcers compared with normal patients.[21]

The possibility that leukocytes were dysfunctional has also been assessed. A significant attenuation in the proliferation of cultured monocytes and lymphocytes taken from the blood of patients with venous ulcers was observed compared with monocytes and lymphocytes extracted from patients with no ulcer. The dysfunction of these cells was associated with clinical venous disease progression.[22] It was therefore hypothesized that altered leukocyte activity and decreased mononuclear cell proliferation contributed to poor wound healing.

CYTOKINES AND GROWTH FACTORS

The changes that have been found in leukocytes and endothelial cells with the environs of venous ulcers and circulating blood have led to speculation that their role in cytokine activation and growth factor production could lead to tissue death and inflammation.

Tumor necrosis factor-α (TNF-α), an important mediator of cell apoptosis, was found to be upregulated in serum during active ulceration, and its levels declined with healing.[23] TNF-α and interleukins (IL) such as IL-1α and IL-6 were found to be reduced in the wound fluid taken from healing chronic venous ulcers when compared with nonhealing ulcers.[24] Transforming growth factor-β (TGF-β) is known to stimulate dermal fibroblast proliferation, which may lead to excessive tissue fibrosis if overexpressed. An excess of TGF-β has been found in some venous ulcers. This has been confirmed at transcriptional and translational levels.[25]

Vascular endothelial growth factor (VEGF) and platelet derived growth factor receptor α and β (PDGFR-α and -β) are well known to be produced by endothelial cells. These factors have been found to be overexpressed in the microenvironment of some venous ulcers.[26,27] It has been suggested that some of these growth factors are ineffectual. Endothelial cell proliferation studies have found that ulcer exudates had no effect on the ulcer healing despite the increase of VEGF in nonhealing ulcers. This suggests that there is posttranslational inhibition of some growth factors in the ulcer environment.[26,28] Ulcer healing appears to be a polygenic complex process involving alterations in cellular function and inflammatory mediators.

It is often difficult to know whether the observed changes are cause or effect or simply the result of local inflammation. They also do not explain why ulcers develop in the gaiter region of the calf or why ulcer healing is achieved by compression treatment. Ideally, leukocyte/endothelial cell interactions and growth factor changes should be studied before skin changes develop in a high-risk group of patients (e.g., following a deep vein thrombosis).

OXIDATIVE STRESS IN VENOUS ULCERS

Reactive oxygen species are present in the ulcer tissue and exudate. Their release is probably mediated by inflammatory cells, fibroblasts, and endothelial cells, which are known to produce superoxidants. These cells are commonly present in chronic venous leg ulcers' environs, leading to hostile oxidative stress and consequent tissue destruction.[29]

Oxidative stress is dependent on the presence of free ion radicals such as iron in the tissue around venous ulcers. Increased iron deposition may cause an elevation of toxic free radicals in venous ulcers' environs. Iron is toxic because of its ability to undergo cyclic oxidation and reduction, generating free radicals via Fenton reaction. Antioxidants have been shown to be upregulated in chronic wounds, perhaps to counterbalance the increased risk of oxidative damage to the tissue.[29,30]

MMP AND VENOUS ULCER FORMATION

Venous ulcers contain a number of protolytic enzymes, especially potent matrix remodeling enzymes, which are capable of degrading the different types of collagen and regulating dermal matrix turnover. The matrix metalloproteinase (MMP) family appears to be involved in different stages of the ulcerative and healing process in chronic venous ulcers.[31] Analysis of venous ulcer exudates found an elevated expression and activation of different subtypes of MMP at different stages of ulcer healing.[32]

Elevated levels of mRNA expression of MMP-1 and MMP-2 have been found in lipodermatosclerotic skin, but only the MMP-2 was found in its active form. These were mainly present particularly in the perivascular regions, basal and suprabasal epidermal layers, and associated with reduced expression of tissue inhibitor metalloproteinase-2 (TIMP-2).[33] Matrix metalloproteinase-2 (MMP-2) digests type I collagen, type III collagen, and fibronectin, all of which are present in the perivascular fibrin cuffs of lipodermatosclerosis. MMP-2 can also degrade elastic fibers that are found fragmented in the reticular dermis of patients with lipodermatosclerosis, indicating perhaps that MMP-2 contributes to the hardening of the skin in this condition. The expression of membrane type metalloproteinases (MT-MMPs) is also elevated in the environs of venous ulcers.[34] These membrane-bound proteinases not only degrade extracellular matrix but are also known to activate other collagenolytic and gelatinolytic enzymes such as MMP-2 and MMP-9. Levels of MMP-9 may also decrease as the wound progresses toward healing.[35]

Venous ulcer fluids, unlike acute wound fluid, have increased proteolytic activity. This encourages the degradation of extracellular adhesion proteins like fibronectin and vitronectin,[36] which are important in the migration of cells such as keratinocytes. MMP-2 and MMP-9 are mainly responsible for degrading fibronectin. They have also been found to be elevated in venous ulcer fluids when compared with acute wound fluids. Altered levels of collagenases MMP-1 and MMP-8 are also present in chronic leg ulcer fluids.[37]

The overall MMP activity in chronic wound fluids is influenced by the types of cells that are present in the ulcer base and by the types and amount of growth factors and cytokines that are present. Ulcers may therefore form because of an enhanced turnover of the dermal extracellular matrix regulated by specific MMPs. It is, however, equally possible that existing local tissue damage and inflammation cause activation of the MMPs.

THE ROLE OF FIBROBLASTS IN VENOUS ULCERS

Fibroblasts are present in abundance in the dermis of patients with lipodermatosclerosis. Dermal fibroblasts are active skin replacement cells and are characterized by an ability to produce collagen and induce fibrosis. Fibroblasts in the dermis of venous ulcers have been found to behave abnormally. Studies have been carried out on their role in venous ulceration.[38,39] The dermal fibroblasts in venous ulcers and chronic wounds appear to have a decreased number of type II receptors to TGF-β and a reduced ability to phosphorylate TGF-β.[38,40] They exhibit a decreased dose-dependent response to TGF-β stimulation compared with fibroblasts taken from normal skin.[38]

Decreased collagen production has been demonstrated in cultured dermal fibroblasts taken from biopsies of the edge of venous ulcers compared with those taken from normal skin, but fibronectin synthesis was not different between both cultured cells.[39] Normal fibroblasts exhibited reduced collagen production although fibronectin production was unchanged when cultured under hypoxic conditions. These findings led Herrick et al. to suggest that fibroblasts' ability to synthesize collagen is probably reduced in the presence of hypoxia, which occurs in the venous ulcer environment. Low fibronectin levels in venous ulcer tissue may be caused by increased degradation rather than poor synthesis.

Slow growth of dermal fibroblasts from the venous ulcer edge and elevated levels of markers of senescence have been demonstrated.[41] It therefore appears that premature "aging" of dermal fibroblasts in venous ulceration is responsible for their poor proliferation and dysfunction.

"Senescence" was observed when neonatal fibroblasts were cultured in a mechanically designed pressure incubator, providing direct *in vitro* evidence of a relationship between elevated pressure and altered fibroblast behavior.[42]

Dermal fibroblast "senescence" may, however, still be the result rather than the cause of ulceration. The role of fibroblast senescence in the development of venous ulcers remains to be confirmed.

SUMMARY

The mechanisms involved in venous ulceration still need to be fully elucidated. The process by which skin changes and venous ulcers develop is complex.

The different hypotheses have generated many further investigations and studies. A better understanding of the cellular and molecular changes in microcirculation of venous ulcers may alter potential therapeutic targets and strategies.

REFERENCES

1. Pratt GH. Arterial varices: A syndrome. *Am J Surg.* 1949;77:456–460.
2. Schalin L. Arteriovenous communications in varicose veins localized by thermography and identified by operative microscopy. *Acta Chir Scand.* 1981;147:409–420.
3. Lofferer O, Mostbeck A, Partsch H. [Arteriovenous anastomosis of the extremities. Nuclear medical examinations with special reference to postthrombotic ulcer of the lower leg]. *Zentralbl Phlebol.* 1969;8:2–20.
4. Lindemayr W, Lofferer O, Mostbeck A, Partsch H. Arteriovenous shunts in primary varicosis? A critical essay. *Vasc Surg.* 1972;6:9–13.
5. Hehne HJ, Locher JT, Waibel PP, Fridrich R. [The importance of arteriovenous anastomoses in primary varicosis and chronic venous insufficiency. A nuclear-medical study]. *Vasa.* 1974;3:396–398.
6. Burnand KG, Clemenson G, Whimster I, Gaunt J, Browse NL. The effect of sustained venous hypertension on the skin capillaries of the canine hind limb. *Br J Surg.* 1982;69:41–44.
7. Burnand KG, Whimster I, Naidoo A, Browse NL. Pericapillary fibrin in the ulcer-bearing skin of the leg: The cause of lipodermatosclerosis and venous ulceration. *Br Med J (Clin Res Ed).* 1982;285:1071–1072.
8. Browse NL, Gray L, Jarrett PE, Morland M. Blood and vein-wall fibrinolytic activity in health and vascular disease. *Br Med J.* 1977;1:478–481.
9. Herrick SE, Sloan P, McGurk M, Freak L, McCollum CN, Ferguson MW. Sequential changes in histologic pattern and extracellular matrix deposition during the healing of chronic venous ulcers. *Am J Pathol.* 1992;141:1085–1095.
10. Stacey MC, Burnand KG, Layer GT, Pattison M. Transcutaneous oxygen tensions in assessing the treatment of healed venous ulcers. *Br J Surg.* 1990;77:1050–1054.
11. Cheatle TR, McMullin GM, Farrah J, Smith PD, Scurr JH. Skin damage in chronic venous insufficiency: Does an oxygen diffusion barrier really exist? *J R Soc Med.* 1990;83:493–494.
12. Falanga V, Eaglstein WH. The "trap" hypothesis of venous ulceration. *Lancet.* 1993;341:1006–1008.
13. Higley HR, Ksander GA, Gerhardt CO, Falanga V. Extravasation of macromolecules and possible trapping of transforming growth factor-beta in venous ulceration. *Br J Dermatol.* 1995;132:79–85.
14. Thomas PR, Nash GB, Dormandy JA. White cell accumulation in dependent legs of patients with venous hypertension: A possible mechanism for trophic changes in the skin. *Br Med J (Clin Res Ed).* 1988;296:1693–1695.

15. Scott HJ, Coleridge Smith PD, Scurr JH. Histological study of white blood cells and their association with lipodermatosclerosis and venous ulceration. *Br J Surg.* 1991; 78:210–211.
16. Coleridge Smith PD, Thomas P, Scurr JH, Dormandy JA. Causes of venous ulceration: A new hypothesis. *Br Med J (Clin Res Ed).* 1988;296:1726–1727.
17. Wilkinson LS, Bunker C, Edwards JC, Scurr JH, Smith PD. Leukocytes: Their role in the etiopathogenesis of skin damage in venous disease. *J Vasc Surg.* 1993;17:669–675.
18. Pappas PJ, DeFouw DO, Venezio LM, et al. Morphometric assessment of the dermal microcirculation in patients with chronic venous insufficiency. *J Vasc Surg.* 1997; 26:784–795.
19. Peschen M, Lahaye T, Hennig B, Weyl A, Simon JC, Vanscheidt W. Expression of the adhesion molecules ICAM-1, VCAM-1, LFA-1 and VLA-4 in the skin is modulated in progressing stages of chronic venous insufficiency. *Acta Derm Venereol.* 1999;79: 27–32.
20. Weyl A, Vanscheidt W, Weiss JM, Peschen M, Schopf E, Simon J. Expression of the adhesion molecules ICAM-1, VCAM-1, and E-selectin and their ligands VLA-4 and LFA-1 in chronic venous leg ulcers. *J Am Acad Dermatol.* 1996;34:418–423.
21. Pappas PJ, Fallek SR, Garcia A, et al. Role of leukocyte activation in patients with venous stasis ulcers. *J Surg Res.* 1995;59:553–559.
22. Pappas PJ, Teehan EP, Fallek SR, et al. Diminished mononuclear cell function is associated with chronic venous insufficiency. *J Vasc Surg.* 1995;22:580–586.
23. Murphy MA, Joyce WP, Condron C, Bouchier-Hayes D. A reduction in serum cytokine levels parallels healing of venous ulcers in patients undergoing compression therapy. *Eur J Vasc Endovasc Surg.* 2002;23:349–352.
24. Trengove NJ, Langton S, Stacey M. Biochemical analysis of wound fluid from nonhealing and healing chronic leg ulcer. *Wound Repair Regen.* 1996;4:234–239.
25. Pappas PJ, You R, Rameshwar P, et al. Dermal tissue fibrosis in patients with chronic venous insufficiency is associated with increased transforming growth factor-beta1 gene expression and protein production. *J Vasc Surg.* 1999;30:1129–1145.
26. Drinkwater SL, Burnand KG, Ding R, Smith A. Increased but ineffectual angiogenic drive in nonhealing venous leg ulcers. *J Vasc Surg.* 2003;38:1106–1112.
27. Peschen M, Grenz H, Brand-Saberi B, et al. Increased expression of platelet-derived growth factor receptor alpha and beta and vascular endothelial growth factor in the skin of patients with chronic venous insufficiency. *Arch Dermatol Res.* 1998;290:291–297.
28. Drinkwater SL, Smith A, Sawyer BM, Burnand KG. Effect of venous ulcer exudates on angiogenesis in vitro. *Br J Surg.* 2002;89:709–713.
29. Wlaschek M, Scharffetter-Kochanek K. Oxidative stress in chronic venous leg ulcers. *Wound Repair Regen.* 2005;13:452–461.
30. Yeoh-Ellerton S, Stacey MC. Iron and 8-isoprostane levels in acute and chronic wounds. *J Invest Dermatol.* 2003;121:918–925.
31. Saito S, Trovato MJ, You R, et al. Role of matrix metalloproteinases 1, 2, and 9 and tissue inhibitor of matrix metalloproteinase-1 in chronic venous insufficiency. *J Vasc Surg.* 2001;34:930–938.
32. Weckroth M, Vaheri A, Lauharanta J, Sorsa T, Konttinen YT. Matrix metalloproteinases, gelatinase and collagenase, in chronic leg ulcers. *J Invest Dermatol.* 1996; 106:1119–1124.
33. Herouy Y, May AE, Pornschlegel G, et al. Lipodermatosclerosis is characterized by elevated expression and activation of matrix metalloproteinases: Implications for venous ulcer formation. *J Invest Dermatol.* 1998;111:822–827.
34. Norgauer J, Hildenbrand T, Idzko M, et al. Elevated expression of extracellular matrix metalloproteinase inducer (CD147) and membrane-type matrix metalloproteinases in venous leg ulcers. *Br J Dermatol.* 2002;147:1180–1186.

35. Trengove NJ, Stacey MC, MacAuley S, et al. Analysis of the acute and chronic wound environments: The role of proteases and their inhibitors. *Wound Repair Regen.* 1999;7:442–452.
36. Drinkwater SL, Smith A, Sawyer BM, Burnand KG. Effect of venous ulcer exudates on angiogenesis in vitro. *Br J Surg.* 2002;89:709–713.
37. Nwomeh BC, Liang HX, Cohen IK, Yager DR. MMP-8 is the predominant collagenase in healing wounds and nonhealing ulcers. *J Surg Res.* 1999;81:189–195.
38. Hasan A, Murata H, Falabella A, et al. Dermal fibroblasts from venous ulcers are unresponsive to the action of transforming growth factor-beta 1. *J Dermatol Sci.* 1997;16:59–66.
39. Herrick SE, Ireland GW, Simon D, McCollum CN, Ferguson MW. Venous ulcer fibroblasts compared with normal fibroblasts show differences in collagen but not fibronectin production under both normal and hypoxic conditions. *J Invest Dermatol.* 1996;106:187–193.
40. Kim BC, Kim HT, Park SH, et al. Fibroblasts from chronic wounds show altered TGF-beta-signaling and decreased TGF-beta Type II receptor expression. *J Cell Physiol.* 2003;195:331–336.
41. Stanley AC, Park HY, Phillips TJ, Russakovsky V, Menzoian JO. Reduced growth of dermal fibroblasts from chronic venous ulcers can be stimulated with growth factors. *J Vasc Surg.* 1997;26:994–999.
42. Stanley AC, Fernandez NN, Lounsbury KM, et al. Pressure-induced cellular senescence: A mechanism linking venous hypertension to venous ulcers. *J Surg Res.* 2005;124:112–117.

3

Epidemiology of Venous Ulcers

Olle Nelzén

A leg ulcer is not a disease in itself but rather a symptom that can be caused by a variety of different diseases with a number of risk factors. Where venous ulcers are concerned, they can be caused by a variety of venous diseases. As it appears today, varicose vein disease seems to be the most common cause, whereas deep vein thrombosis and primary deep vein disease are becoming less frequent in most studies. In addition, combinations of various venous diseases are far from uncommon, which underlines the necessity for careful diagnostic measures. Venous ulcers are not a new problem, since they are well known to have plagued humankind since ancient times. The first written description of treatment of probable venous ulcers was found in the Papyrus of Eber originating from Egypt around 1550 BC. In the Milan Cathedral in Italy, four big paintings show miracles performed by Saint Carlo for patients with leg ulcers in the beginning of the 17th century, clearly indicating that at that time leg ulcers were already a huge problem within the population (Figure 3-1). It appears that leg ulcers have been a voluminous problem for generations, and this problem still remains. The greatest problem with leg ulcers is that so few people realize that they really are a problem, and that is why epidemiological studies are needed.

WHAT IS EPIDEMIOLOGY?

Epidemiology is the science dealing with the distribution and frequency of diseases or disorders in the population. To have knowledge of the epidemiology of a certain disease is of fundamental importance in order to

FIGURE 3-1 The miracle of Aurelia Degli Angeli. She had suffered from a smelly, painful ulcer for 3 years when, in 1601, she called on the Holy Archbishop Saint Carlo, who prayed, and her sores instantly closed and she returned to full health. Painting by Giovanni Battista Crespi, known as Il Cerano, 1610. Milan Cathedral, Italy. Copyright © 1995, Veneranda Fabbrica del Duomo di Milano.

be able to plan appropriate actions to counteract the disease and to improve management, including care and treatment. A good epidemiological survey will form a very valuable baseline to calculate the magnitude of needed treatment changes, both monetary and workload for the carers, as well as serve as a useful comparison resource for measuring any outcome changes as a result of performed treatment changes. To be able to do that, a repeat epidemiological survey has to be undertaken with a similar methodology. Further epidemiological research can be used to assess or detect possible risk factors for a certain disease, but to finally prove a risk factor, longitudinal studies are generally necessary. It is of vital importance to be in control of confounding factors, such as age distributional changes, that may very well be responsible for observed changes of prevalence for a certain disease rather than treatment changes. Many new treatments are made available only for subgroups of patients with a disease. Will the result of such changes make a difference in the occurrence of this specific disease within the total population? An epidemiological study of that population will probably give you the answer. Furthermore, epidemiological studies are used to evaluate the natural history of diseases and can ascertain the characteristics of healthy persons compared with diseased patients.

WHAT IS THE DIFFERENCE BETWEEN INCIDENCE AND PREVALENCE?

To be able to read and to understand epidemiological papers, you have to understand the terminology (Table 3-1). The *incidence* of a disease or

TABLE 3-1   Epidemiological Terminology

Incidence	Number of new cases per time unit and population, usually 1 year
Point prevalence	Proportion with a certain disease at any point of time—time period usually shorter than 3 months
Period prevalence	Proportion with a certain disease within a longer period of time—usually 1 year or more
Overall prevalence	Proportion that have ever had a certain disease—lifetime period = lifetime prevalence

disorder is the number of new cases appearing per time unit, usually 1 year. It is generally presented as a proportion of the total population studied. The *prevalence* means the number of people with a certain disease or disorder within the population studied, also given as a proportion. The prevalence measurement includes both newly diseased and old cases in contrast to the incidence measurement, which includes only new cases. For so-called chronic diseases, such as venous leg ulcers, the prevalence is generally much higher than the incidence, whereas flu, for example, would probably yield more equal proportions. Unfortunately, there are various ways of assessing the prevalence, so usually prevalence is divided into point prevalence or period prevalence. A *point prevalence* measurement is a sample taken within a fairly narrow time frame that, for chronic diseases, generally is less than 3 months. A *period prevalence* is used for longer time periods, usually 1 year or longer. An *overall prevalence*, or if you will, a lifetime prevalence, will give you the proportion of a population that has ever suffered a venous leg ulcer, for example, which, by definition, also includes people with a history of previous ulceration in addition to people with open ulcers.

WHAT IS A LEG ULCER?

What may be obvious to some is not always clear to others, even for people within the medical profession. It is even more difficult for the average patient or for the general public to say what is or is not a leg ulcer. Thus, it is extremely important to define what you mean by a leg ulcer, or a venous leg ulcer, to be more specific. Such a definition has been commonly forgotten in many previous epidemiological studies, which will affect the credibility of such studies negatively. The definition of a *chronic leg ulcer* that we used in our Swedish studies[1-6] was as follows: any ulcer below the knee level that did not heal within a 6-week period—a definition also used by others. To be able to diagnose a venous ulcer, you also need proven venous incompetence or obstruction shown by additional noninvasive or invasive investigations, in addition to clinical signs. By relying on clinical signs alone, we have shown that you only get ideally 75% accuracy compared with the outcome when you add information from additional investigations.[3] This means that you risk misdiagnosing 1 out of 4 by relying on clinical inspection only.

LEG ULCER PREVALENCE

Through epidemiological research, we know that leg ulcers are encountered worldwide.[7,8] The most common way of estimating the size of the problem is to assess the point prevalence of leg ulcers in a defined population. Usually this has been undertaken through cross-sectional studies identifying all patients receiving professional care for leg ulcers at a specific point of time. The most widely used definition of a chronic leg ulcer is an ulcer anywhere below the knee that has been present for a period of 4–6 weeks or longer. These estimates usually range from 0.1% to 0.3% of the total population having open ulcers at any one time.[7,8] There are, in addition, people who do not seek professional help, and in Sweden we found that group to be as big as the number of people receiving professional care,[4,7] Thus, we found a point prevalence of about 0.6% representing around 50,000 people with open leg ulcers in Sweden,[4] with a total population of about 9 million. Since leg ulcer is a chronic condition, periods of healing are often followed by recurrences unless the causes of the ulcer are permanently treated.[6,7] Thus, the total population that has a history of ulceration is about three times as great, around 150,000, but 100,000 are, for the moment, healed in Sweden.[7]

It is important to realize that different studies can reach different conclusions regarding leg ulcer point prevalence. The observed level is dependent on whether all leg ulcer patients are identified, the methodology, the quality of leg ulcer treatment in the area, and the proportion of people who take care of their ulcers by themselves,[7,8] Knowledge of the true size of the problem in any country cannot really be anticipated until an assessment is made of whether self-care is common. Epidemiological research is a valuable tool to improve leg ulcer management and to assess the effect of changed management strategies.

ETIOLOGIC SPECTRUM OF LEG ULCERS

To diagnose the ulcer is today of vital importance not only for epidemiological research reasons but, more importantly, to be able to treat the ulcer appropriately. Tailored treatment has become increasingly important since there are now a variety of conservative treatment options as well as a number of surgical or minimally invasive alternatives, which alone or in combination will give the patient the most benefit in terms of rapid ulcer healing and prevention of recurrence. In some epidemiological studies, the etiological spectrum has been validated. Venous leg ulcers are the most common, and ulcers of major venous causes constitute around 50% of all ulcers.[2,7] Since about 20% to 25% of all ulcers are located on the foot, the proportion is around 70% if foot ulcers are excluded.[2,9] About half of these are caused by varicose veins (superficial venous insufficiency +/− perforators),[2,3,7,10] and there are reasons to believe that varicose ulcers will become even more predominant in future years. The most common diagnosis for foot ulcers is arterial insufficiency, which is responsible for about half of all foot ulcers.[2] The detailed etiologic spectrum is fairly complex, and many ulcers have a mixed etiology.[2,7] In more than one-third of patients, combinations of causative factors are likely to be responsible for the ulcers. Mixed venous and arterial etiology is a commonly encountered combination seen in about every fifth ulcer with dominating venous etiology.[2,3] Multifactorial ulcers, in which venous incompetence may be involved, are common, with no obviously dominating causative factor; therefore, they are more difficult to treat.[2,7,11] These ulcers are most often seen among the elderly. From our latest studies and from data from other researchers, it appears that multifactorial ulcers are increasing in number, which poses new challenges to the health care system.[11,12]

VENOUS ULCER PREVALENCE

Very few studies can actually give you details regarding venous ulcer prevalence, since only a few have validated the diagnoses, in contrast to the many studies assessing leg ulcers regardless of etiology.[7,8] One of the most uniform prevalence estimations regards overall prevalence of venous ulcers. Most studies have shown that around 1% of the adult population has a history of healed or open venous leg ulcers, and that number seems stable over the years and in many different countries (Table 3-2). Based on data from the Skaraborg study in Sweden, we also found that roughly an equal proportion had chronic lower limb ulceration of other causes than venous. Thus, around 2% of populations can be expected to have a history of chronic lower limb ulceration.[4,7]

There are only a few point prevalence estimates regarding venous ulcers published (Table 3-2). In the five latest studies, the diagnosis of venous ulcer was validated with noninvasive methods.[3,4,5,8,11] The lowest point prevalence, 0.024%, was found in the latest published study from the UK.[11] That prevalence is, however, based on a study with a questionable methodology using an extensive questionnaire, which is likely to bias recruitment of patients negatively. The highest prevalence was found in the two studies that were based on large random population samples and thus also included people in self-treatment. Both in Germany[15] and Sweden,[4] the point prevalence was 0.29%. The other remaining results were based on patients receiving professional care and range from 0.06% to 0.20%. It

TABLE 3-2 Prevalence Estimates of Venous Leg Ulcers

Authors (Publ. Year)	Country	Method	All Known to Health Care	Noninvasive Diagnosis	Prevalence % Overall	Point
Bobek et al. (1966)[13]	Czechoslovakia	Pop. study n = 15,060 adults >15 yrs	No	No	1.0 adult pop.	—
Widmer (1978)[14]	Switzerland	Selected sample n = 4,529 industrial workers 25–74 yrs	No	No	1.0 adult pop.	—
Fischer (1981)[15]	West Germany	Random pop. sample n = 4,260 adults 20–74 yrs	No	No	*2.7 in sample* 2.3 based on examined	*0.44 in sample* 0.29 based on examined
Nelzén et al. (1994)[3]	Sweden	Cross-sectional study Pop. 270,800 n = 387/827 ulcer pat. validated (randomly selected)	Yes	Yes Bidirectional Doppler arterial and venous	n.a.	0.16 total pop. 0.22 adult pop.

Baker et al. (1991)[8]	Australia	Cross-sectional study Pop. 238,000 n = 246/259 ulcer pat. validated	Yes	Yes Doppler + photoplethysmography	n.a. 0.06 total pop.
Nelzén et al. (1996)[4]	Sweden	Random pop. sample n = 12,000 people 50–89 yrs	No	Yes Bidirectional Doppler arterial and venous	0.8 total pop. 1.0 adult pop. 0.29 total pop.
Nelzén et al. (1996)[5]	Sweden	Selected sample n = 2,785 industrial workers 30–65 yrs	No	Yes Bidirectional Doppler arterial and venous	0.8 in sample 0.2 in sample
Moffatt et al. (2004)[11]	UK	Cross-sectional study in health care Pop. 252,000 n = 113 ulcer pat. validated	Yes	Yes Doppler + photoplethysmography	n.a. 0.024 total pop.

Adult pop. = population above the age of 15.
n.a. = not assessed.

TABLE 3-3 Combined Prevalence Estimates from Skaraborg County of Venous Ulcers[7]

A. Point Prevalence Open Venous Ulcers

	Known to Health Care	Self-Care Included
Total population	0.16	0.3
Adult population (>15 years)	0.22	0.4
Retired population (>65 years)	0.76	1.0

B. Overall Prevalence Healed and Open Venous Ulcers

	Known to Health Care	Self-Care Included
Total population	0.5	1.0
Adult population (>15 years)	0.6	1.3
Retired population (>65 years)	2.3	3.0

Figures given as percentages.

seems reasonable to expect the point prevalence in the total population to be somewhere in the region of 0.1% to 0.3% in most Western populations, although local variations are likely to exist. The weighted results from Skaraborg County (population 270,800) that were based on the results from three separate studies[3,4,5] are shown in Table 3-3.

INCIDENCE OF VENOUS ULCERS

There is not much information on the incidence of venous ulcers published. In one study from the United States,[16] the yearly incidence had remained unchanged during a 25-year period; the overall age- and sex-adjusted incidence was 18 per 100,000 person years. From New Zealand,[17] the annual cumulated incidence rate of lower limb ulceration was 32 per 100,000. We do not know the proportions having venous ulcers in that study, but a reasonable guess would be around 50%, leaving an incidence figure similar to the U.S. estimate. Based on our studies from Skaraborg, we could retrospectively estimate the yearly incidence to one-tenth of the point prevalence, giving an incidence of 16 per 100,000 based on patients receiving professional care and a total figure of around 30 per 100,000 if people in self-treatment are included.[7] In summary, the yearly incidence of venous ulcers is likely to be in the region of 15 to 30 per 100,000.

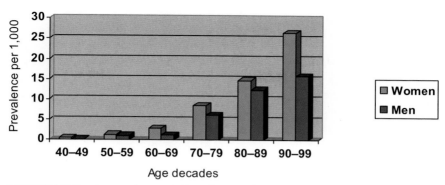

FIGURE 3-2 Age- and sex-specific point prevalence of venous ulcer based on patients receiving professional care in Skaraborg.

AGE AND SEX DISTRIBUTION

Patients with venous ulcers who are taken care of by health care professionals are old. The median age was about 75 years in Australia[8] and 77 years in Sweden.[3] It appears that females outnumber men only above retirement age.[3,8] The female predilection for venous ulcers has, however, been overestimated in the past as a result of observed frequencies not being adjusted for age.[3] Female longevity has been the cause of such misinterpretation.[7] The age-adjusted male:female ratio in the Skaraborg study was 1 : 1.6.[3] In Figure 3-2, the age- and sex-specific point prevalence of venous ulcers is shown.[3] Only a slight female predominance was reported from Australia.[8] There are not enough data on people in self-treatment to assess the sex distribution, but it is clear the age distribution is shifted toward people of working age.[4,5]

Two recent population studies that have assessed venous disease thoroughly within random population samples have both shown surprisingly similar occurrence for varicose veins for men and women. In the Edinburgh vein study from Scotland,[18] there were actually four times more men with a history of venous ulcers, but the numbers were small and therefore somewhat uncertain. Nevertheless, these numbers question the female predilection. In Bonn, Germany,[19] the researchers found more equal figures of ulceration between genders. In a previous U.S. study from San Diego,[20] varicose veins were still found more commonly among women, while trophic skin changes were more common among men.

Thus, it is obvious that increasing age is a risk factor for venous ulcers, but today it is highly questionable if female sex truly is. It appears men tend to feel more ashamed of ulcers and thus avoid contact with the health care system. This leads to an overrepresentation of females within studies based on patients treated by health care professionals, not only because of female longevity.[7] Thus, female sex is probably not a true risk factor for venous leg ulcer as supported by earlier cited studies. The latter is also supported by preliminary results from the Skaraborg leg ulcer study in 2002, where males and females were close to equal in numbers after correcting for age. One other reason, we believe, is that men now have become more positive in seeking professional help for leg ulcers.

VENOUS ULCER HISTORY

Most venous ulcers develop for the first time in middle age,[3] but, as shown, patients have a tendency not to seek professional help until after retirement (Figure 3-3). This means that a large number of younger people in the community have venous ulcers that are not taken care of professionally.[7] A large number of those (50% to 75%) have only superficial and/or perforator incompetence (SVI/PVI) and could thus benefit from varicose vein surgery.[4,5] In the Skövde study[5] in which only people below retirement age were included (age 30–65 years), only 25% of patients with venous ulcers had detectable deep venous incompetence (DVI) in contrast to the Skaraborg study[2] in which the corresponding figure was 60%. This may indicate that venous ulcer caused by SVI/PVI is more common among younger people and that the insufficiency may progress with time, if left untreated, to also involve the deep veins.[21,22] Unfortunately, these younger patients with first-time venous ulcers rarely seek medical help at an early stage, when their venous incompetence could easily be cured by surgery (Figure 3-3).

The chronicity of venous leg ulcers is empirically well known. Venous ulcers are more difficult to heal and have a greater tendency to recur than other types of chronic ulcers.[3] The median duration of current ulceration was 26 weeks in Australia,[8] and in Sweden 54% of patients had had their

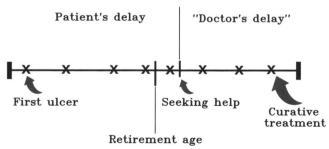

FIGURE 3-3 The typical scenario and history for a patient with a venous ulcer.[7] X represents ulcer episodes.

ulcer for more than 1 year.[3] Ulcers caused by DVI are the most difficult to heal,[4,23] with 64% having a duration of more than 1 year.[3] A venous leg ulcer has a tendency to recur, with more than 70% of open venous ulcers being recurrent.[3,8] Ulcers caused by varicose veins have a similar tendency to recur, as do ulcers caused by deep vein involvement.[3,6] In Sweden, 61% of patients with venous leg ulcers had had their first ulcer episode before the age of retirement, and 37% even before the age of 50.[3] The median duration of the ulcer history was 13.5 years, in Skaraborg as compared with 2.5 years for patients with ulcers from other causes.[3]

OBESITY

It is a well-known fact that obesity is quite common among patients with venous ulcers, and there is an ongoing discussion regarding whether obesity is a risk factor for venous disease and ulcers. In the Skaraborg study, we found that patients with venous ulcers were more obese than patients with nonvenous ulcers and also more obese than a matched Swedish population.[3] However, when we looked at the same problem in a large population sample, also including people in self-care, we did not find that patients with ulcers were more obese than age- and sex-matched controls.[7] This finding indicated that the ulcer causes obesity rather than the opposite.

HISTORY OF DEEP VEIN THROMBOSIS (DVT)

Following a clinically apparent DVT, about 5% of patients will develop leg ulcers.[24,25] A direct history of previous DVT is not common among patients with venous leg ulcers. In Sweden, only 37% had such a history,[3] and in Australia, even fewer (17%).[8] A positive history of DVT is more frequently found among patients with deep venous insufficiency than among those with varicose veins only (54% versus 14%, $p < 0.0001$).[3] If all potential factors predisposing for DVT are included, such as pregnancy, limb fracture, major accidents, and general anesthesia, only 4% of Australian patients had no evidence of a predisposing factor.[8] Still, it seems likely that post-thrombotic ulcers comprise less than 50% of all venous ulcers, with the true figure being in the region of 25% to 50%.[7] The latter is supported by the fact that in a U.S. study,[20] only 22% of legs with trophic changes showed signs of deep vein impairment. There seems to be a trend in which post-thrombotic ulcers are becoming less frequent, probably because of better DVT treatment and frequent use of DVT prophylaxis. Very few of the patients who had their DVTs before the era of Heparin treatment are still alive today, which in part also explains this trend toward superficial cases.

HEALING AND SURVIVAL

The healing prognosis for patients in the community is rather poor, with expected 12 weeks healing results often lower than 40%. In Scotland, a nationwide randomized study[26] failed to detect any benefit from a nurse-training program. The 12 weeks of healing remained virtually unchanged. Only around 28% of patients healed their ulcers during a follow-up of 21 months.

As a part of the Skaraborg study, we made a 5-year follow-up of all patients to assess healing and survival.[6] Venous ulcers had a bad outcome: Only 54% had healed, 44% without recurrence and 10% with a recurrence, during the period. At that time most venous ulcers were treated with conservative compression bandaging usually by district nurses in the community. Patients with deep vein incompetence had a significantly worse outcome compared with patients with superficial venous incompetence alone. Regarding patient survival, we found that the patients with venous ulcers had a 5-year survival equal to an age- and sex-matched control population (Figure 3-4).

FUTURE TRENDS

A more holistic approach in Skaraborg, Sweden, with a multidisciplinary approach including interventions such as intense use of varicose vein

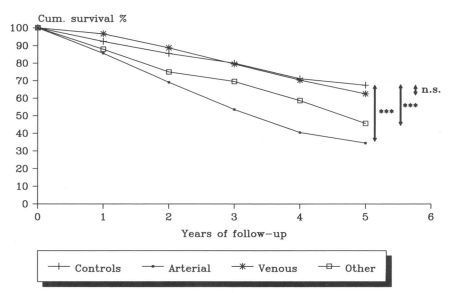

FIGURE 3-4 Life table of the 5-year survival for patients with leg ulcers of various origin. Wicoxon (Gehan) test *** p * .014 0.001, n.s. p = 0.53.[6]

surgery and arterial reconstruction, has resulted in lowered point prevalence both for venous and arterial ulcers and has created a marked shift of the etiologic spectrum, according to preliminary data.[12,27] These data were the result of a repeat of a cross-sectional study in 2002 comparing the data with a similar study performed in the area in 1988,[1,2,3] showing a 46% reduction in patients with venous ulcers.[27] The point prevalence for venous ulcers has been reduced from 0.16% down to 0.09%. Preliminary data from a repeat large size (10,000 participants) study validated a population questionnaire study also pointing strongly in the same positive direction. It appears that men now, to a greater extent, are seeking professional help and that varicose ulcers now outnumber ulcers with deep vein incompetence, leaving room for further use of surgical treatment. There still appears to be room for further reductions of venous ulcer prevalence, at least within the Swedish population. It appears that a radical change in management strategy does result in improved healing and a lowered rate of recurrent ulcers, especially when venous and arterial ulcers are concerned. Changing the course of ulcers of multifactorial or other origins seems, however, to be more difficult.

Leg ulcers are common in most populations, and the prevalence rises with age. Due to the expected increasing proportion of elderly in most populations, an increased prevalence of chronic venous leg ulcers is expected, unless radical changes of management are undertaken. If we continue to treat the majority conservatively with compression regimes, few, if any, will be cured, and the majority will experience recurrent ulceration over time and some will never heal. We have strong evidence that superficial venous surgery reduces the risk of recurrence of venous ulcers significantly within 1 year.[28] Surgically untreated superficial venous and incompetence has been shown to be a risk factor for venous ulcer recurrence,[29] which is in line with our experience in Skaraborg. So, by increased use of leg ulcer diagnosis and tailored treatment, including superficial and perforator surgery, for each individual it is likely that the scenario can be substantially improved, and the prevalence of leg ulcers can actually be lowered worldwide, not only in Sweden.

REFERENCES

1. Nelzén O, Bergqvist D, Hallböök T, Lindhagen A. Chronic leg ulcers: An underestimated problem in primary health care among elderly patients. *J Epidemiol Community Health.* 1991;45:184–187.
2. Nelzén O, Bergqvist D, Lindhagen A. Leg ulcer etiology—A cross-sectional population study. *J Vasc Surg.* 1991;14:557–564.
3. Nelzén O, Bergqvist D, Lindhagen A. Venous and non-venous leg ulcers: Clinical history and appearance in a population study. *Br J Surg.* 1994;81:182–187.
4. Nelzén O, Bergqvist D, Lindhagen A. The prevalence of chronic lower-limb ulceration has been underestimated: Results of a validated population questionnaire. *Br J Surg.* 1996;83:255–258.

5. Nelzén O, Bergqvist D, Fransson I, Lindhagen A. Prevalence and aetiology of leg ulcers in a defined population of industrial workers. *Phlebology.* 1996;11:50–54.
6. Nelzén O, Bergqvist D, Lindhagen A. Long term prognosis for patients with chronic leg ulcers: A prospective cohort study. *Eur J Vasc Endovasc Surg.* 1997;13:500–508.
7. Nelzén O. Patients with chronic leg ulcers: Aspects of epidemiology, aetiology, clinical history, prognosis and choice of treatment. Comprehensive Summaries of Uppsala dissertations from the Faculty of Medicine 664. *Uppsala: Acta Universitatis Upsaliensis,* 1997;1:1–88.
8. Graham ID, Harrison MB, Nelson EA, Lorimer K, Fisher A. Prevalence of lower-limb ulceration: A systematic review of prevalence studies. *Advances in Skin & Wound Care.* 2003;16:305–316.
9. Baker SR, Stacey MC, Jopp-McKay AG, Hoskin SE, Thompson PJ. Epidemiology of chronic venous ulcers. *Br J Surg.* 1991;78:864–867.
10. Cornwall JV, Dore CJ, Lewis JD. Leg ulcers: Epidemiology and aetiology. *Br J Surg.* 1986;73:693–696.
11. Moffatt CJ, Franks PJ, Doherty DC, Martin R, Blewett R, Ross F. Prevalence of leg ulceration in a London population. *Q J Med.* 2004;97:431–437.
12. Jakobsson A, Nelzén O, Fransson I. (Abstract) Dramatic changes in the aetiological spectrum of leg ulcers: Results from a cross-sectional based study. *Int Angiol.* 2005;24(Suppl. 1):92.
13. Bobek K, Cajzl L, Cepelak V, Slaisova V, Opatzny K, Barcal R. Étude de la frequénce des maladies phlebologiques et de l'influence de quelques facteurs étiologiques. *Phlebologie.* 1966;19:227–230.
14. Widmer LK. *Peripheral Venous Disorders. Basle Study III.* Bern: Hans Huber, 1978.
15. Fischer H. *Venenleiden: Eine Repräsentative Untersuchung in der Bevölkerung der Bundesrepublik Deutschland (Tübinger-studie).* München: Urban Schwartsenberg, 1981.
16. Heit JA, Rooke TW, Silverstein MD, et al. Trends in the incidence of venous stasis syndrome and venous ulcer: A 25-year population-based study. *J Vasc Surg.* 2001;33:1022–1027.
17. Walker N, Rodgers A, Birchall N, Norton R, MacMahon S. The occurrence of leg ulcers in Auckland: Results of a population-based study. *New Zealand Med J.* 2002;1151:159–162.
18. Evans CJ, Fowkes FGR, Ruckley CV, Lee AJ. Prevalence of varicose veins and chronic venous insufficiency in men and women in the general population: Edinburgh Vein Study. *J Epidemiol Community Health.* 1999;53:149–153.
19. Rabe E, Pannier-Fischer F, Bromen K, et al. Bonner Venenstudie der Deutschen Gesellschaft für Phlebologie. *Phlebologie.* 2003;32:1–14.
20. Criqui MH, Jamosmos M, Fronek A, et al. Chronic venous disease in an ethnically diverse population: The San Diego Population Study. *Am J Epidemiol.* 2003;158:448–456.
21. Walsh JC, Bergan JJ, Beeman S, Comer TP. Femoral venous reflux abolished by greater saphenous vein stripping. *Ann Vasc Surg.* 1994;8:566–570.
22. Magnusson M, Nelzén O, Volkmann R. Leg ulcer recurrence and its risk factors: A duplex ultrasound study before and after vein surgery. *Eur J Vasc Endovasc Surg.* 2006;32:453–461.
23. Skene AI, Smith JM, Dore CJ, Charlett A, Lewis JD. Venous leg ulcers: A prognostic index to predict time to healing. *Br Med J.* 1992;305:1119–1121.
24. Milne AA, Ruckley CV. The clinical course of patients following extensive deep venous thrombosis. *Eur J Vasc Surg.* 1994;8:56–59.
25. Saarinen J, Sisto T, Laurikka J, Salenius J-P, Tarkka M. Late sequelae of acute deep venous thrombosis: Evaluation five and ten years after. *Phlebology.* 1995;10:106–109.

26. Scottish Leg Ulcer Trial Participants. Effect of a National Community Intervention Programme on Healing Rates of Chronic Leg Ulcer: Randomized Controlled Trial. *Phlebology.* 2002;17:47–53.
27. Jakobsson A, Nelzén O, Fransson I. (Abstract) Leg ulcer point prevalence in a defined geographical population: A repeat cross-sectional study. *Int Angiol.* 2005;24(Suppl. 1): 91–92.
28. Barwell JR, Davies CE, Deacon J, et al. Comparison of surgery and compression with compression alone in chronic venous ulceration (ESCHAR study): Randomised controlled trial. *Lancet.* 2004;363:1854–1859.
29. Gohel MS, Taylor M, Earnshaw JJ, Heather BP, Poskitt KR, Whyman MR. Risk factors for delayed healing and recurrence of chronic venous leg ulcers. An analysis of 1,324 legs. *Eur J Vasc Endovasc Surg.* 2005;29:74–77.

4

LOWER EXTREMITY ULTRASOUND EVALUATION AND MAPPING FOR EVALUATION OF CHRONIC VENOUS DISEASE

STYLIANOS K. TSINTZILONIS AND NICOS LABROPOULOS

Duplex ultrasound scanning is considered the test of choice among different diagnostic tools for detailed evaluation of chronic venous insufficiency. It combines color flow imaging with B-mode and pulsed Doppler. It is very well documented that duplex ultrasound scanning is the most reliable, cost-effective, quick, and noninvasive method to identify venous reflux. This type of scanning is also able to detect venous obstruction and assess recurrence of both reflux and thrombosis.

Most recent studies have shown that the superficial and perforating veins are involved in the majority of patients.[1] However, several reports document that the type and duration of treatment and the rate of recurrence of venous ulcers are significantly different when the deep venous system is involved.[2,3] Therefore, knowledge of the venous anatomy, pathology, and their variations is an important prerequisite for the person who performs the duplex ultrasound in order to identify precisely the etiology and the anatomic distribution of the venous malfunction in each patient.

REQUIREMENTS AND SETTINGS

The test starts by having the patient stand with support of his/her bodyweight on the contralateral limb. The limb under examination should be relaxed with the knee slightly flexed. It has been shown that this is the best position for testing reflux, and the supine position is no longer recommended.[4] If the patient cannot stand, the sitting position is preferred.

Superficial veins can be imaged with different transducers based on the depth of their location, especially in patients with obese and edematous limbs. Superficial veins within 1 cm in the subcutaneous fat can be evaluated with a 10-MHz transducer. A 3-MHz transducer is used for evaluation of veins deeper than 6 cm. The veins to be examined are identified with B-mode imaging. Many of the modern ultrasound machines have transducers with a range of frequencies that can adjust according to the depth, so it is not necessary to change the transducer often.

Manual compression is used to demonstrate reflux by rapidly squeezing and suddenly releasing different muscle beds above and below the vein in question. At the groin, Valsalva maneuver can also be used. In obese patients or in patients with edema, dorsi/plantar flexion may be used if the compression is insufficient to augment the flow. However, when precise measurements are needed regarding flow or duration of reflux, then automatic rapid inflation-deflation cuffs should be used.

All main veins should always be imaged along their entire length. Nonsaphenous, accessory veins and major tributaries when incompetent should be followed, and their course and connections should be noted. It is important, for the study to be reliable and accurate, that the examiner is aware of the venous anatomy and the different variations such as duplications, hypoplasia, and segmental aplasia (Figure 4-1). Duplication is usually seen in the popliteal vein in up to 40% of limbs and in 25% to 30% in the femoral vein. Pathologic findings may be detected in one of the two veins. Duplications of the GSV and SSV are found in <3%, while hypoplasia and aplasia are frequently seen.

Throughout the exam, color flow imaging is used on top of the B-mode ultrasound. Initially, the artery is identified by a red color on the monitor. Then and while applying compression, the adjacent veins are identified by a blue color. Blue color indicates that flow is in the opposite direction than the flow in the artery. If no color is noticed in the vein after the release of the compression, then no reflux is present, implying that the valves are competent. The appearance of red color in the vein indicates the presence of reflux. The Doppler waveform will document the presence of reflux. Retrograde flow with short duration is seen behind competent valves, and it is normal.[4,5,6] This is explained as a result of reverse flow just before valve closure, and this should not last more than 0.5 second. Reverse flow greater than 0.5 second in the vein under investigation is considered abnormal.[4] Several studies were done to define significant reflux on different veins. The results of the most recent and complete study defined reflux as a retrograde flow lasting >1,000 ms in the common femoral, femoral, and pop-

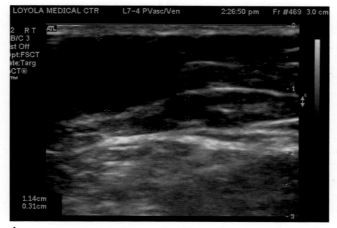
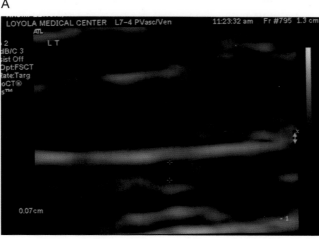

FIGURE 4-1 Anatomic variations in the superficial veins. **A.** Dilated accessory saphenous vein and normal GSV in the upper thigh. The accessory vein extended to the midcalf and had prolonged reflux. **B.** Hypoplasia of the GSV in the lower thigh that measured only 0.7 mm. A dilated accessory vein starting at midthigh had replaced the GSV to the calf where they rejoined.

liteal veins; >500 ms in the superficial veins, deep femoral veins, and deep calf veins; and >350 ms in the perforating veins (Figure 4-2).[4]

Reflux also can be distinguished into *segmental* when it is confined to a single venous segment and *multisegmental* when it is confined to more than one venous segment.[7] The location of reflux and etiology are described with the CEAP classification.[8]

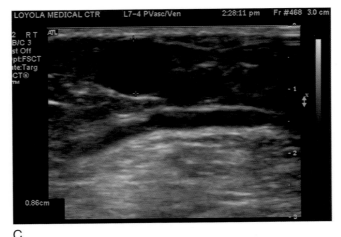

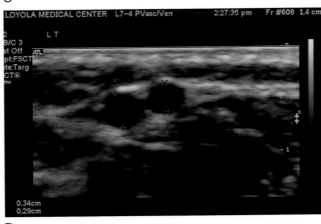

FIGURE 4-1 *Continued*
C. Aplasia of the GSV at the knee. An incompetent dilated accessory vein is seen outside the saphenous canal where the GSV is absent. **D.** Duplication of the SSV from the upper calf to midcalf. Both veins are seen inside the saphenous canal.

SUPERFICIAL VEINS

Imaging of the GSV starts at the groin where the saphenofemoral junction (SFJ) is identified medial to the common femoral artery. The terminal valve in the GSV is located 1–2 mm distal to the SFJ; the preterminal valve is located 2 cm more distally. Between those two valves, there are tributaries that join the GSV, which are often associated with pathology. They can have reflux together with GSV or be independent to it. In cases of deep

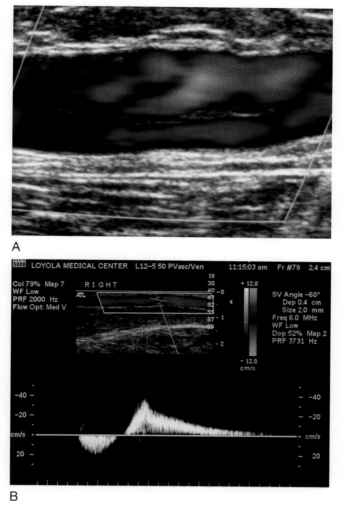

FIGURE 4-2 Reflux in different veins in the lower extremity. **A.** The GSV in the thigh is dilated, has old thrombosis, is partially recanalized, and has reflux. **B.** The anterior accessory saphenous vein is dilated and has reflux lasting about 3 seconds.

vein obstruction, they are dilated and act as collateral pathways. For example, reflux in the accessory saphenous vein without the involvement of GSV occurs in about 9% of patients. This is important because only the accessory vein should be treated, and the GSV can be spared.[9] In the thigh, GSV is always identified in the "saphenous eye," which the fascial sheets create.[10,11] The upper echogenic layer seen in a transverse scan is the super-

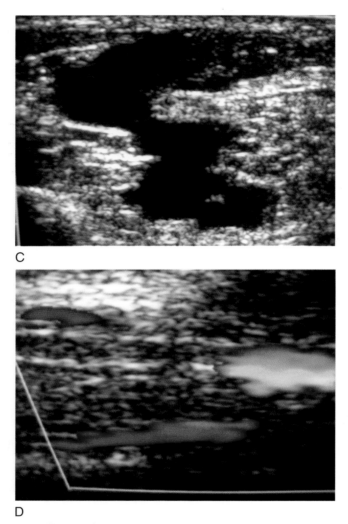

FIGURE 4-2 *Continued*
C. A lower calf medial perforator vein joins a tributary of the posterior arch vein and the posterior tibial vein. It is very dilated and has reflux. Its largest diameter was 6.7 mm.
D. The femoral vein in the midthigh is partially recanalized and has reflux.

ficial fascia, and the lower echogenic layer is the muscular fascia. Any vein outside the saphenous eye is not the GSV, but an accessory saphenous, or a tributary. In the upper thigh, there is an accessory saphenous vein that creates a second "eye" and can easily be distinguished from the GSV because it lies anteriolateral and parallels the course of the femoral artery and vein. In the knee area, it is difficult to recognize the "saphenous eye," and the GSV is usually identified in a triangle formed by the tibia, medial gastrocnemius muscle, and fascial sheet. Occasionally, a large tributary arises above the knee and may be mistakenly identified as the GSV. In general, the GSV in the thigh and the leg is often followed by parallel veins that may be confused with the GSV. These veins lie subcutaneously and then travel through the superficial fascia to enter the saphenous compartment. Most of the patients have reflux in GSV and its tributaries. The anterior and medial accessory veins in the thigh and the posterior and anterior arch vein in the calf are the tributaries most commonly incompetent.[12]

The SSV is found at the popliteal fossa within the triangular fascia over the medial and lateral heads of the gastrocnemius muscle.[13] The vein lies in the leg in a fascial compartment that is identified with the ultrasound as an "eye" similar to the GSV eye. The SSV may terminate at different levels of the popliteal vein, may extend to the thigh, or may join the GSV via the Giacomini vein. Termination of SSV above the popliteal vein is now collectively termed as "thigh extension of SSV."[14] Knowledge of the variable anatomy of the SSV is very important, as the intervention may be modified according to the patterns of reflux.[15,16] There is a tributary called "popliteal fossa perforating vein" that may be misidentified as the SSV because it runs parallel to the SSV and joins separately with the popliteal vein.[17]

Reflux in nonsaphenous veins is found in 10% of patients with CVD.[18] Most of these patients are usually multiparous women. The location and origin of these veins are variable. Often they are found as an extension of the pelvic veins in the thigh (vulvar and gluteal), lateral and posterior thigh (lateral thigh system, posterolateral perforator veins, and sciatic nerve veins), popliteal fossa (vein of the popliteal fossa, tibial nerve veins, atypical perforators, and tributaries), and lateral knee and calf (atypical tributaries and perforators). About 90% of these patients present with CVD classes 1 to 3, and skin damage is present in 10%. Inaccurate clinical evaluation has been found in one-third of cases with this type of reflux. Therefore, DU is very important in identifying reflux in these veins because there are many different ones and there is great variability in the patterns of reflux.

PERFORATING VEINS

The role of the perforator veins is controversial in the development of venous ulcers in patient with CVD. They are seen in different planes with the ultrasound because the direction of flow is different in those veins. Transverse and oblique scanning along the course of the superficial veins is more appropriate. The fascia has to be identified by the examiner. The fascia appears white due to high collagen content, and the perforators are identified as they travel through the fascia. About 150 perforating veins have been identified, and they are grouped based on their location.[19] However, only about 20 of them have been shown to become incompetent and are frequently found in patients with ulceration. Most of these veins are found in the medial and posterior calf, and they are connecting tributaries of the GSV and SSV with muscular and deep axial veins.

The examination starts from the medial malleolus going upward to the knee. The posterior arch and GSV are scanned on a transverse plane for incompetent perforator veins. Localized varicose veins are highly suspicious for an underlying incompetent perforator, and that region requires detailed evaluation. Then, the SSV is tested starting from the lateral malleolus and ending at its termination in the fossa or higher up. In order to determine normal valvular function, distal compression is applied for augmentation of flow, and the perforators are tested in combination with the deep and superficial veins. The flow in the calf perforator veins may be bidirectional. The net outward flow from deep to superficial veins is measured to determine reflux.[20] In patients with significant edema, dorsi/plantar flexion is performed to determine reflux in these veins. This is important because manual compression alone may not show reflux in dilated perforator veins.

DEEP VEINS

The common femoral vein is tested from the inguinal ligament to its union with the femoral and the deep femoral veins. Any evidence of continuous flow, poor augmentation, and asymmetrical waveforms to the contralateral CFV should alert the examiner to extend the imaging in the iliac veins and the inferior vena cava. The CFV should also be examined above and below the SFJ in order to determine the presence or absence of reflux at the SFJ level and test the effect of the GSV on the CFV.[21] The femoropopliteal veins are examined along their length from the groin to the tibioperoneal trunk. Duplications and any pathology in them are noted. In the popliteal fossa also, the gastrocnemial veins are being evaluated, particularly those in the medial head because they are more often implicated in CVD. Unusual veins if present—such as the popliteoprofunda vein, persistent sciatic vein, and veins from the sciatic nerve and other muscular veins—are being evaluated as well. The popliteal vein is examined carefully above and below its union with the SSV and gastrocnemius veins for the reasons mentioned previously for the CFV.

In the calf, the posterior tibial, peroneal, and soleal veins are imaged because the frequency of deep vein thrombosis in these veins is high, and this in turn can lead to post-thrombotic reflux. The anterior tibial veins are imaged only in the presence of CVD signs and symptoms in the anterior aspect of the calf.

ULCER BED

The ulcer bed should always be evaluated for incompetent veins including the area surrounding the ulcer as far as 2 cm from the periphery of the ulcer. In ulcers over the medial malleolus, the veins that need to be tested are the GSV, the posterior arch, posterior tibial, and the peroneal veins. In ulcers at the lateral malleolus, the veins are the SSV and the peroneal veins. In ulcers on the anterior aspect of the leg, the anterior arch of the GSV and the anterior tibial veins need to be assessed. Sterile technique can be used for the evaluation of the ulcer bed. The surface of the ulcer may be covered with a transparent sterile dressing, or the probe can be covered with a sterile glove that is filled with ultrasonic gel.[22,23] The prevalence of deep venous reflux has been found to be higher in patients with venous ulcers when compared to patients with less severe forms of CVD. In addition, it was found that the ration of outward flow in these veins is longer in patients with ulcers compared to patients without ulcers.

ULTRASOUND FINDINGS AND IMPLICATIONS FOR TREATMENT

The use of DU has contributed to a better understanding of the mechanisms of CVD and has changed dramatically the management of patients. The overall involvement of the superficial veins was 88% in patients with venous ulceration.[24] Reflux confined to the superficial and perforator veins alone was found in 40% to 50%,[24] and treatment of these veins was found to be successful in healing the ulcers in almost 90% of patients.[25-27] When the deep veins are involved, particularly in patients with previous thrombosis, the ulcer healing rate is lower, and the recurrence rate is higher.[2,3] The presence of deep vein reflux is higher in patients with venous ulcers compared with patients with less severe symptoms of CVD.[7,28,29] Overall, the involvement of deep venous insufficiency in venous ulcer disease has been shown to be up to 56%.[24] Also, a documented episode of thrombosis is found in one-third of patients with ulcers, which is significantly higher compared to those without an ulcer (Figure 4-3).[24,30] Reflux confined to the deep veins alone is found in 12%, while obstruction alone is detected in about 2%.[24,30] In a recent prospective multicenter study, of the 2.6% (21/799) of limbs that appeared to be venous in origin, no venous

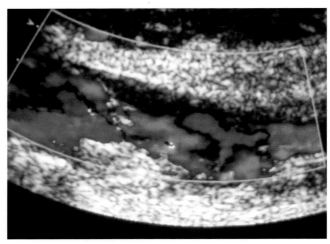

FIGURE 4-3 Femoral and popliteal vein old thrombosis (two previous episodes in the same site) with partial recanalization and severe reflux (>5 seconds). The patient had an ulcer in the lateral malleolus and also reflux in the SSV from the popliteal fossa to the ulcer. Ligation and stripping of the SSV healed the ulcer in 2 months. The duration of reflux in the femoral and popliteal veins was reduced to 2 seconds. However, the ulcer recurred after 5 months.

abnormality was found.[31] These patients had many different pathologies such as carcinomas, infections, vasculitis, hematologic abnormalities, etc.

DU MAPPING FOR GUIDING TREATMENT

Recently, with the evolution of endovenous ablation techniques such as the radiofrequency, laser, and foam sclerotherapy, DU is of major importance, since these methods are being performed with DU guidance.[32–34] DU is used to select the patients during the procedures and follow-up. In this way, only the involved veins are being treated under direct vision. The outcome of the treatment in these veins and its effect in the untreated veins are evaluated. Residual from recurrent disease is identified, and also thrombosis and neovascularization are detected as well.[35,36]

REFERENCES

1. Bergan JJ, Schmid-Schonbein GW, Smith PD, Nicolaides AN, Boisseau MR, Eklof B. Chronic venous disease. *N Engl J Med.* 2006;355:488–498.
2. Gloviczki P, Bergan JJ, Rhodes JM, et al. Mid term results of endoscopic perforator vein interruption for chronic venous insufficiency: Lessons learned from the North American

subfascial endoscopic perforator surgery registry. The North American Study Group. *J Vasc Surg.* 1999;29:489–502.
3. Scriven JM, Bianchi V, Hartshorne T, et al. A clinical and hemodynamic investigation into the role of calf perforating vein surgery in patients with venous ulceration and deep venous incompetence. *Eur J Endovasc Surg.* 1998;16:148–152.
4. Labropoulos N, Tiongson J, Pryor L, et al. Definition of venous reflux in lower extremity veins. *J Vasc Surg.* 2003;38:793–798.
5. Vasdekis SN, Clarke GH, Nicolaides AN. Quantification of venous reflux by means of DU. *J Vasc Surg.* 1989;10:670–677.
6. Van Bemmelen PS, Bedford G, Beach K, Strandness DE. Quantitative segmental evaluation of venous valvular reflux with duplex ultrasound scanning. *J Vasc Surg.* 1989;10:425–431.
7. Labropoulos N, Delis K, Nicolaides AN, Leon M, Ramaswami G. The role of the distribution and anatomic extent of reflux in the development of signs and symptoms in chronic venous insufficiency. *J Vasc Surg.* 1996;23:504–510.
8. Eklof B, Rutherford RB, Bergan JJ, et al. American Venous Forum International Ad Hoc Committee for Revision of the CEAP Classification. Revision of the CEAP classification for chronic venous disorders: Consensus statement. *J Vasc Surg.* 2004;40:1248–1252.
9. Labropoulos N, Leon L, Engelhorn CA, et al. Sapheno-femoral junction reflux in patients with a normal saphenous trunk. *Eur J Vasc Endovasc Surg.* 2004;28:595–599.
10. Caggiati A, Ricci S. The long saphenous vein compartment. *Phlebology.* 1997;12:107–111.
11. Caggiati A. Fascial relationships of the long saphenous vein. *Circulation.* 1999;100:2547–2549.
12. Labropoulos N, Kang SS, Mansour MA, Giannoukas AD, Buckman J, Baker WH. Primary superficial vein reflux with competent saphenous trunk. *Eur J Vasc Endvasc Surg.* 1999;18:201–206.
13. Caggiati A. Fascial relationships of the short saphenous vein. *J Vasc Surg.* 2001;34:241–246.
14. Caggiati A, Bergan JJ, Gloviczki P, Eklof B, Allegra C, Partsch H. International Interdisciplinary Consensus Committee on Venous Anatomical Terminology. Nomenclature of the veins of the lower limb: Extensions, refinements, and clinical application. *J Vasc Surg.* 2005;41:719–724.
15. Labropoulos N, Giannoukas AD, Delis K, et al. The impact of isolated lesser saphenous vein system incompetence on clinical signs and symptoms of chronic venous disease. *J Vasc Surg.* 2000;32:954–960.
16. Delis KT, Knaggs AL, Khodabakhsh P. Prevalence, anatomic patterns, valvular competence, and clinical significance of the Giacomini vein. *J Vasc Surg.* 2004;40:1174–1183.
17. Delis KT, Knaggs AL, Hobbs JT, Vandendriessche MA. The nonsaphenous vein of the popliteal fossa: Prevalence, patterns of reflux, hemodynamic quantification, and clinical significance. *J Vasc Surg.* 2006;44:611–619.
18. Labropoulos N, Tiongson J, Pryor L, et al. Nonsaphenous superficial vein reflux. *J Vasc Surg.* 2001;34:872–877.
19. Limborgh J van. L'anatomie du système veineux de l'extremité inférieure en relation avec la pathologie variqueuse. *Folia Angiol.* 1961;8:240–257.
20. Labropoulos N, Mansour MA, Kang SS, Gloviczki P, Baker WH. New insights into perforator vein incompetence. *Eur J Vasc Endovasc Surg.* 1999;18:228–234.
21. Labropoulos N, Tassiopoulos AK, Kang SS, Mansour MA, Littooy FN, Baker WH. Prevalence of deep venous reflux in patients with primary superficial vein incompetence. *J Vasc Surg.* 2000;32:663–668.

22. Hanrahan LM, Araki CT, Rodriguez AA, Kechejian GJ, LaMorte WW, Menzoian JO. Distribution of valvular incompetence in patients with venous stasis ulceration. *J Vasc Surg.* 1991;3:805–812.
23. Labropoulos N, Giannoukas AD, Nicolaides AN, Leon M, Ramaswami G, Volteas N. New insights into the pathophysiologic condition of venous ulceration with color flow duplex imaging: Implications for treatment? *J Vasc Surg.* 1995;22:45–50.
24. Tassiopoulos AK, Golts E, Oh DS, Labropoulos N. Current concepts in chronic venous ulceration. *Eur J Vasc Endovasc Surg.* 2000;20:227–232.
25. Bello M, Scriven M, Hartshorne T, et al. Role of superficial venous surgery in the treatment of venous ulceration. *Br J Surg.* 1999;86:755–759.
26. Darke SG, Penfold CAD. Venous ulceration and saphenous ligation. *Eur J Vasc Surg.* 1992;6:4–9.
27. Burnand KG, Thomas ML, O'Donnell TF Jr, Browse NL. Relation between postphlebitic changes in the deep veins and results of surgical treatment of venous ulcers. *Lancet.* 1976;1:936–938.
28. Labropoulos N. CEAP in clinical practice. *Vasc Surg.* 1997;31:224–225.
29. Labropoulos N. Clinical correlation to various patterns of reflux. *Vasc Surg.* 1997;31: 242–248.
30. Labropoulos N, Patel PJ, Tiongson JE, Pryor L, Leon LR, Tassiopoulos AK. Patterns of venous reflux and obstruction in patients with skin damage due to chronic venous disease. *Vasc Endovasc Surg.* 2007;41:33–40.
31. Labropoulos N, Manalo D, Patel N, Tiongson JE, Pryor L, Giannoukas AD. Uncommon leg ulcers. *J Vasc Surg.* 2007;45:568–573.
32. Lurie F, Creton D, Eklof B, et al. Prospective randomised study of endovenous radiofrequency obliteration (closure) versus ligation and vein stripping (EVOLVeS): Two-year follow-up. *Eur J Vasc Endovasc Surg.* 2005;29:67–73.
33. Almeida JI, Raines JK. Radiofrequency ablation and laser ablation in the treatment of varicose veins. *Ann Vasc Surg.* 2006;20:547–552.
34. Guex JJ. Foam sclerotherapy: An overview of use for primary venous insufficiency. *Semin Vasc Surg.* 2005;18:25–29.
35. Perrin MR, Labropoulos N, Leon LR Jr. Presentation of the patient with recurrent varices after surgery (REVAS). *J Vasc Surg.* 2006;43:327–334.
36. van Rij AM, Jones GT, Hill GB, Jiang P. Neovascularization and recurrent varicose veins: More histologic and ultrasound evidence. *J Vasc Surg.* 2004;40:296–302.

5

Hypercoagulable States Associated with Chronic Venous Insufficiency

Jose L. Trani Jr. and
Jeffrey H. Lawson

INTRODUCTION

Venous disorders affect a significant percentage of the U.S. population, as many as one in 1,000 persons annually.[1] For many patients with deep venous thrombosis (DVT), the clinical course following an initial thrombotic event may be complicated by further manifestations of venous disease formerly referred to as the *post-thrombotic syndrome.* In patients who progress to *chronic venous insufficiency (CVI)*, the currently accepted terminology, approximately 4% will progress to develop venous ulceration.[2] Chronic venous insufficiency is a direct result of sustained venous hypertension, as described elsewhere in this volume. Over time, the persistence of venous hypertension leads to the most severe form of chronic venous insufficiency: chronic venous ulceration. A combination of factors including high thrombotic recurrence rates, ineffective anticoagulant treatment, and a poorly understood disease process makes venous ulceration difficult to both prevent and manage.[3] Identified causes in the development of chronic venous insufficiency include a family history of venous insufficiency, especially of maternal origin, and a history of vigorous exercise. Subsequent chronic venous ulceration includes persistent untreated primary venous insufficiency and repetitive post-thrombotic episodes of recurrent DVT.[3,4]

The CVI syndrome is a combination of sequelae of a deep venous thrombosis. Studies demonstrate that this syndrome may occur in more than 50% of patients who have symptomatic deep venous thrombosis.[5] The underlying pathophysiology of the thrombotic event leading to post-thrombotic syndrome is the resultant scarring from thrombus resolution that leads to valve incompetence and subsequent distal venous hypertension.[6] At least one risk factor is identified in the majority of patients who have a symptomatic deep venous thrombosis. Many patients have multiple risk factors that contribute to their CVI.[7] The inciting procoagulant etiologies for developing thromboses may be divided into inherited and acquired thrombophilias (Table 5-1). The overall prevalence of thrombophilia, both inherited and acquired, in patients who proceed to develop CVI is reported to be as high as 41%. This is similar to the reported prevalence of thrombophilia in patients with an antecedent deep vein thrombosis, making thrombophilias an important contributors to CVI.[8] The remainder of the chapter will be devoted to a discussion of the major inherited and acquired thrombophilias, as well as their relationship to CVI in general and venous ulceration in particular.

INHERITED THROMBOPHILIA CONDITIONS

Inherited thrombophilias are characterized by a genetic predisposition to venous thrombotic events. They are frequently the cause of vascular thrombosis in younger individuals (less than 45–50 years of age) and may contribute to deep venous thrombosis at any age. Attempts have been made to quantify the risk of developing a deep venous thrombus in patients with inherited thrombophilias. Between 8% and 13% of patients in an unselected patient population who present with a first deep venous thrombus will have a genetic anticoagulant deficiency. Early studies took into account all major coagulation disorders with the exception of factor V

TABLE 5-1 Inherited and Acquired Thrombophilias That Increase the Risk of Developing a Deep Vein Thrombosis

Inherited	Acquired
Factor V Leiden mutation	antiphospholipid syndrome
Activated protein C mutations	anticardiolipin antibodies
Antithrombin mutations	Lupus anticoagulant antibodies
Protein C mutations	Hyperhomocysteinemia
Protein S mutations	Activated protein C resistance
Elevated factor VIII	Malignancy
Prothrombin gene mutation	Autoimmune disorders, etc.
Dysfibrinogenemia	

Leiden mutation.[9] For this particular disorder, two separate studies have determined that in patients with a first-time deep venous thrombosis or pulmonary embolism, 12% to 19% of patients were positive for the factor V Leiden mutation.[9]

Factor V Leiden Mutation

Factor V Leiden mutation is the most common prothrombotic genetic defect. It is present in approximately 5% of all Caucasians, and it accounts for 40% to 50% of all cases of inherited thrombophilia. Biochemically, the etiology of the factor V Leiden mutation is a point mutation resulting in an arginine to glutamine substitution at position 506 in coagulation factor V. This alteration results in an inability of activated protein C to cleave factor Va, promoting coagulation via continued thrombin production.[10] The factor V Leiden mutation is among the most prevalent inherited thrombophilia condition identified in patients with a first deep venous thrombosis. MacKenzie and colleagues also identified the factor V Leiden mutation as a specific thrombophilia condition associated with chronic venous ulceration in the setting of a previous deep vein thrombosis.[8] Separate studies have determined that this mutation is present in 36% of all patients with post-thrombotic leg ulcers.[10] Post-thrombotic venous ulceration accounts for 30% to 60% of all lower extremity venous ulcerations. Initial screening for factor V Leiden mutations requires measurements of the aPTT with and without the addition of a standard amount of activated protein C in order to produce an activated protein C ratio. Second-generation tests included diluting the patient plasma in order to produce a more standard assay. In the current era, most authorities recommend using a PCR-based screen to identify the factor V Leiden mutation.

As indicated previously, the factor V Leiden mutation has been detected in up to 20% of individuals with a venous thrombotic event.[11] Also, a separate study that evaluated patients with chronic venous ulceration determined that 13% of patients with a venous ulcer also had a factor V Leiden mutation.[8] As this mutation is present in 5% of the population, a 4–5-fold risk of developing a deep venous thrombosis is conferred among carriers (Table 5-2).

Although the factor V Leiden mutation is the most common form of activated protein C resistance, other mutations in factor V may produce similar procoagulant effects. In addition, activated protein C resistance may be an acquired condition, most commonly associated with the antiphospholipid syndrome.

Antithrombin Deficiency

Antithrombin (formerly known as antithrombin III) is a vitamin K-independent glycoprotein that functions to inhibit thrombin as well as activated factors of the intrinsic clotting cascade (including factors IXa and

TABLE 5-2 Comparison of the Prevalence of Various Thrombophilic Conditions in the General Population, and Among Patients with Deep Vein Thrombosis and Chronic Venous Ulcers. Data Represents a Compilation of Statistics from Multiple Sources.

	General Population	Deep Venous Thrombus	Relative Risk of Developing a DVT	Chronic Venous Ulcer
FVL	2–10%	10–20%	2.2–5.0	8–22%
AT	0.2%	0.5–1%	5.0–8.1	4%
PC	0.2–0.3%	3–3.2%	6.5–7.4	6%
PS	0.2–1%	3–7.3%	2.5–10.4	6%
LA	2–5%	10–15%	3.0	9%
ACL	5–10%	15–20%	3.0	12%
HH	4.5–5%	11–11.7%	2.5–3.0	*

Key: *FVL*, factor V Leiden; *AT*, antithrombin; *PC*, protein C; *PS*, protein S; *LA*, lupus anticoagulant; *ACL*, anticardiolipin antibody; *HH*, hyperhomocysteinemia, * No data available.

Xa). The addition of heparin as a cofactor enhances the activity of antithrombin, causing rapid inactivation of both thrombin and factor Xa. Hereditary antithrombin deficiencies occur in two flavors. In type I, an overall reduction in the production of antithrombin occurs. There have been more than 80 mutations reported that produce type I mutations with most interfering with protein synthesis. Type II deficiencies result in a functional reduction in antithrombin activity through heparin binding site defects, thrombin binding site defects, and pleiotropic defects that can affect both heparin binding and antithrombin activity.[12]

Antithrombin deficiencies may be detected through immunoassays to quantify the amount of antithrombin antigen present. This will establish a type I deficiency. Functional assays are required to detect type II deficiencies. An antithrombin-heparin cofactor assay will measure binding of heparin to lysyl residues on antithrombin through the neutralization of coagulation enzymes. Depending on the substrate, either thrombin or factor Xa, this test detects defects in the catabolism of the respective target. In current clinical practice, the favored test of choice is the antithrombin-heparin cofactor assay that detects inhibition of factor Xa due to its increased specificity.[13]

The risk of developing a deep vein thrombus in patients with an antithrombin deficiency has been prospectively evaluated. It is estimated that this deficiency conferred a risk ratio of 8.1 when compared to patients with no thrombophilia defect.[14] A history of antithrombin deficiency was present in three times as many patients with a history of a deep venous thrombus and presence of chronic venous ulcer than in patients with a chronic venous ulcer and without a history of a deep vein thrombus. Overall, antithrombin deficiency has been reported to be present in approximately 5% of all cases of chronic venous ulcers.[8]

Protein C and S Abnormalities

Proteins C and S are vitamin K–dependent glycoproteins that are synthesized in the liver. Activated protein C functions to inhibit coagulation factors Va and VIIIa. Protein S serves as a cofactor to enhance the functionality of activated protein C. Similar to antithrombin deficiencies, type I and type II mutations may occur for both proteins C and S. Immunoassays are used to quantify the amount of protein C or S present, thereby delineating a type I mutation. Functional studies are required to determine type II deficiencies. Over 160 different mutations have been reported for protein C and over 70 mutations are described for protein S.[11]

The number of patients who have an identified protein C or S deficiency in an unselected group of patients who develop a deep venous thrombotic event is between 6% and 17%. In contrast, the prevalence of these diseases is between 0.5% and 2% in an unselected group of control patients.[8,14,15] Thirteen percent of chronic venous ulcer patients studied by MacKenzie and colleagues were found to have a protein C or S deficiency (6% and 7%, respectively).[8] Although this study represents a single group's experience with a small sample size, protein C and S defects demonstrated the most disproportionate increase in prevalence from patients with a first episode of venous thrombotic event to the development of a chronic venous ulcer, suggesting that patients affected by these diseases may be at greatest risk to progressing to the post-thrombotic syndrome and eventually to a venous ulceration.

Other Inherited Conditions

Elevated factor VIII levels are associated with an approximately 5-fold risk of developing a deep vein thrombosis. Although this is thought to be a heritable condition, no identifiable genetic link has been demonstrated. Factor VIII levels can arise as part of an acute phase response.[9] Prothrombin, also known as factor II, is a vitamin K–dependent glycoprotein that is the precursor to thrombin. A guanine to adenine mutation at position 20210 confers a prothrombotic effect possibly due to increased circulating prothrombin levels. This mutation is associated with an almost 3-fold risk of developing a deep venous thrombosis.[11] Twice as many patients with chronic venous ulcers were positive for this mutation as was detected in the general population without a venous ulcer.[8] Dysfibrinogenemia conditions are ones in which structural defects cause alterations in the cleavage of fibrinogen to fibrin. Although greater than 300 mutations have been identified, half are silent mutations. The remaining conditions are approximately evenly divided between a hypercoagulable and a coagulopathic state.[9]

ACQUIRED THROMBOPHILIA CONDITIONS

A variety of acquired thrombophilia conditions exist that may lead to the venous thrombotic disease. They include processes such as malignancy, surgery, trauma, pregnancy, and oral contraceptive use. Despite the frequency with which these processes contribute to deep vein thrombosis, no evidence is present in the literature to connect a resultant deep vein thrombosis to the pathogenesis of a chronic venous ulcer. In contrast, there are data to implicate acquired conditions such as the antiphospholipid syndrome, hyperhomocysteinemia, and a history of a previous thromboembolism in the development of venous thrombosis and ultimately venous ulceration.

Antiphospholipid Syndrome

The antiphospholipid syndrome is characterized by vascular thrombosis and/or recurrent fetal loss in association with medium to high titers of antibodies to plasma proteins. These antibodies may be directed against phospholipid-dependent cofactors in the coagulation cascade. The three main tests for antiphospholipid antibodies include Lupus Anticoagulant activity, which may be suspected based on an unexplained prolongation of the PTT that is not reversed when the patient's plasma is diluted 1 : 1 with normal plasma; the anticardiolipin antibody; and antib2-glycoprotein I antibody testing. The antiphospholipid syndrome was first described in patients with systemic lupus erythematosus, but has since been shown to occur independent of any underlying disease process. In addition to venous thrombosis, other manifestations of the syndrome include arterial thrombosis, including stroke, transient ischemic attack, multi-infarct dementia, cardiac complications, pulmonary effects, and intra-abdominal pathology, among other processes.

Venous thrombosis is the most common initial clinical presentation of the antiphospholipid syndrome. In a series of 1,000 patients with diagnosed antiphospholipid syndrome, a history of DVT was present in 32% of patients.[16] Bradbury et al. noted in their review of the literature that anticardiolipin antibodies are prevalent in 5% of the Western population, while they were detected in 15% of an unselected group of patients who developed a first-time venous thrombosis, thus instilling a 3-fold increased risk in this population.[11] MacKenzie and colleagues in their examination of the prevalence of thrombophilia in patients with chronic venous ulceration determined that anticardiolipin antibodies were present approximately twice as often in patients with a deep vein thrombosis than in the population in general (20% vs. 10%). Furthermore, in the CVU patient population, anticardiolipin antibodies were present in 14% of their patients. This same study demonstrated a 10% incidence of patients with positive Lupus Anticoagulant activity and a deep venous thrombus. The Lupus Anticoagulant antibody was present in 9% of patients with a chronic venous ulcer. Together, these two antiphospholipid antibodies were present in the greatest number of patients with both a first deep venous thrombosis and venous ulcer.[8]

Hyperhomocysteinemia

Hyperhomocysteinemia may be either a genetic or an acquired characteristic. Homocysteine is an intermediary in the breakdown pathway from methionine to cysteine. Homocysteine may be converted to cysteine, a process known as transsulfuration, through the use of vitamin B6 as a cofactor. Alternatively, homocysteine may also be reconverted, through remethylation, to produce methionine in a process that requires vitamin B12 as a cofactor. Elevated levels of homocysteine may result in thromboembolic complications, osteoporosis, and premature atherosclerosis, among other pathologic processes. The proposed mechanisms by which hyperhomocysteinemia produces a prothrombotic effect include impairing endothelial cell anticoagulant function, increasing procoagulant factors in the clotting cascade, and inhibiting cofactors required for anticoagulant function.[17-19]

The most common genetic form of hyperhomocysteinemia results from the production of a thermo-labile variant of methylene tetrahydrofolate reductace, an enzyme vital to the remethylation pathway. Additional pathway defects of genetic origin have also been identified. The prevalence of any genetic defect leading to hyperhomocysteinemia is approximately 1%.[11] The far more common cause of hyperhomocysteinemia results from a deficiency in B6, B12, or folate, thus restricting cofactors necessary for transsulfuration or transmethylation. Overall, hyperhomocysteinemia is estimated to increase the risk of developing a DVT 2.5- to 2.95-fold.[11,20]

Recurrent Deep Venous Thrombosis

Recurrent deep venous thrombosis is associated with venous ulcerations. It is estimated that the recurrence rate for deep venous thrombosis is between 12.4% and 25% at 5 years. In the largest population of patients to be evaluated, Christiansen and colleagues prospectively followed 474 patients with heterogeneous etiologies of an initial deep vein thrombosis. Within this patient population, 67% had at least one prothrombotic laboratory abnormality during their initial work-up. The risk of recurrence was increased 1.4-fold in this population over those patients without a prothrombotic abnormality. In contrast to other reports, no increased risk of recurrent deep vein thrombosis was bestowed upon patients with either a factor V Leiden mutation or hyperhomocysteinemia. A 1.8-fold risk was seen in patients with protein C, S, and antithrombin deficiencies. Patients with multiple prothrombotic abnormalities demonstrated a higher rate or recurrence than patients with a single abnormality (1.6-fold vs. 1.2-fold).[21]

As indicated previously, the increased risk of recurrent deep venous thrombosis has important implications for the development of CVI. One study found that in an unselected population of patients with a symptomatic DVT, 28% went on to develop CVI, and 9.3% developed major complications. Of all factors analyzed in this study, only a recurrent ipsilateral DVT significantly correlated with development of CVI.[22] Further analysis of this patient population determined that patients with either an underlying cancer or coagulation defect were at greatest risk for the development of recurrent DVT. A history of deep venous thrombosis has been reported to be found in patients with chronic venous ulceration between 17% and 45% of the time.[1] Elsewhere in this volume, Bergan reports only 30% of his ulcer population as having previous DVT. It is this population, however, that experiences recurrent deep venous ulceration.

TREATMENT STRATEGIES

The role of anticoagulation strategies in treating deep venous thrombosis has been extensively evaluated. Most studies intend to use anticoagulation as a means of preventing thrombus extension, a recurrent deep vein thrombosis, or, more importantly, a pulmonary embolism. Current recommendations include anticoagulation with heparin and warfarin in all patients with a deep venous thrombosis. The role of adequate anticoagulation for the prevention of the post-thrombotic syndrome has been evaluated in a limited number of studies and is discussed by Caprini in Chapter 10 of this volume. One study evaluated the quality of anticoagulation and its relationship to the development of the post-thrombotic syndrome. Univariate analysis demonstrated that ipsilateral recurrence, body mass index, and greater than 50% of time spent below the therapeutic target were the only variables that correlated with the development of the post-thrombotic syndrome.[23]

The duration of anticoagulation in patients with heritable thrombophilia conditions but not acquired conditions has been addressed by Greaves, who noted that the risk of subsequent events in patients with inherited thrombophilias should not be considered greater than in patients with no attributable etiology of first venous thrombotic event because both have shown an increased tendency toward thrombosis. The benefit of anticoagulation in preventing subsequent events over time is diminished by the risk of bleeding complications while receiving anticoagulation therapy. There is currently no data to support a prolonged course of anticoagulation for either prevention of a recurrent deep venous thrombosis or prevention of the post-thrombotic syndrome in the setting of an insolated thrombophilia factor. Patients who display two or more inherited thrombophilia defects may derive benefit from anticoagulation for a greater length of time as the risk of either a second thrombotic event or sequellae of thrombosis are so much greater.[24]

The use of elastic compression stockings for the prevention of post-thrombotic syndrome following a deep venous thrombosis has been addressed by three separate studies and is discussed by Partsch in Chapter 7. A meta-analysis of these three studies determined that the use of compression stockings conferred a 54% reduction in the development of the post-thrombotic syndrome. Information on the length of stocking compression required was not adequately addressed in these trials. Finally, the length of time required to obtain this benefit was also not definitively stated strongly enough to make adequate recommendations regarding these garments.[6]

SUMMARY

Thrombophilia conditions can be identified in one-quarter to one-third of patients with a deep venous thrombosis in Western countries. The incidence of CVI in patients with a deep venous thrombosis approaches 30% at 5 years, with those bearing the most severe changes of chronic venous insufficiency, including venous ulceration, reaching 9% during the same time period. Post-thrombotic disease accounts for a significant percentage of patients who develop venous ulcers. Between 25% and 35% of patients with chronic venous ulceration have post-thrombotic disease (Figure 5-1).

Despite the high incidence of thrombophilia in patients who present with a new deep venous thrombosis as well as in patients who ultimately

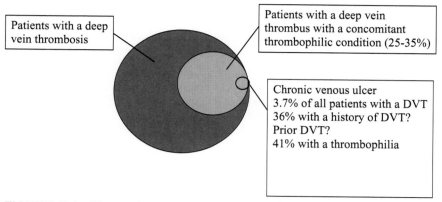

FIGURE 5-1 Diagram demonstrating the contribution of thrombophilia to deep vein thrombosis and chronic venous ulceration.

progress to develop venous ulceration, prolonged anticoagulation is not recommended. Screening patients for thrombophilia conditions should be undertaken only under specific circumstances. Young patients (under 50) without any other identifiable risk factors, such as oral contraceptive use or recent surgery; patients with a family history of deep venous thrombosis; and patients with recurrent venous thrombosis should all receive screening for a hypercoagulable state should receive additional screening consideration. Patients who have more than one thrombophilia condition identified by screening may benefit from prolonged anticoagulation.

REFERENCES

1. Abbade LP, Lastoria S. Venous ulcer: Epidemiology, physiopathology, diagnosis and treatment. *Int J Dermatol.* 2005;44:449–456.
2. Mohr DN, Silverstein MD, Heit JA, Petterson TM, O'Fallon WM, Melton LJ. The venous stasis syndrome after deep venous thrombosis or pulmonary embolism: A population-based study. *Mayo Clinic Proceed.* 2000;75:1249–1256.
3. Pappas P, Lal B, Cerveira J, Duran W. The pathophysiology of chronic venous insufficiency. In: Rutherford R, ed. *Vascular Surgery*, 6th ed. Two vols. Elsevier Saunders, New York, 2005: Vol. 2, 2220–2229.
4. Berard A, Abenhaim L, Platt R, Kahn S, Steinmetz O. Risk factors for the first-time development of venous ulcers of the lower limbs: The influence of heredity and physical activity. *Angiology.* 2002;53:647–657.
5. Schulman S, Lindmarker P, Holmstrom M, et al. Post-thrombotic syndrome, recurrence, and death 10 years after the first episode of venous thromboembolism treated with warfarin for 6 weeks or 6 months. *J Thromb Haemost.* 2006;141:249–256.
6. Kahn SR. The post-thrombotic syndrome: Progress and pitfalls. *Br J Haematol.* 2006;134:357–365.
7. Bertina RM. Genetic approach to thrombophilia. [see comment]. *Thromb, Hemost.* 2001;86:92–103.
8. Mackenzie RK, Ludlam CA, Ruckley CV, Allan PL, Burns P, Bradbury AW. The prevalence of thrombophilia in patients with chronic venous leg ulceration. *J Vasc Surg.* 2002;35:718–722.
9. Bauer K, Lip G. Overview of the causes of venous thrombosis. In: Rose B, Rush J, eds. *UpToDate*, Waltham, MA: UpToDate.com, 2006.
10. Gaber Y, Siemens HJ, Schmeller W. Resistance to activated protein C due to factor V Leiden mutation: High prevalence in patients with post-thrombotic leg ulcers. *Br J Dermatol.* 2001;144:546–548.
11. Bradbury AW, MacKenzie RK, Burns P, Fegan C. Thrombophilia and chronic venous ulceration. *Eur J Vasc, Endovasc Surg.* 2002;24:97–104.
12. Michiels JJ, Hamulyak K. Laboratory diagnosis of hereditary thrombophilia. *Sem Thromb, Hemostasis.* 1998;24:309–320.
13. Demers C, Henderson P, Blajchman MA, et al. An antithrombin III assay based on factor Xa inhibition provides a more reliable test to identify congenital antithrombin III deficiency than an assay based on thrombin inhibition. *Thromb, Hemost.* 1993;69:231–235.
14. Martinelli I, Mannucci PM, De Stefano V, et al. Different risks of thrombosis in four coagulation defects associated with inherited thrombophilia: A study of 150 families. *Blood.* 1998;92:2353–2358.

15. Heijboer H, Brandjes DP, Buller HR, Sturk A, ten Cate JW. Deficiencies of coagulation-inhibiting and fibrinolytic proteins in outpatients with deep-vein thrombosis. [see comment]. *New Eng J Med.* 1990;323:1512–1516.
16. Cervera R, Piette JC, Font J, et al. Antiphospholipid syndrome: Clinical and immunologic manifestations and patterns of disease expression in a cohort of 1,000 patients. *Arthritis Rheum.* 2002;46:1019–1027.
17. Rodgers GM, Kane WH. Activation of endogenous factor V by a homocysteine-induced vascular endothelial cell activator. *J Clin Invest.* 1986;77:1909–1916.
18. Hajjar KA. Homocysteine-induced modulation of tissue plasminogen activator binding to its endothelial cell membrane receptor. *J Clin Invest.* 1993;91:2873–2879.
19. Lentz SR, Sadler JE. Inhibition of thrombomodulin surface expression and protein C activation by the thrombogenic agent homocysteine. *J Clin Invest.* 1991;88:1906–1914.
20. Ray JG. Meta-analysis of hyperhomocysteinemia as a risk factor for venous thromboembolic disease. *Arch Intern Med.* 1998;158:2101–2106.
21. Christiansen SC, Cannegieter SC, Koster T, Vandenbroucke JP, Rosendaal FR. Thrombophilia, clinical factors, and recurrent venous thrombotic events. [see comment]. *JAMA.* 2005;293:2352–2361.
22. Prandoni P, Lensing AW, Cogo A, et al. The long-term clinical course of acute deep venous thrombosis. [see comment]. *Ann Intern Med.* 1996;125:1–7.
23. van Dongen CJ, Prandoni P, Frulla M, Marchiori A, Prins MH, Hutten BA. Relation between quality of anticoagulant treatment and the development of the postthrombotic syndrome. *J Thromb, Hemost.* 2005;3:939–942.
24. Greaves M, Baglin T. Laboratory testing for heritable thrombophilia: Impact on clinical management of thrombotic disease annotation. *Br J Haematol.* 2000;109:699–703.
25. Mateo J, Oliver A, Borrell M, Sala N, Fontcuberta J. Laboratory evaluation and clinical characteristics of 2,132 consecutive unselected patients with venous thromboembolism—Results of the Spanish Multicentric Study on Thrombophilia (EMET-Study). *Thromb, Hemost.* 1997;77:444–451.

6

THE CHRONICALLY SWOLLEN LEG: FINDING THE CAUSE:

THEORY AND PRACTICE

WARNER P. BUNDENS

INTRODUCTION

Physicians are usually much better at diagnosing the cause of an acutely swollen limb than they are a chronically swollen leg or legs. In acute cases the temporal relationship of a likely inciting cause to the development of limb swelling usually leads rapidly to the diagnosis. In contrast, many physicians are at a relative loss in diagnosing the cause of chronic leg swelling, and many patients are simply given the diagnosis of "edema." Edema is not, however, a diagnosis. It is a physical finding that is caused by excess fluid in tissues. The excess fluid is the result of an ongoing pathologic process. That pathologic process and not the edema is the diagnosis. Vascular specialists are frequently referred patients with a history of months, or even years, of a swollen leg or legs.

Physicians who were cognizant of the underlying mechanisms of chronic edema and the patterns of clinical presentation can make an accurate diagnosis as to the etiology of the chronically swollen limb in 90% or more of cases.[1] Also, in the majority of cases, the diagnosis can be made rapidly based on history and physical findings. The remaining cases, perhaps 10%, require more extensive investigation, including laboratory tests and imaging studies. The correct diagnosis as to the cause of a chronically swollen/edematous leg(s) is obviously important if the proper treatment is to be instituted. Referrals commonly occur following unsuccessful and often inappropriate treatment with diuretics and/or compression stockings.

The need for treatment is more than one of cosmesis. The reasons include the fact that edema impedes wound healing. Oxygen and other nutrients move through the interstitial fluid largely by diffusion. Diffusion is an efficient transport system over short distances (micrometers), but very inefficient at large distances (millimeters). Also, edema fluid is a good culture medium, and swollen limbs are prone to infections that can be difficult to eradicate and are prone to recurrence. Lastly, patients want treatment. Their symptoms may be minimal, often none or only a vague feeling of heaviness. However, they know the swelling indicates that something is wrong, and they want something done about it.

FLUID EXCHANGE IN THE MICROCIRCULATION

The exchange of fluids between the blood and interstitium takes place at the capillary level of the microcirculation. A diagrammatic microcirculation unit is shown in Figure 6-1. The exchange of fluid between the intravascular and interstitial spaces is the result of forces in the form of hydraulic and osmotic pressures. Systemic arterial pressure is reduced by the arterioles and precapillary sphincters to approximately 30 mmHg. The systolic-

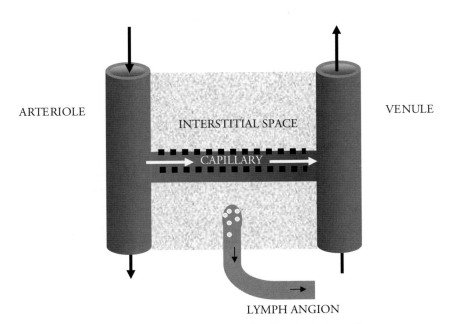

FIGURE 6-1 Diagrammatic microcirculation unit.

diastolic variation of pressure is typically only 1–2 mmHg or 2–4 mmHg with dilation of the precapillary sphincters. The resistance of the microcirculation vessels causes a further reduction in pressure such that in a supine patient, the blood enters the arterial end of the capillaries with a pressure of approximately 30 mmHg and leaves the venular end with a pressure of 10 mmHg. The fluid/hydraulic pressure of the interstitial space in a normal subject is approximately 0 mmHg. Thus, the net effect of hydraulic pressure is to force fluid out of the capillaries and into the interstitium.

Opposing these "outward" forces is an osmotic pressure gradient, which is the result of the proteinaceous colloids in the blood and interstitium. These colloids make up only a small percentage of the total plasma osmotic pressure, but because they are the main molecules that cannot cross the capillary wall, they are responsible for the osmotic pressure gradient between the plasma and the interstitium. This colloid osmotic pressure is usually referred to as the oncotic pressure. The oncotic pressure of the plasma is typically 28 mmHg. Only a small fraction of the plasma is filtered during a single transit through the microcirculation, and the oncotic pressure remains at approximately 28 mmHg throughout the circulatory system. The oncotic pressure of the interstitium is typically 8 mmHg. Thus, the effect of the osmotic forces across the capillary walls is to cause fluid to flow from the interstitium into the capillary lumen.

The net effects of the hydraulic and osmotic pressures are shown in Figure 6-2. At the arterial end of capillaries the net forces of 30 mmHg hydraulic versus 20 mmHg oncotic causes fluid to leave the capillary and flow into the interstitium. As the blood transits the capillary, the hydraulic forces diminish and the net effect at the venular end of the capillary is for fluid to pass from the interstitium into the capillary.

The net effect of these forces and the resulting fluid flux were quantified by Starling in 1896.[2] The equation, known as Starling's hypothesis, is shown in Figure 6-3. Several modifications have been made to Starling's hypothesis to adjust for experimental observations and other factors; however, it is still a very good working model.

Normally, the above-described forces maintain a good, but not perfect, balance of fluid between the circulation and the interstitial tissues. There is usually a slight excess of fluid flowing into the interstitium with time. This excess is normally removed by the lymphatics. Total lymphatic flow in the body is, however, only about 2 liters per day.

MECHANISMS OF EDEMA

Examination of the Starling equation and Figure 6-2 reveals there are four possible mechanisms of edema formation. They are (1) an increase in

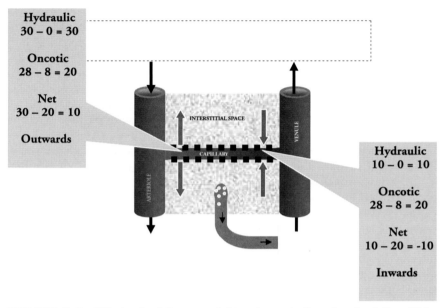

FIGURE 6-2 Effects of net forces on fluid exchange in the microcirculation. Water normally leaves the capillary on the arteriolar side and flows back into the capillary on the venular side.

$$J = k[(Pc - Pi) - \sigma(\Pi c - \Pi i)]$$

J – flux
k – capillary filtration coefficient
Pc – intraluminal hydraulic pressure
Pi – interstitial hydraulic pressure
σ – reflection coefficient (the effectiveness of the semipermiable membrane; 1 = perfect)
Πc – intraluminal oncotic pressure
Πi – interstitial oncotic pressure

FIGURE 6-3 Starling's hypothesis.

the hydraulic pressure gradient, (2) a decrease in the oncotic pressure gradient, (3) increased capillary permeability, and (4) decreased lymphatic flow.

The result of an increased hydraulic pressure gradient (an elevation of Pc in the Starling equation) can occur on either the arterial or venous end of the capillary. On the arterial end, the mechanism is usually relaxation of the arteriolar sphincters. This can be the result of inflammation (from any cause) and medications. Calcium channel blockers, which cause smooth muscle relaxation, are often involved. Nifedipine is the most common of the calcium channel blockers that can cause edema with an incidence of 10%. Hydralazine, also a vasodilator, is also known to cause edema. Cilostazol, a phosphodiesterase inhibitor used in the treatment of claudication, has vasodilatory properties and an 8% incidence of peripheral edema.

It is also possible to raise the hydraulic pressure gradient on the venous side of the microcirculation. Simple dependency of the limbs, without movement, raises the venous pressure at the ankle level to approximately 90 mmHg when standing and 60 mmHg in the sitting position. This is due to the vertical column of blood between the heart and the ankle. In normal subjects, even minimal movements can reduce the venous pressure to 20 mmHg as the valve leaflets close and negate the effects of a long column of blood when the limbs are below heart level. In contrast, people with chronic venous insufficiency (i.e., chronic obstruction or more commonly valvular insufficiency) cannot relieve this venous hypertension with movement. So-called fluid retention is also a cause of venous hypertension and edema. Causes of excess venous volume include congestive heart failure, renal disease, hormones (especially estrogen and progesterone), secondary aldosteronism, and drugs such as the nonsteroidal anti-inflammatories. Naprosyn has a 1% to 3% incidence of peripheral edema. Recently, oral hypoglycemics such as Actos and Avandia have been shown to cause fluid retention by a renal mechanism.

A decrease in the oncotic pressure gradient, the force that "pulls" fluid back into the intravascular space, can cause edema. A decrease in plasma oncotic pressure (Π_c in the Starling equation) can be caused by malnutrition, protein losing enteropathies such as Chron's or celiac disease, protein losing nephropathy, or hepatic disease with decreased production of plasma proteins. Interstitial oncotic pressure (Πi) can be increased as a result of decreased lymphatic flow, which leads to decreased clearing of macromolecules from the interstitium, or by a disease that causes the deposition of osmotically active substance in the tissues. An example would be the deposition of mucopolysaccharides that occurs in hypothyroidism leading to myxedema.

An increase in capillary permeability will also cause edema. The hydraulic pressure gradient becomes more "effective" (increased k in the Starling equation), and the increased permeability causes the capillary wall to be less effective as a semipermeable membrane (lower σ) decreasing the oncotic gradient. The causes of increased capillary permeability include endothelial damage, allergy, and inflammation. Endothelial damage can be caused by burns and toxins. Allergic reactions cause the release of histamine. Histamine causes endothelial cells to shrink and increases the distance between cell margins, thus increasing capillary permeability. Inflammation can be from any cause such as infection, or a systemic inflammatory condition such as lupus or vasculitis. It is also believed that cyclic idiopathic edema, sometimes seen in premenstrual females, is the result of increased capillary permeability on a hormonal basis.

Decreased lymphatic flow, either primary or secondary, will lead to edema. Primary or idiopathic lymphedema is classified as congenital, praecox (before age 35), or tarda (after age 35). Secondary lymphedema can be due to invasion of the lymphatic system by parasites (filiariasis); lymph node excision and/or radiation; tumor invasion by malignancy; or damage to lymphatic vessels or nodes by surgery, trauma, or infection.

CLINICAL PRESENTATION

The theoretical causes of edema that have been listed will simply lead to a long list of possible etiologies. Fortunately, recognizable clinical patterns usually point an experienced clinician to a rapid diagnosis.

Edema caused by venous pathology, usually venous valvular insufficiency, may involve one or both legs. If both legs are affected, it is usually not equal. The edema is very well relieved by overnight elevation. The feet are usually spared. The edema may or may not pit. The symptoms range from simple heaviness to possibly an aching pain. The presence of other stigmata of venous disease such as varicosities and skin changes also points to a possible venous etiology.

Lymphedema, in contrast to venous edema, is worse distally. The foot and toes are involved and are often described as "pillow" feet and "sausage" toes. The relief by overnight elevation is minimal to mild. The edema typically pits early in the course of lymphedema, but later, as the tissues become more fibrotic, pitting is absent. Symptoms are often absent, or the patient complains of only heaviness of the leg. Lymphedema can be bilateral, but is usually unequal in such cases.

Edema-caused cardiac, renal, endocrine, GI, and hepatic etiologies are referred to as edema with a "central" cause. Both legs are involved equally, and the edema is very well relieved by elevation. The distribution is similar to lymphedema (i.e., worst distally); however, often the toes are not involved. The edema typically pits. The symptoms are none or minimal. Dependent edema of a limb is often indistinguishable from centrally caused edema by simple clinical appearance. The clue in these cases is the patient's activities. It is seen in subjects with minimal activities who sit most of the day, often in a wheelchair.

Edema caused by medication is virtually always bilateral and equal. Relief by elevation and the distribution in the legs, however, can be variable but usually acts and looks like central or dependent edema. Symptoms are little or none. Common medications that can cause edema are listed in Table 6-1.

TABLE 6-1 Commonly Used Drugs That Can Cause Edema

Calcium channel blockers	Clonidine
Hydralazine	Minoxidil
Estrogen	Cilostazol
Progesterone	Gabapentin
Nonsteroidal anti-inflammatories	Glucocorticoids
COX2 inhibitors	

TABLE 6-2 Clinical Patterns of Lower Extremity Edema

Cause	Venous	Lymph	Central* Depend.	Medication
Bilateral	Occasional / ≠	Occasional / ≠	Always	Always
Relief by Elevation	Complete	Minimal	Complete	Varies
Distribution	Foot usually spared	Worst distally	Worst distally	Varies
Pitting	±	−	+	+
Symptoms	Heavy/ache	None or heavy	Little or none	Little or none

* Central causes include cardiac, renal, hepatic, GI, and endocrine etiologies.

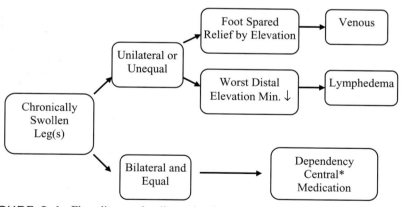

FIGURE 6-4 Flow diagram for diagnosis of common causes of chronic leg swelling. *Central causes include cardiac, renal, hepatic, GI, and endocrine etiologies.

The summary of the clinical appearance of the limbs of various etiologies is shown in Table 6-2. A flow diagram for the likely etiology of a chronically swollen limb or limbs is shown in Figure 6-4. This flow diagram works in approximately 90% of cases. The other 10% require a more careful history with possible laboratory and imaging studies.

Lastly, vascular specialists may occasionally be presented with patients with swollen legs who do not have true edema. These patients have an unusual fat distribution in the legs called lipedema. The patients are female, and the condition often runs in families. These patients will describe having developed "tree trunk" legs or "thick ankles" in their teenage years. The feet are not involved. Elevation, diuretics, and compression have minimal effects on limb size. The tissues have a doughy consistency distinct from true edema. The symptoms are minimal, though patients may describe increased skin sensitivity to touch.

REFERENCES

1. Rutherford RB. Initial patient evaluation: The vascular consultation. In: Rutherford RB, ed. *Vascular Surgery*, 5th ed. Philadelphia: W.B. Saunders Company, 2000:9–11.
2. Starling EH. Physiologic forces involved in the causation of dropsy. *Lancet.* 1896;I; 1267–1270.

SECTION

II

NONOPERATIVE MANAGEMENT OF CHRONIC VENOUS INSUFFICIENCY

7

Compression Therapy in Venous Leg Ulcers

Hugo Partsch

INTRODUCTION

Compression therapy is the basic treatment modality in venous leg ulcers that has been shown to be effective for healing ulcers and also for keeping them healed.[1,2] In every single case, the underlying venous pathology has to be ascertained, preferably by Duplex investigation, and methods to correct pathophysiology by surgery or sclerotherapy have to be considered.

COMPRESSION DEVICES

Different devices can be used for compression therapy of venous ulcers (Table 7-1).

MEDICAL COMPRESSION STOCKINGS

Medical compression stockings exerting interface pressures in the gaiter area up to 40 mmHg are effective in preventing and reducing edema.[3]

For treating venous ulcers, the use of compression stockings may be considered, if the ulcers are not too large (less than 5 sq cm) and not too long-standing (history shorter than 3 months).[4] Even unskilled patients are able to handle this kind of treatment, producing a constant degree of pressure and allowing the patient to change the dressing, clean the ulcer, and have a shower. One of the disadvantages is that the exudates may soil the

TABLE 7-1 Types of Compression Devices

- Graduated compression stockings
    - Custom made
    - Standard size
        - Knee length
        - Thigh length
        - Compression tights
- Bandages
    - Single component—multiple components
    - Inelastic—elastic
- Intermittent pneumatic compression
    - Single chamber
    - Sequential chambers
        - Foot-pump
        - Lower leg
        - Full leg, trunk

stocking, which then has to be frequently washed, causing a weakening of the stocking fibers.

Recently, some randomized controlled trials have reported favorable results of ulcer healing using different kinds of compression stockings.[5-7]

Light compression stockings may be used for keeping the local ulcer dressing in place. A class II compression stocking placed on top will not only add its pressure to that of the underlying light stocking but will also increase the stiffness of the final kit.[8] Several ulcer stockings consisting of two layers have been introduced (Venotrain®, Bauerfeind, Medi Ulcertec®, Ulcerkit® Gloria, Saphenamed UCV®, Hartmann). The basic layer keeping the ulcer dressing in place stays on the leg overnight, while the second stocking is donned over the basic liner during the daytime. This regime allows the patient to clean the ulcer and change the dressing whenever necessary. This treatment modality is very cost effective, since no specialized medical personnel are required to perform the dressing changes.

Stockings with a zip have been introduced in the United States (Ulcercare®, Jobst-Baiersdorf). A ready-made tubular device, which can be washed and reused, was introduced in some European countries under the name Tubulcus® or Rosidal mobil®.[5]

Usually, below-knee stockings are prescribed. Strong compression stockings exerting a pressure of more than 30 mmHg at ankle level are recommended.

After the ulcers are healed, compression stockings are essential in order to prevent recurrence.[2,9] Compliance with the daily use may be a problem, especially in elderly patients who have difficulties applying elastic stockings. In order to facilitate the application of a stocking, some helpful

TABLE 7-2 Pressure Ranges of Medical Compression Stockings According to Different National Regulations*

Compression Class	EU (CEN)	USA	UK	France	Germany
A	10–14 (light)				
I	15–21 (mild)	15–20 (moderate)	14–17 (light)	10–15	18–21 (light)
II	23–32 (moderate)	20–30 (firm)	18–24 (medium)	15–20	23–32 (medium)
III	34–46 (strong)	30–40 (extra firm)	25–35 (strong)	20–36	34–46 (strong)
IV	>49 (very strong)	40+		>36	>49 (very strong)

* The values indicate the pressure (mmHg) that should be exerted by the hosiery at a hypothetical cylindrical ankle.

devices have been introduced. Nylon or silk socks (e.g., Easy slide ®) help to slide the stocking over the foot.

Medical compression stockings are differentiated into several compression classes according to the pressure they are supposed to exert on the leg. Unfortunately, there are considerable discrepancies between different national standards, as shown in Table 7-2. A class II stocking is defined by a pressure range at ankle level between 15 and 20 mmHg in France, between 18 and 24 mmHg in the UK, between 20 and 30 mmHg in the United States, and between 23 and 32 mmHg in Germany and other European countries.

Also, the descriptive terms vary considerably: A stocking called "medium or moderate" in European countries may be called "strong" in the United States.

For a better international understanding, it has therefore been recommended to use the pressure range in mmHg rather than the compression class and also to specify the description of "mild, light," "medium, moderate," and "strong, firm, or extra firm" by adding the pressure range.[10]

COMPRESSION BANDAGES

For the routine management of venous ulcers, stockings cannot replace compression bandages, which may exert much higher pressures.

Table 7-3 shows values of interface pressure of bandages measured on the distal leg according to the British standard, which is the only existing standard for bandages[11] in comparison to new proposals from a consensus conference on compression bandaging.[12]

TABLE 7-3 Interface Pressure Ranges (mmHg) Exerted by Compression Bandages on the Distal Leg in the Supine Position*

	Consensus Group	BS 7505
Light compression	<20	<20
Moderate compression	20–40	21–30
Strong compression	40–60	31–40
Very strong compression	>60	41–60

* The definitions proposed by a consensus group[12] are compared with those of the British Standard 7505.[11]

The pressure values indicated by the consensus group are clearly higher than those from the British Standard, which come closer to the stocking regulations.

It has to be stressed that all pressure values from Table 7-3 are entirely based on experiments performed in the laboratory and not measured on the human leg.

Interface Pressure and Stiffness Measured on the Leg

The pressure on the leg is one deciding parameter that determines the efficacy of a bandage. Another important feature is stiffness, which characterizes the elastic property of the material and which is defined by the increase of pressure due to an increase of the circumference of the leg during movement.[13]

Both parameters can be assessed by measurements on the individual leg using simple, battery-powered, portable transducers like the Kikuhime tester® (MediTrade; Soro, Denmark) or the Sigat-tester® (Ganzoni; Winterthur, Switzerland). Such measurements will be indispensable in future trials comparing different compression products and for training purposes.

A preferred measuring point is the area on the leg that shows the most pronounced changes of curvature and of circumference by active standing and by walking, which is about 8–12 cm above the inner malleolus.[13,14] A comparison between *in vivo* and *in vitro* measurements of pressure and stiffness of compression stockings revealed a good correlation.[8]

ELASTIC AND INELASTIC BANDAGES

Table 7-4 shows a differentiation of compression material based on the elastic properties of single layers.

TABLE 7-4 Elastic Property of Single-Layer Compression Material

	Inelastic/Nonstretch	Inelastic/Short Stretch	Elastic/Long Stretch
Stretch	0	<100%	>100%
Stiffness	Very high	High	Low
Application	Trained staff	Trained staff	Every patient
Stays on the leg	Day and night	Day and night	Daytime

- **Elastic compression bandages:** Bandages that incorporate materials which exert pressure when applied with stretch.
- **Inelastic compression bandages:** Bandages that exert pressure which increases when movement causes the calf muscle to contract.

Elastic, long stretch material is relatively easy to handle and can be used also by the patients. In contrast to the inelastic material, these bandages produce an active force by the elastic constriction of their fibers. The pressure drop after some hours is minimal. Therefore, such bandages may cause pain and discomfort when the patient sits or lies down, especially when they have been applied too tightly. Single-component elastic bandages or compression stockings are therefore applied in the morning, preferably before getting up and are removed before going to bed at night.

During walking, the peak pressure waves are lower than with inelastic material.

Inelastic material produces a much higher increase when the patient is standing up and performing dorsiflexion than does the elastic material (Figure 7-1). The pressure increase from standing up and the amplitudes during ankle movement are useful parameters for characterizing stiffness.

For a practical differentiation between elastic and inelastic material, it had been proposed to define the difference between standing and supine pressure measured with a small pressure transducer at the medial gaiter region as the so-called Static stiffness index (SSI).[13] Values higher than 10 are indicative of inelastic material, while values lower than 10 are typical for elastic material.[15] This parameter signifies the range between an effective working pressure and a tolerable resting pressure.

A good compression bandage is characterized by a well-tolerated resting pressure and high pressure peaks during walking. An inelastic bandage achieving pressure peaks of 80 mmHg will compress superficial and deep veins intermittently. This can be demonstrated by compressing the leg with blood pressure cuffs containing an ultrasound permeable window. Using a Duplex instrument, it can be demonstrated that in the upright position, a pressure of 40–60 mmHg will narrow the leg veins.[16] Dorsiflexion leading

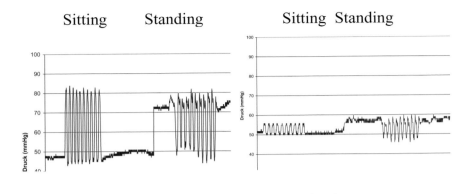

FIGURE 7-1 Sub-bandage pressure (mmHg) with inelastic and elastic material, both applied with a resting pressure of 50 mmHg on the distal lower leg in the sitting position. During up and down movement of the feet, the pressure gradient between muscle systole (peak values) and muscle diastole (lowest values) is much higher for the inelastic than for the elastic material. When the patient is standing up, the pressure rises with the inelastic bandage by 22 mmHg and only by 8 mmHg with the elastic bandage.

to intermittent pressure peaks of 60–80 mmHg causes intermittent occlusions of the veins.

Two main disadvantages of inelastic bandages have to be considered:

- One is the loss of bandage pressure starting immediately after bandage application. After only 1 hour, the initial resting pressure will drop by about 25%, mainly due to an immediate decrease of the volume of the limb.
- The second disadvantage is the fact that the application of a good inelastic compression bandage is not easy and should be learned and trained. Due to the fast loss of pressure, bandages with inelastic material should be applied with much higher initial tension than elastic bandages. An inadequate bandaging technique with inelastic bandages is likely the main reason for the poorer clinical outcome described in some studies.[1]

SINGLE-LAYER AND MULTILAYER BANDAGES

Usually, single-layer bandages are applied with an overlap of about 50%. Single-layer bandages (e.g., one Ace bandage) are insufficient to treat a venous leg ulcer.

Some randomized controlled trials have clearly shown that multilayered compression was more effective in healing venous ulcers than single-layer compression.[1]

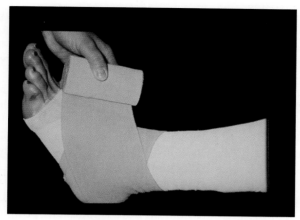

FIGURE 7-2 Unna boot bandage (10 meters applied under considerable tension) and wrapped over by an inelastic bandage (Rosidal K®). In order to avoid wrinkles and folds during application, the zinc paste bandage has to be cut and molded to the leg.

A multilayer bandage may consist of one or several components of different compression materials.

SINGLE-COMPONENT AND MULTIPLE-COMPONENT BANDAGES

Bandages consisting of different materials are called multiple-component bandages or compression bandaging systems. They may be provided by companies in one package, so-called kits.[11]

It has to be stressed that multiple layers will not only increase the pressure but will also change the elastic property toward a stiffer final bandage. The same is true if multiple layers of different materials are used. Such multicomponent bandages will act like an inelastic bandage even when the single components are elastic and may also stay on the leg overnight.

EXAMPLES OF COMPRESSION BANDAGING SYSTEMS

Zinc paste (unna boot) bandages are examples of completely nonstretch material with high stiffness that may remain on the leg for several days up to 2 weeks (Figure 7-2).

In contrast to many other institutions, we prefer to apply the unna boot layers with considerable tension and cover the zinc paste with an outer layer of an inelastic cotton wool bandage (Rosidal® or Comprilan®), thereby producing a multilayer, multicomponent inelastic bandage.

The force used during bandaging depends on the amount of edema, the walking ability of the patient, and the leg circumference. In contrast to bandages containing elastic fibers, the resulting interface pressure is solely produced by the strength during application of the bandage, which should be much higher when used with elastic material.

In a patient with severe edema, the inelastic bandage should be modeled with a pressure of about 60 mmHg to the distal portion of the lower leg. As with every compression treatment, it is mandatory to exclude arterial occlusive disease before such high pressure bandages are applied.

An even pressure distribution can be achieved by applying several layers. Due to the increasing circumference of the leg toward the calf, the applied pressure will decrease even when the bandager applies the bandage with the same strength as at the ankle. This is explained by Laplace's law, stating that the pressure on a cylinder is indirectly proportional to its radius.

Velcro-band gaiters like *Circ-Aid®*,[17,18] consisting of completely non-stretch material, can be applied by instructed patients themselves and can be readjusted when the bandages get loose due to edema reduction.

The Pütter bandage, consisting of several layers of inelastic cotton bandages (e.g., Rosidal K®, Comprilan®), is an example of a multilayer, single-component compression bandage. Usually, two 5-m long bandages are used to cover a lower leg. More important than the details of application in terms of spirals and the amount of overlap is the high resting pressure on the distal lower leg, which should diminish over the proximal parts in order to create a pressure gradient from distal to proximal.

Rosidal sys® is another inelastic bandaging system offered as kit; it contains a foam-padding layer, two Rosdal K® bandages, and a hose for protection.

Such bandages are frequently used in continental Europe. Routinely, they are changed every week; in the case of strong exudation in the initial phase, they are changed after a few days. The bandages can be washed and reused. Thereby, they increasingly lose their elastic properties, which make the bandages more and more inelastic.

A new inelastic two-component system was recently developed, consisting of a basic layer and an adhesive bandage (*Coban two-layer bandage®*, 3M).

New materials with special elastomeric properties (Vari-Stretch system used in *Pro-Guide®*) try to overcome the problem of inadequate pressure produced by an inexperienced bandager.[19] The sub-bandage pressure generated by this system remains relatively constant as the tension with which the bandage is applied increases. In order to provide guidance to the user on the optimal tension and ensure that two bandage-turns applied with a 50% overlap achieve a satisfactory pressure, the outer bandage layer contains ovals that change into circles when the level of tension is applied to the fabric.

Several elastic bandages have been supplied with cohesive (self-adhering) or *adhesive* basic sticky layers in order to prevent slipping down of the bandage. Due to the high friction between several layers, such bandages act like inelastic material. They are also able to maintain prolonged compression.[20]

The *four-layer bandage* (e.g., *Profore®*) contains several layers of four materials: cotton wool, crepe, elastic, and self-cohesive. These bandages are relatively easy to apply and have gained considerable widespread use, especially in the UK, where ulcers are treated mainly by nurses.[21] The final bandage that should exert a pressure of about 40 mmHg on the distal leg shows a relatively high stiffness. Therefore, such multilayer bandages may also stay on the leg for several days. In centers that are experienced with applying inelastic material, no significant superiority of four-layer bandages could be found.[22]

INTERMITTENT PNEUMATIC COMPRESSION

The application of intermittent pneumatic compression in addition to the use of compression bandages is able to accelerate ulcer healing.[23] This adjunctive treatment may be very helpful in patients who have edema and are unable to walk or who suffer from a stiff ankle, especially when concomitant arterial disease is also present. In such cases, there is often a vicious circle starting with pain and inability to walk, sitting for many hours, and progression of edema, which reduces the arterial skin perfusion, which in turn will worsen the pain. Bed rest with leg elevation will not be tolerated in this situation. Intermittent pneumatic compression is able to reduce edema and to enhance arterial blood flow.[24,25]

One deciding point that intermittent compression may be beneficial in patients with arterial occlusive disease while sustained compression with bandages or stockings is contraindicated is the pressure-free interval between the pressure cycles, in which short periods of reactive hyperemia develop.

PELOTTES AND PADS

Venous ulcers are frequently localized behind the inner malleolus or on the flat areas of the medial lower leg. The pressure of a bandage or a stocking will be low in these areas due to Laplace's law, which states that the pressure is indirect proportional to the radius. A local increase of pressure can be achieved by applying rubber foam pads over the ulcer region, thereby decreasing the radius of the leg segment (Figure 7-3). Care should be taken that the edges of such pelottes should be flattened in order to

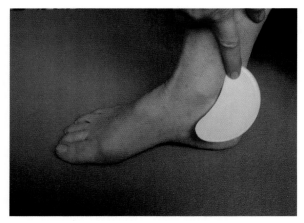

FIGURE 7-3 Rubber foam pad to increase the local bandage pressure over an ulcer in the retromalleolar region.

avoid sharp impressions into the skin by theses devices (Figure 7-4a and Figure 7-4b).

PREVENTION OF ULCER RECURRENCE

Management of leg ulcers consists of two phases:

1. The *healing phase* until epithelialization is obtained;
2. The *maintenance phase* after ulcer healing in which a frequently occurring recurrence should be prevented.

In general, it is often easier to heal a venous ulcer than to keep it healed.

To keep the ulcer healed, continuous compression is essential ("maintenance phase"). Medical compression below-knee stockings, 30–45 mmHg, are the preferred method of choice. Patients who are unable to put on the stockings may use elastic bandages instead. Eradication of venous refluxes by surgery or sclerotherapy should be considered in every single patient.

COMPRESSION TECHNIQUES

Many different techniques of compression bandaging have been described. Some general rules are as follows:

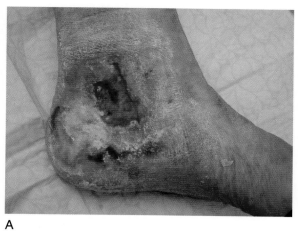

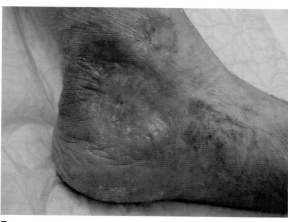

FIGURE 7-4 **A.** Venous ulcer due to a post-thrombotic syndrome behind the inner malleolus.
B. After 16 weeks of compression therapy using Unna boot bandages and a rubber foam pad, the ulcer is healed. (The impression of the pad can still be seen.) Consideration to abolish venous refluxes by surgery or by sclerotherapy and wearing of compression stockings is essential to keep the ulcer closed.

- Elastic bandages are easier to handle than inelastic bandages and may also be applied by staff that is not specifically trained and by the patients themselves. This is also true for compression stockings.
- Inelastic material like zinc paste should be applied with much higher resting pressure in a kind of modeling work, pressing the bandage roll toward the leg. To obtain a homogenous pressure

distribution without constricting bands or folds, it is advisable to cut the zinc bandage when it does not exactly follow the cone-shaped leg surface during application. A 10-m bandage is recommended for one lower leg. After the lower leg is covered with several layers, a 5-m-long short stretch bandage is wrapped over, and the patient is encouraged to walk around immediately for 30 minutes at least. This short stretch bandage can be washed and reused with each change of the bandage.

- After some minutes, the pressure will drop to around 40 mmHg due to the immediate removal of a considerable amount of edema. When in the edematous phase, the bandage will get loose after a few days; it should be renewed or wrapped over with a short stretch bandage. The same is advisable when exudates from the ulceration penetrate the bandage. This may occur especially during the initial treatment phase, and the patient should be informed to come back if this happens. Thereafter, the bandage is changed every 7 days on average.
- Bandaging should cover the foot and go up to the capitulum fibulae. The initial turn may be placed around the ankle or between the heel and the dorsal tendon to fix the bandage. Then the bandage is taken down to the foot to the base of the toes. The ankle joint is always bandaged with maximal dorsal extension of the foot.
- Taking the bandage up to the leg and overlapping can be done in a spiral fashion or with figure eights.
- Graduated compression is achieved by exerting higher pressure on the distal lower leg than on the proximal calf. Local pressure over ulcers or firm lipodermatosclerotic areas can be increased by pads and pelottes. Tendons and the shin should always be protected by cotton wool.
- There should be an overlapping of the layers between 30% and 50%.
- The proximal end of the bandage should cover the capitulum fibulae.
- Bandaging of the lower leg is sufficient for the majority of patients. Only in cases with extensive swelling or phlebitis of the thigh are compression bandages reaching up to the inguinal fold advisable. The flexor tendons in the knee-hole are protected by cotton wool. Thigh bandages can best be applied using adhesive material starting from the proximal lower leg and going up to the proximal thigh. In order to narrow the veins, the sub-bandage pressure at mid-thigh level should amount to 40 mmHg at least.
- Highly exudative ulcers may need frequent dressing changes in the initial phase. However, exudation will subside after several days of firm compression.

Walking exercises are essential to optimize the effect of compression therapy.

REFERENCES

1. Cullum N, Nelson EA, Fletcher AW, Sheldon TA. Compression for venous leg ulcers (Cochrane Review). Cochrane Library, Issue 3. Oxford: Update Software, 2003.
2. Nelson EA, Bell-Syer SE, Cullum NA. Compression for preventing recurrence of venous ulcers. Cochrane Library, Issue 3. Oxford: Update Software, 2003.
3. Partsch H. Do we still need compression bandages? Haemodynamic effects of compression stockings and bandages. *Phlebology.* 2006;21:132–138.
4. Horakova MA, Partsch H. Ulcères de jambe d'origine veineuse: Indications pour les bas de compression? *Phlébologie.* 1994;47:53–57.
5. Jünger M, Partsch H, Ramelet AA, Zuccarelli F. Efficacy of a ready-made tubular compression device versus short-stretch compression bandages in the treatment of venous leg ulcers. *Wounds.* 2004;16:313–320.
6. Jünger M, Wollina U, Kohnen R, Rabe E. Wirksamkeit und Verträglichkeit eines Ulkus-Kompressionsstrumpfes zur Therapie des Ulcus cruris venosum im Vergleich zu einem Unterschenkelkompressionsverband-Resultate einer prospektiven, randomisierten, multizentrischen Studie. *Current Med Res Opin.* 2004;20:1613–1624.
7. Koksal C, Bozkurt AK. Combination of hydrocolloid dressing and medical compression stockings versus Unna's boot for the treatment of venous leg ulcers. *Swiss Med Wkly.* 2003;133:364–368.
8. Partsch H, Partsch B, Braun W. Interface pressure and stiffness of ready made compression stockings. Comparison of in vivo versus in vitro measurements. *J Vasc Surg.* 2006;44:809–814.
9. Mayberry JC, Moneta GL, Taylor LM, Porter JM. Fifteen-year results of ambulatory compression therapy for chronic venous ulcers. *Surgery.* 1991;109:575–581.
10. Comerota A, Delis K, Guex JJ, Partsch H, Ruckley V. International Compression Club, Hawaii, January 19, 2006. *J Vasc Surg.* In press.
11. British Standard. Specifications for the elastic properties of flat, non-adhesive, extensible fabric bandages. BS 7505, 1995;1–5.
12. Partsch H, Clark M, Mosti G, et al. Classification of Compression bandages. Practical aspects. Eur J Vasc Endovasc Surg. Submitted.
13. Partsch H, Clark M, Bassez S, et al. Measurement of lower leg compression in vivo: Recommendations for the performance of measurements of interface pressure and stiffness. *Dermatol Surg.* 2006;32:229–238.
14. Stolk R, Wegen van der-Franken CPM, Neumann HAM. A method for measuring the dynamic behavior of medical compression hosiery during walking. *Derm Surg.* 2004;30:729–736.
15. Partsch H. The static stiffness index. A simple method to assess the elastic property of compression material in vivo. Dermatol Surg. 2005;31:625–630.
16. Partsch B, Partsch H. Calf compression pressure required to achieve venous closure from supine to standing positions. *J Vasc Surg.* 2005;42:734–738.
17. DePalma RG, Kowallek D, Spence RK, et al. Comparison of costs and healing rates of two forms of compression in treating venous ulcers. *Vasc Surg.* 1999;33:683–689.
18. Blecken SR, Villavicencio JL, Kao TC. Comparison of elastic versus nonelastic compression in bilateral venous ulcers: A randomized trial. *J Vasc Surg.* 2005;42: 1150–1155.

19. Moffatt CJ on Behalf of the EXPECT Trial. A multi-centre randomised trial comparing a Vari-stretch compression system with Profore. Abstract. 13th Conference European Wound Management Association, Pisa, Italy. May 2003.
20. Travers JP, Dalziel KL, Makin GS. Assessment of a new one-layer adhesive bandaging method in maintaining prolonged limb compression and effects on venous ulcer healing. *Phlebology.* 1992;7:59–63.
21. Blair SD, Wright DDI, Backhouse CM, Riddle E, McCollum CN. Sustained compression and healing of chronic venous ulcers. *Br Med J.* 1988;297:1159–1161.
22. Partsch H, Damstra RJ, Tazelaar DJ, et al. Multicentre, randomised controlled trial of four-layer bandaging versus short-stretch bandaging in the treatment of venous leg ulcers. *VASA.* 2001;30:2108–2113.
23. Coleridge Smith P, Sarin S, Hasty J, Scurr JH. Sequential gradient pneumatic compression enhances venous ulcer healing: A randomised trial. *Surgery.* 1990;108:971–975.
24. Eze AR, Comerota AJ, Cisek PL, et al. Intermittent calf and foot compression increases lower extremity blood flow. *Am J Surg.* 1996;172:130–134.
25. Delis KT, Nicolaides AN, Wolfe JHN, Stansby G. Improving walking ability and ankle brachial pressure indices in symptomatic peripheral vascular disease using intermittent pneumatic foot compression: A prospective controlled study with one year follow-up. *J Vasc Surg.* 2000;31:650–661.

8

WOUND HEALING: ADJUVANT THERAPY AND TREATMENT ADHERENCE

DALE BUCHBINDER AND SHARON B. BUCHBINDER

Although the interventions are well known and documented in the literature, venous leg ulcers are a costly, chronic, and recurrent source of frustration to both patients and health care providers.[1-3] It has been reported that the average venous leg ulcer patient has a mean duration of ulceration of 95 months (about 8 years) during the course of her/his lifetime.[4]

Over half a million people in the United States are affected by venous leg ulcers, and most of the time this condition is associated with chronic pain, loss of productive work time, loss of leisure and social activities, poor self-rated heath, depression, and a decreased quality of life.[5-7] This chronic disease has even greater importance in light of the fact that the prevalence of venous leg ulcers increases with age, and a large segment of the world population is aging.[3,8-10]

MANAGEMENT OF VENOUS LEG ULCERS

Currently, there are no nationally agreed upon clinical practice guidelines for the treatment of venous leg ulcers in the United States. Among the published guidelines are those put out by Grace, the Registered Nurses Association of Ontario (RNAO), the Royal College of Nursing (RCN) in the United Kingdom, and the Wound Ostomy and Continence Nurses Society (WOCN).[11-15]

Despite the lack of one national set of guidelines for venous ulcer control, experts in the field almost unanimously agree the main method of treatment and management of venous leg ulcers is edema control through patient self-care (i.e., compression stockings, ambulation, and leg elevation) for healing as well as preventing recurrences.[5,12,14–16] Over 50% of venous leg ulcer patients will heal within 6 months with compression and edema therapy alone.[17] This means that a large proportion of patients will either take an excessive amount of time to heal, or will not heal with standard compression therapy alone. A recent study in the UK examining risk factors for delayed healing and recurrence of venous leg ulcers in 1,324 legs concluded that "elderly patients with longstanding ulcers... may benefit from adjunctive treatments to improve clinical outcomes."[18] The authors recommended a holistic approach to wound healing with careful, individualized assessment of risk factors among patients with venous ulcers.

At the Greater Baltimore Wound Care Center, we have treated over 12,000 patients with over 30,000 wounds between 1991 and 2006. Chronic venous insufficiency is associated with 30%, or approximately 10,000 of these wounds. The philosophy of our center is to begin with a comprehensive patient evaluation, checking for systemic diseases, such as diabetes mellitus, systemic lupus erythematosis, congestive heart failure, arterial insufficiency, nutritional deficiency, etc. These conditions must be addressed, and the patient needs optimal medical management. In our center, any patient with nonpalpable pulses receives an arterial Doppler exam, and all patients with chronic venous insufficiency receive venous duplex scans. Often we find treatable conditions, such as superficial venous incompetency or iliac vein stenosis. The wounds are treated at the same time that we are optimizing the patient's other conditions.

A complete wound evaluation is conducted, all wounds are measured and photographed, and aggressive debridement is performed to remove all necrotic tissue. Since these wounds are often exquisitely tender, adequate pain control is essential. Debridement is carried out using either topical analgesia in the center, or in the operating room with conscious sedation or general anesthesia. Wounds are evaluated for the presence of infection, and if we believe there is local infection or cellulitis, we begin antibiotics and topical treatments, such as silver alginates, antibiotic creams (silver sulfadine), or Bacitracin® ointment. In addition, compression dressings are applied, and patients are placed in changeable compression garments. Initially, we follow patients on a weekly basis, as they often require compression wraps (i.e., unna boots, or multilayer wraps because they are unable to perform self-care with daily or twice daily dressings or are unable to wear compression garments.) These patients are given a trial of this type of treatment for a minimum of 4 weeks. This conservative therapy is continued for those patients whose wounds show rapid healing (i.e., excellent granulating tissue and a decrease in wound volume by 50% in the first 4 weeks). Patients with correctable venous pathology are appropriately treated.

If the wounds are slow to progress (i.e., less than 50% volume reduction at week 4), or if they are large (i.e., >200 cm^3), then we start adjuvant treatments. The adjuvant therapies we will discuss in this chapter include xenografts, allografts, collagen dressings, growth factor therapy, human skin equivalents, and vacuum-assisted closures. We will also review the necessity for patient adherence to the wound care protocol to increase healing rates and prevent venous ulcer recurrence.

ADJUVANT THERAPIES

Adjuvant therapies should be reserved for those patients whose wounds are difficult to heal because these treatments are time consuming and expensive. Adjuvant therapies are only cost effective in the healing of difficult wounds and should not be used in place of appropriate conservative therapy. However, in selective patients, they will help heal wounds that are not progressing, decrease healing times, and improve outcomes.

XENOGRAFTS

Xenografts serve as biologic dressings and protect wounds from bacterial and physical trauma, reduce pain, and increase moisture and heat retention.[19] While not a replacement for human skin, xenografts do serve as a barrier until the wound is able to heal. They go back as far as the 16th century and have been obtained from species as varied as dogs, cats, rats, chickens, bullfrogs, sheep, cows, and pigs. Porcine xenografts are used most commonly because they are cheap and readily available. The wound should be debrided and, if infected, antibiotics should be initiated. The porcine skin is more likely to adhere if the wound is clean than if it is infected. Porcine xenografts can be used in conjunction with meshed allografts; however, they need to be inspected and changed weekly before rejection occurs. Compression therapy should be applied to reduce edema, and the patient should be seen at least once a week, at a minimum, to assess wound healing.

ALLOGRAFTS

An allograft is donated cadaver skin, usually cryopreserved, and readily available through tissue banks.[20] Early on, this type of therapy was used as a biological dressing to cover large burn wounds. It has also been used in treatment of chronic wounds, including venous stasis ulcers.[3] Cadaver allografts act as a matrix for growth of granulation tissue. After the wound bed is prepared as previously noted, the cadaveric skin is put in place, an antibiotic/antimicrobial dressing is applied over the graft, and a compression dressing is used to reduce edema. In our practice, patients are seen weekly. The allografts are debrided at 1 week. They may be reapplied or an alternate dressing may be used (i.e., silver alginate). Between 1 and 2 weeks, patients will reject the grafts if they are not removed, and there will be increased inflammation in the wound. Allografts control pain, decrease risk of infection, stimulate neovascularization, reduce healing time, cover tendon or bone, and provide good preparation for split thickness skin grafting, or use of human skin equivalents.[20] Allograft skin has the capability to adhere to the wound bed, thus controlling the bacterial count. In addition, this biological dressing functions as a "bag of growth factors," bathing the wound with essential cytokines and proliferating at the wound bed.[21] Allografts are usually available, moderately priced, and have a beneficial effect on wound healing.

HUMAN SKIN EQUIVALENTS (HSES)

The introduction of human skin equivalents (HSEs) is the latest revolution in biologic dressings for wounds. One of the most widely used and researched HSEs is a cultured, allogenic, bilayered product called Apligraf™, which is FDA approved for the treatment of venous ulcers. A randomized clinical trial (RCT) of 283 patients demonstrated that Apligraf™ plus compression therapy healed difficult venous ulcers faster than compression therapy alone.[22] In addition, no rejection or sensitivity was seen during or after the intervention. The key objection to Apligraf™ is that it is very expensive at about $1,000 USD per graft. However, when the costs of venous ulcers with respect to society, health care, the government, private plans, and the patient were compared to the costs of Apligraf™, researchers found that speeding up healing saved time and money. Apligraf™ was more cost effective over a longer time frame and when used selectively among patients with long-standing, difficult-to-heal wounds.[23] Proper handling of Apligraf™ is critical. It is provided in a Petri dish on a porous membrane. After the wound bed is prepared, the fenestrated Apligraf™ is put into place, making sure the dermal glossy layer is in direct contact with the wound bed. This graft may be meshed to provide coverage for large areas. As much graft as possible should be placed in the wound, and a nonadherent dressing should be placed over it. Antibiotics (e.g., Bacitracin®) should go on the nonadherent dressing, and a folded piece of sterile gauze should go on top of that and compression wraps should be applied.[24] Patients should be seen weekly, at a minimum, to assess wound healing. Rejection is uncommon. Clinicians should wait at least 6 weeks, per product description, prior to reapplying Apligraf™, as healing may not be obvious to the naked eye.

COLLAGEN DRESSINGS AND GROWTH FACTOR THERAPY

There are a myriad of collagen dressings available for the treatment of chronic wounds. They are usually derived from bovine or porcine collagen and come in a variety of preparations, including collagen particles (Medifil™), matrix of either pure collagen (Oasis™), or collagen with other products, such as Promogran Prisma™ matrix, which contains a combination of collagen; oxidized, regenerated cellulose; and silver.[25,26] These dressings can be applied as powder or sheets to wounds. In a randomized clinical trial, Oasis™ collagen product was found to be significantly better than Allevyn™ alone.[27] Collagen dressings are moderately expensive and may be promising adjuvants.

Growth factor therapy has been advocated; however, there are only a few readily available products such as Regranix®, recombinant platelet derived growth factor. This therapy is not approved for venous wounds and is quite expensive. One tube costs between $300 and $500 USD, and there are no randomized controlled trial (RCT) findings that show improved healing in venous ulcers. Platelet releasate is also available to produce autologous growth factor in wounds, but there are no RCTs showing that these treatments are effective in chronic venous wounds. These treatments are also expensive and involve drawing blood from the patient to produce the releasate. DaCosta found that granulocytic macrophage colony stimulating factor administered by periulcer injection improved healing rates of chronic venous ulcers.[28] This treatment, however, is not readily available and can be done only in a research setting.

VACUUM-ASSISTED CLOSURE (VAC)

VAC has been shown to be effective in increasing healing rates, especially in large wounds.[29] A major RCT in the Netherlands demonstrated VAC improved wound bed preparation rates in hospitalized patients when compared to standard compression therapy alone.[30] Venous ulcers were included in this study with decreased healing times, as well. The VAC can be applied to chronic leg wounds, and there is a portable pump available so the patient can use it on an outpatient basis. Also, it has been used in conjunction with HSE and growth factor therapy. This treatment is expensive and requires home health professionals to change the dressing several times a week. Compression needs to be applied over the VAC dressing, as well. Since these studies do not show total healing, only improved wound bed preparation for skin grafting and improved skin graft outcomes, further studies should be performed to show total healing. However, with large wounds that may require split thickness skin grafting, the VAC should be considered.

TREATMENT ADHERENCE

The role of adherence and self-care has received more attention lately due to the potential for reduced utilization of the health care system.[31] The increasing importance of chronic conditions in an aging population has created a greater emphasis on individual responsibility for health maintenance and care.[32–34] Estimates of patient adherence to a medication regimen vary from 30% to 78% depending on the type of protocol. When medications are involved, adherence can be monitored by counting prescription refills, blood levels, and number of pills.[35]

Behavioral adherence (e.g., smoking cessation, good nutrition, leg elevation, exercise, skin care, and use of compression stockings) is not as easily monitored as medication therapy. The deleterious effects of nonadherence to these recommendations are insidious and can lead to increased healing times and recurrent leg ulcers.[36] Patients who are nonadherent are sometimes labeled as "noncompliant," "problem patients," "dependent," "helpless," "pitiful," "worrying," or "difficult personalities."[37,38] However, when patients are educated about their disease, given treatment choices, and empowered to become partners in their wound care, treatment adherence improves.[37–42]

Negotiated responsibility, collaboration, and patient empowerment occur when the health care providers and patient have thoroughly discussed the options available to the patient for ambulatory compliance and self-care. The patient must have the ability, will, motivation, acknowledgment of responsibility for life and decisions, along with the determination to achieve self-care as a goal.[37,38] The provider needs critical information to assess the ability of the patient to comply with the treatment regimen and to provide self-care or to obtain ambulatory care. The patient needs to be able to understand the level of involvement that the provider is able to give. The patient also needs to be educated about the wound healing process.

Some of the questions the health care provider should discuss with the patient are

- What is the patient's need for help and capacity to look after self with respect to physical needs, psychosocial needs and social network, dwelling, and local environment?
- What Activities of Daily Living (ADLs) is the patient able to complete?
- Is informal care available from people in the house, family, neighbors, friends, or volunteers?
- Where are the gaps in care?

Additionally, the health care provider and the patient need to have a conversation about what types of changes need to be made to meet the patient's needs. These changes may be relative to informal versus formal care, the physical home environment, changes in the employment environment, and changes in external cues.

External cues from the health care provider include, but are not limited to, the discussion of negotiated responsibility, provider-patient interactions, and patient and family education materials. Patient education should be positive, respectful, nonjudgmental, compassionate, and clear.[43] The goal of these discussions should be to ensure that the patient and family are motivated to collaborate in working to heal the wound. If there are gaps in the care, they must be identified and addressed. Unless venous ulcer care is provided in a multidisciplinary wound care center where providers all adhere rigorously to a treatment algorithm, care can be inconsistent and fragmented. Enabling the patient to become an expert in her/his own health care empowers the patient to become proactive, take ownership of care, and ask better questions.[44] Provider-initiated materials can be in the form of patient education seminars, pamphlets, videotapes, or DVDs describing the disease process and treatment regimen to be followed. Careful documentation on the part of wound care providers is critical to evaluation of the impact of these strategies on wound healing and other patient outcomes.

External cues from the patient can be patient-initiated tools that take the form of notes, diaries, journals, or logs to record the daily at-home treatment regimen. Diaries have been found to be useful adjuncts to treatment for arthritis, headaches, urinary incontinence, pain, and asthma.[45,46] These patient-initiated tools can be brought to appointments for review and discussion with the wound care providers. However, patient-initiated cues are not enough to ensure adherence to treatment. The interplay or "dance" between the patient and the provider using an array of patient-centered tools, along with provider-initiated cues and documentation, can lead to improved adherence and self-care, as well as greater patient satisfaction with medical care.

Between August 2000 and July 2001, we conducted a randomized clinical trial (RCT) at the Greater Baltimore Wound Care Center to evaluate the utility of combining wound care provider and patient cues. After approval from both the Towson University and Greater Baltimore Medical Center Institutional Review Boards, verbal and written informed consents for patient interviews and chart reviews were obtained. Patients who chose to participate in the study were asked to complete an SF-36 Health Survey, a paper and pencil quality of life instrument.[47] This survey was administered at the time of registration and at the clinical endpoint (i.e., venous leg ulcer healing). Healing rates were assessed by physicians and were based on size of wound and length of time to complete healing. If the wound did not heal, the predetermined study endpoint was 120 days.

Through use of a random number table, 33 patients were assigned to either a daily diary intervention or no daily diary. Patients in the daily diary arm were asked to complete the paper and pencil data collection tool, which contained questions about treatment features clinicians indicated were critical to good wound management (i.e., use of compression stockings, leg elevation, an index of pain, and an index of interference with each day's Activities of Daily Living). Patients were asked what pain medications they took each day and how often. An open-ended statement asked patients to "Write about how you are feeling about taking care of your wound." In addition, a satisfaction with the wound care center survey was mailed to the participants' homes at the end of the study.

At baseline, per the center's nurses' notes, all patients were ambulatory and mentally competent; capable of physically providing self-care; moderately to highly prepared for self-care; and free of barriers (physical, mental, religious, language, cultural) to self-care. Assessment of patient adherence to treatment protocol was also noted. Of the 15 patients who started in the daily diary intervention, 12 patients did not continue to keep the daily diary; thus complete in-depth data were available for only three patients who finished both the SF-36 Health Survey and the daily diary intervention over a range of 32 to 120 days.

The first patient, a white female, age 55, arrived at the center with systemic lupus erythematosis, and 4-cm long ulcers on both legs. Almost daily, she experienced moderate to severe pain and interference with activities. Her wound did not heal by the 120-day study endpoint. Over the 120 days, according to the baseline and endpoint SF-36 Health Survey scores, her physical functioning remained the same and her pain control improved. However, her mental health score declined. She often wrote that she felt frustrated, angry, and helpless. But she was very happy with her wound care and would definitely recommend the wound center to others. She also continued to smoke cigarettes but never noted that in her diary. That appeared only in the nurses' notes.

The second patient, a white female, age 69, arrived at the wound center with a 6-cm long venous leg ulcer. In slightly over a month (32 days), her wound healed. Her SF-36 physical functioning score declined, but her pain and mental health scores improved dramatically. She was happy with the care received and would recommend the wound center to others. However, she was nonadherent with her diet and with leg elevation. Once again, that information was not in the patient diary; it was in the nurses' notes.

The third patient was an African American male, age 78, who had a swollen, discolored leg, a 1-cm long venous leg ulcer, and diabetes mellitus. Over the 42 days that it took his wound to heal, not once did he indicate that he was in pain, but that his legs were "weak," most likely due to neuropathies. He, too, was happy with the care received and would recommend the wound center to others.

Self-documentation of the patient experience is potentially a powerful therapeutic tool.[48,49] However, the patients had to be very committed to completing the daily diaries, one of the major reasons our n was so small. Our research demonstrated that diaries are best used in conjunction with physician and nursing observations as well as other, more objective measures, such as the SF-36.

EMERGING COLLABORATIVE MODELS FOR VENOUS ULCER CARE

Loneliness, isolation, and depression are often associated with chronic venous ulcers and have been determined to be a barrier to wound healing.[5–7,50] Pain and wound odors can exacerbate these feelings due to interference with ADLs, poor body image, and decreased social contacts.[51,52] A UK study of a nurse-led education program to increase patient concordance (adherence) was found to be effective at increasing the time patients spent with their legs elevated at heart level and in preventing venous ulcer occurrence.[39] Patients who were 18 years of age or older, were able to understand the content of an education program, and had a venous ulcer that had healed within the past 2 years were eligible for inclusion in the experimental group. The intervention was composed of a 3-hour session educating nurses on enhancing patient concordance. Patients were given educational pamphlets, and nurses reinforced the prevention strategies described in the pamphlets every 3 months. A year later, only one leg ulcer recurred in the experimental group compared with 15 in the control group. Brooks and co-authors concluded that regular patient support and reinforcement of patient education were critical for concordance to be sustained.

In another model, also from the UK, a Leg Club was established to (1) deliver holistic empathic care in a nonthreatening environment, (2) encourage compliance with social interaction and peer support, and (3) become a local center for excellence in wound care.[40] Nurses partnered with a church and community volunteers helped with fund-raising to equip the club. Sponsorship was also obtained from the health care industry. Two years later, the author conducted a qualitative research study among patients attending the Leg Club using Becker's Health Belief Model. The themes that emerged from the research were enthusiasm for the friendly environment, motivation to attend for treatment, positive attitude toward the management of their care, belief in the efficacy of the treatment due to increased knowledge and support, and a sense of ownership in the club.

Members of the health care team felt that the informal clinic provided a forum for health promotion and education and an opportunity for team-building outside a traditional health care setting. In the United States, a support group utilizing the Leg Club model could provide a useful complement to wound care center protocols.

CONCLUSION

With the aging of the world's population, health care providers should be prepared to see a rise in the prevalence of venous ulcers. While almost half of the venous ulcers seen will heal with conservative therapy, a significant proportion of hard-to-heal wounds will remain and frustrate both patient and wound care provider. Adjuvant therapies, in conjunction with a collaborative relationship between the clinician and the patient, offer opportunities for greater adherence to the treatment protocol, improved healing rates, increased patient satisfaction, and decreased recurrence of venous ulcers. However, clinicians should be mindful that these therapies are constantly evolving, and new products are brought to market almost daily. These products should be evaluated with randomized clinical trials for efficacy and cost effectiveness. Expensive treatments should be used only in patients with wounds not responsive to conventional therapies, and adjunctive treatments should not be used in lieu of conventional therapy.

REFERENCES

1. Kerstein M, Gemmen E, van Rijswijk L, et al. Cost and cost effectiveness of venous and pressure ulcer protocols of care. *Dis Manage Health Outcomes.* 2001;9:651–663.
2. Kerstein M. Economics of quality ulcer care. *Dermatol Nursing.* 2003;15:59–61.
3. Rojas AI, Phillips TJ. Venous ulcers and their management. In: Falanga V, ed. *Cutaneous Wound Healing.* London: Martin Dunitz, Ltd, 2001:263–286.
4. Menzoian JO, LaMorte WW, Woodson J. Nonoperative treatment of venous ulcers In: Ernst CB, Stanley JC, eds. *Current Therapy in Vascular Surgery,* 3rd ed. St. Louis, MO: Mosby, 1995:924–927.
5. Brem H, Kirsner RS, Falanga V. Protocol for the successful treatment of venous ulcers. *Am J Surg.* 2004;188:S1–S8.
6. Cooper SM, Hofman D, Burge SM. Leg ulcers and pain: A review. *Lower Extremity Wounds.* 2003;2:189–197.
7. Mantyselka PT, Turunen JHO, Ahonen RS, Kumpusalo EA. Chronic pain and poor self-rated health. *JAMA.* 2003;290:2435–2442.
8. De Araujo T, Valencia I, Federman DG, Kirsner R. Managing the patient with venous ulcers. *Ann Intern Med.* 2003;138:326–334.
9. Kinsella K, Velkoff V. An aging world: 2001, U.S. Census Bureau, Series P95/01–1, Washington, DC: US Government Printing Office. Available at: http://www.census.gov/prod/2001pubs/p95-01-1.pdf. Accessed December 5, 2006.

10. Wipke DD, Sae-Sia W. Management of vascular leg ulcers. *Advances in Skin & Wound Care.* 2005;18:437–445.
11. Grace P, ed. Guidelines for the management of leg ulcers in Ireland Dublin (Ireland): Smith & Nephew Ltd. Available at: http://www.guideline.gov/summary/summary.aspx?doc_id=3616&nbr=2842&string=Venous+AND+leg+AND+ulcers. Accessed December 5, 2006.
12. Registered Nurses Association of Ontario (RNAO). Assessment and management of venous leg ulcers Toronto (ON): RNAO. Available at: http://www.rnao.org/bestpractices/PDF/BPG_venous_leg_ulcer.pdf. Accessed February 7, 2007.
13. Royal College of Nursing (RCN). The management of patients with venous leg ulcers: Recommendations London (UK): RCN. Available at: http://www.rcn.org.uk/publications/pdf/guidelines/venous_leg_ulcers.pdf. Accessed December 5, 2006.
14. Royal College of Nursing (RCN). Management of patients with venous leg ulcers: Audit protocol London (UK): RCN. Available at: http://www.rcn.org.uk/publications/pdf/guidelines/leg_ulcer_audit_protocol.pdf. Accessed December 5, 2006.
15. Wound, Ostomy, and Continence Nurses Society (WOCN). Guideline for management of wounds in patients with lower-extremity venous disease Glenview (IL): Wound, Ostomy, and Continence Nurses Society (WOCN); 2005 (WOCN clinical practice guideline; no. 4). Available at: http://www.guideline.gov/summary/summary.aspx?doc_id=7485. Accessed November 28, 2006.
16. Hess CT. Putting the squeeze on venous ulcers. Travel. *Nursing.* 2004;34(Suppl 1):8–13. Available at: http://www.nursing2004.com/pt/re/nursing/pdfhandler.00152193-200411001-00008.pdf;jsessionid=CtAMitDrGlHs24t2fGyL1D8JzcN253EQL9gR1h7x8GDXDAU5zkUm!430004150!-949856032!9001!-1. Accessed December 5, 2006.
17. Falanga V, Sabolinski M. A bilayered living skin construct (APLIGRAF®) accelerates complete closure of hard-to-heal venous ulcers. *Wound Repair and Regeneration.* 1999;7:201–207.
18. Gohel MS, Taylor M, Earnshaw JJ, Healther BP, Poskitt KR, Whyman MR. Risk factors for delayed healing and recurrence of chronic venous ulcers—An analysis of 1,324 legs. *Eur J Vasc Endovasc Surg.* 2004;29:74–77.
19. Chiu T, Burd A. "Xenograft" dressing in the treatment of burns. *Clin Dermatol.* 2005;23:419–423.
20. Snyder RJ. Treatment of nonhealing ulcers with allografts. *Clin Dermatol.* 2005;23:388–395.
21. Wang J, Jin Y, Guo Z. Expression of basic fibroblast growth factor and fibronectin in tissue engineering and skin allograft during healing process *Hua Xi Qiang Yi Xue Za Zhi.* 2003;21:41–43.
22. Falanga V, Margolis D, Alvarez O, et al., and the Human Skin Equivalent Investigators Group. Rapid healing of venous ulcers and lack of clinical rejection with an allogenic cultured human skin equivalent. *Arch Dermatol.* 1998;134:293–300.
23. Sibbald RG, Torrance GW, Walker V, Attard C, MacNeil P. Cost-effectiveness of Apligraf™ in the treatment of venous leg ulcers *Ostomy/Wound Manage.* 2001;47:36–46.
24. Dolynchuk K, Hull P, Guenther L, et al. The role of Apligraf™ in the treatment of venous leg ulcers *Ostomy/Wound Manage.* 1999;45:34–43.
25. Morris J, Dowlen S, Cullen B. Early clinical experience with topical collagen in vascular wound care. *JWOCN.* 1994;21:247–250.
26. Chronic wound dressings: New matrix restores a more normal healing environment Available at: http://www.medicalnewstoday.com/medicalnews.php?newsid=25458. Accessed January 15, 2007.
27. Mostow EN, Haraway GD, Dalsing M, Hodde JP, King D, Oasis Venous Ulcer Study Group. Effectiveness of an extracellular matrix graft (OASIS Wound Matrix) in the

treatment of chronic leg ulcers: A randomized clinical trial. *J Vasc Surg.* 2005;41: 837–843.
28. DaCosta RM, Ribeiro JRM, Aniceto C, Mendes M. Randomized, double-blind, placebo-controlled, dose-ranging study of granulocyte-macrophage stimulating factor in patients with chronic venous leg ulcers. *Wound Repair and Regeneration.* 1999;7: 17–25.
29. Moues CM, Vos MC, van de Bend GJ, Stijnen T, Hovius SE. Bacterial load in relation to vacuum-assisted closure wound therapy: A prospective randomized trial. *Wound Repair and Regeneration.* 2004;12:11–17.
30. Vuerstaek JDD, Vainas T, Wuite J, Nelemans P, Neumann MHA, Veraart JCJM. State-of-the-art treatment of chronic leg ulcers: A randomized controlled trial comparing vacuum-assisted closure (VAC) with modern wound dressings. *J Vasc Surg.* 2006;44: 1029–1038.
31. Wykle ML, Haug MR. Multicultural and social class aspects of self-care. *Generations.* 1993;17:25–29.
32. Becker M, Radius S, Rosenstock I, Drachman R, Schuberth K, Teets K. Compliance with a medical regimen for asthma. *Public Health Reports.* 1978;93:268–277.
33. Lee JK, Grace KA, Taylor AJ. Effect of a pharmacy care program on medication adherence and persistence, blood pressure, and low-density lipoprotein control: a randomized controlled trial. *JAMA.* 2006;296:2563–2571.
34. Simpson RJ. Challenges for improving medication adherence. *JAMA.* 2006;296: 2614–2615.
35. Osterberg L, Blaschke T. Adherence to medication. *N Engl J Med.* 2005;353:487–497.
36. Peters J. A review of the risk factors influencing nonrecurrence of venous leg ulcers. *J Clin Nurs.* 1998;7:3–9.
37. Sparks J. Venous leg ulcers—Why patients won't comply. *ACCNS Journal for Community Nurses (Australia).* 2002;7:15–16.
38. Van Agthoven WM, Plomp HN. The interpretation of self-care: A difference in outlook between clients and home nurses. *Soc Sci Med.* 1989;29:245–252.
39. Brooks J, Ersser SJ, Lloyd A, Ryan TJ. Nurse-led education sets out to improve patient concordance and prevent recurrence of leg ulcers. *J Wound Care.* 2004;13: 111–116.
40. Lindsay E. Leg clubs: A new approach to patient-centered leg ulcer management. *Nurs Health Sci.* 2000;2:139–141.
41. Lindsay E, Hawkins J. Care study: The leg club model and the sharing of knowledge. *Br J Nurs.* 2003;12:784, 786, 788–790.
42. Vowden K, Vowden P. Bridging the gap: The impact of patient choice on wound care. *J Wound Care.* 2006;15:143–145.
43. Ayello EA. Twenty years of wound care: Where have we been, where are we going? *J Wound Care.* 2006;19:28.
44. Bethel E, Cadogan J, Charles H, et al. Issues in wound care: Empowering patients—Developing a framework for practitioners. *J Wound Care.* 2006;15:62–63.
45. Burman ME. Health diaries in nursing research and practice. *Image: J Nurs Scholarship.* 1995;27:147–152.
46. Winkler R, Underwood P, Fatovich B, James R, Gray D. A clinical trial of a self-care approach to the management of chronic headache in general practice. *Soc Sci Med.* 1989;29:213–219.
47. Ware JE, Kosinski M, Keller SD. *SF-36: Physical and Mental Health Summary Scales: A User's Manual.* Boston, MA: Health Assessment Lab, 1994.
48. Hopkins A. Disrupted lives: Investigating coping strategies for non-healing leg ulcers. *Br J Nurs.* 2004;13:556–563.

49. Smyth J, Stone A, Hurewitz A, Kaell A. Effects of writing about stressful experiences on symptom reduction in patients with asthma or rheumatoid arthritis: A randomized trial. *JAMA*. 1999;281:1304–1309.
50. Fraser M. Psychosocial needs of the wound care patient. Presented at: Art and Science of Wound Management Meeting; September 4, 1997; Orlando, FL.
51. Doughty DB. Strategies for minimizing chronic wound pain. *Advances in Skin & Wound Care*. 2006;19:82–85.
52. Young CV. The effects of malodorous fungating malignant wounds on body image and quality of life. *J Wound Care*. 2005;14:359–362.

9

NEGATIVE PRESSURE DRESSINGS IN VENOUS ULCERS

BRIAN D. LEWIS AND
JONATHAN B. TOWNE

An estimated 80% of lower extremity ulcers are venous in origin, of which many are chronic and have a protracted course with associated morbidity. These ulcers have a negative impact on the patient's quality of life and are an economic burden to the patient and society. Some estimate that the cost in the United States may be between $1.9 billion to $2.5 billion annually.[1]

Management of these chronic wounds has been problematic in medicine for many years. Some mainstays of therapy over time have been debridement of necrotic tissue, utilization of saline gauze, leg elevation, compression, medicated wraps, various wound care solutions, topical treatments, and intermittent compression. All of these modalities have been utilized in the management of chronic venous wounds, which remain problematic and difficult to heal.

One of the newer modalities in managing both acute and chronic wounds is the use of topical negative pressure dressings. The topical negative pressure dressing in its most common form consists of an open cell polyurethane sponge with 400–600 µm pores, a transparent adhesive covering, noncollapsible tubing, and a vacuum-generating device with a collection reservoir. It should be noted that there is also a polyvinyl alcohol sponge with 60–270-µm open cell design that can be utilized in certain circumstances. The device has a set of extensive controls that will maintain a wide range of negative pressure settings and is capable of maintaining continuous or intermittent suction.[2]

Normal wound healing goes through many stages: migration of cells, such as macrophages, fibroblasts, and epithelial cells; removal of debris; decrease in bacterial burden; angiogenesis; formation of granulation tissue; wound contraction; production of connective tissue; and ongoing remodeling.[3] The topical negative pressure dressing can enhance many of these stages, which are needed for the wound healing process to occur.

The exact mechanism by which topical negative pressure dressings lead to improvement in wound healing is unknown, but several theories have been presented. The topical negative pressure dressing decreases local edema, results in an increase in local blood flow, may decrease wound's bacterial burden, and causes mechanical deformation of the wound bed, leading to improved wound healing. It also decreases chronic wound fluid and contributes to the maintenance of the moist wound healing environment, which is necessary for wound healing. All of these effects promote an increase in cell division and proliferation, an increase in matrix synthesis, and an increase in wound protein synthesis.[4]

CHRONIC WOUND FLUID

The open cell sponge utilized in the topical negative pressure dressing allows for an even application of subatmospheric pressure to the wound bed.[5] This allows for removal of excess fluid from the chronic wound bed and aids in wound healing by removing substances that are inhibitory to the wound healing process. It has been demonstrated that matrix metalloproteinases and the products of their breakdown that accumulate in the chronic wound fluid impair wound healing.[6] The removal of these products advances the wound healing process. Wound fluid from chronic venous ulcers may also inhibit cell growth by having fibroblasts enter a quiescent phase of the cell cycle; additionally, the chronic wound fluid also suppresses epithelial cells and endothelial cells, further inhibiting the wound healing process.[7]

LOCAL EDEMA

Local edema accompanies chronic venous wounds. This local interstitial edema may act as a physical barrier to local blood and lymphatic flow. The topical negative pressure dressings will decrease the local interstitial edema and improve lymphatic flow, resulting in a lower interstitial pressure. This allows for better flow of blood and nutrients through the capillaries and improves perfusion to the granulating bed, overall improving the local environment and aiding in the wound healing process.[2]

Improved perfusion of the wound bed may additionally aid in bacterial clearance.[5,6,7,9]

LOCAL BLOOD FLOW

In a study by Morykwas et al., using a Doppler probe in a porcine model, they were able to demonstrate that when subatmospheric pressure was decreased to −125 mmHg, there was optimization of local blood flow. The increase in flow at this level was approximately 400% over baseline. As the pressure was further decreased, there proved to be no additional increase in local blood flow.[8] It should also be noted that the increase in blood flow was present for only a limited period of time. The investigators have determined that the intermittent application of negative pressure for 5 minutes at −125 mmHg followed by 2 minutes with no negative pressure would be the preferred treatment modality. In a similar study, also using a porcine model, granulation tissue formation significantly increased using a topical negative pressure dressing when compared to saline gauze. The rate of granulation tissue formation was improved when either intermittent or continuous negative pressure was utilized; however, there was a trend toward more rapid granulation tissue formation during intermittent therapy, thus supporting the idea that intermittent cycling as opposed to continuous application of topical negative pressure is more efficacious in the wound healing process.[9]

BACTERIAL BURDEN

Another benefit of the topical negative pressure treatment is a potential decrease in wound bacterial burden. In wound studies done by Morykwas et al., 10^8 bacteria were inoculated into a fresh wound. The wounds were then treated with a topical negative pressure dressing at −125 mmHg pressure. The investigators were able to demonstrate that there was a decrease in the bacterial burden to approximately 10^5 bacteria by days 4 or 5 after treatment.[9] Moües et al. found that the bacterial burden may actually increase for some bacteria, in particular *S. aureus*. However, there was a decrease in the concentration of other bacteria, mainly aerobic gram-negative rods. They noted that even with the elevation in the number of *S. aureus* colonies in the wound, there was no impairment in the wound healing process.[10] Further supporting the decrease in bacterial burden, DeFranzo compared the bacterial burden in wounds treated with a topical negative pressure dressing to those managed with wet-dry dressings and found that the bacterial burden decreased from 10^7 to 10^2 or 10^3 by day 4 or 5 with the topical negative pressure dressing while it took until approximately day 11 to achieve the same results with wet-dry dressings.[5]

WOUND BED DEFORMATION

Another explanation for the potential benefits of topical negative pressure dressing in accelerated wound healing is the fact that the open cell sponge induces mechanical deformations of the wound base when negative pressure is applied. The walls of each open cell anchor portions of the wound bed, while the intervening segments are not anchored in the same way. Topical negative pressure then stretches and deforms these intervening areas. This induces deformation of the cells and stimulates cellular proliferation and migration.[2,11] This is similar to the cellular proliferation that occurs with the use of tissue expanders in plastic surgery and also is present with the use of bone distraction devices in orthopedics and oral surgery.[11] *In vitro* studies have revealed that cells allowed to stretch tend to proliferate, while unstretched cells are less likely to divide.[11]

This effect is thought to be modulated through cytoskeletal and indirect connections between the cell wall and the nucleus. These connections transfer mechanical forces that the cell has experienced to the nucleus, leading to a series of events resulting in expression of growth factor, protein formation, and gene expression, which induces cellular conformational changes and increases cellular proliferation.[2]

DIRECT WOUND CLOSURE

The direct effect of negative pressure on the wound and wound edges also exerts a constant effect on the wound edges contracting the ulcer. This effect directly decreases the size of the wound independent of any cellular proliferation that may be occurring.[7]

WOUND HEALING ENVIRONMENT

One of the mainstays of therapy through time has been wet-to-dry dressings. It is well known that a moist wound healing environment in the wound bed is beneficial in the wound healing process. The moist wound environment creates an environment that stimulates angiogenesis, encourages expression of normal growth factors, and increases fibrinolysis.[14] Typically, wet-to-dry dressings are changed multiple times throughout the day. The theory is that moistened dressings slowly dry out during the intervening time period, and when removed, there is mechanical debridement of any necrotic tissue from the wound bed; however, there is also the potential for removal of healthy granulation tissue.

Other modalities are also employed, which further maintain a moist wound healing environment. In addition to simple saline gauze, various hydrocolloid gels as well as occlusive dressings are used to maintain a moist wound healing environment. A properly used topical negative pressure dressing will maintain a moist wound healing environment. It should, however, be noted that when a topical negative pressure dressing has a large persistent air leak, the wound may desiccate, which will be detrimental to wound healing.[8]

WOUND HYPOXIA

The direct effects of negative pressure on the wound bed lead to a localized decrease in the partial pressure of the oxygen within the wound. This may stimulate the formation of new blood vessels and further result in improvement in the granulation bed.[12] This results in a healthier microenvironment within the wound bed.

ADJUNCT TO ASSISTED CLOSURE

Despite continued progress in the management of wounds, some wounds continue to prove extremely difficult to heal with standard techniques and often require other surgical modalities such as flap closure and skin grafting in order to obtain final coverage. The topical negative pressure dressing, through promotion of an adequate granulation bed, increases success of local flap closures. Because this type of dressing decreases overall wound size, the extent of definitive procedures may also be lessened.[2]

The topical negative pressure dressing may be effective in bolstering split thickness skin grafts after they are placed on a granulating wound bed. The negative pressure aids in the removal of excess fluid from the wound bed, thus allowing for good apposition of the split thickness skin graft to the granulation tissue. It also accentuates angiogenesis, which would also potentially lead to improvements in the outcomes from split thickness skin grafting. The even negative pressure applied through the open cell design allows for even apposition of the skin over the wound surface. When topical negative pressure dressing is used for split thickness skin graft coverage, it is most often recommended that subatmospheric pressure be utilized at −50 to −75 mmHg instead of −125 mmHg. Most individuals recommend placement of a nonadherent barrier between the open cell sponge and the split thickness skin graft to avoid ingrowth of healthy tissue into the sponge and inadvertent damage to the graft with dressing changes. The dressing is left in place for up to approximately 4 days prior to removal. When used to bolster a split thickness skin graft, the negative pressure is usually applied in a continuous, not intermittent, fashion.[6]

INDICATIONS AND CONTRAINDICATIONS

Some of the common indications for use of the topical negative pressure dressing as approved by the FDA include "diabetic foot wounds, pressure ulcers, chronic wounds, acute and traumatic wounds, dehisced surgical wounds, partial-thickness wounds, flaps and grafts" (VAC® Therapy Guidelines). Contraindications for use of the topical negative pressure dressing include "Malignancy in the wound, untreated osteomyelitis, non-enteric and unexplored fistula, necrotic tissue and eschar presenting the wound, exposed organs and blood vessels" (VAC® Therapy Guidelines). There are few reported complications attributed to the use of topical negative pressure dressings. The most common reported complications are pain, typically resolved when the negative pressure is decreased, and excessive tissue ingrowths into the dressing sponge.[8] This may lead to bleeding when the dressings are changed. Other reported complications have included erysipelas, damage to surrounding skin, and failure to heal ulcers.[13]

CONCLUSION

In conclusion, although the exact mechanism of action for topical negative pressure dressing in the management of chronic venous wounds is unknown, numerous theories exist. The utility of a topical negative pressure dressing may be as simple as removal of chronic wound fluid, increasing local blood flow; decrease in wound bacterial burden; effects of mechanical deformation of the wound bed; decrease in interstitial edema; and the maintenance of a moist wound healing environment. Some of these processes, all of these processes, or a combination of them may account for the topical negative pressure dressing's beneficial effects.

In a search of the literature completed by Shirakawa,[7] six randomized controlled trials were found to have been completed comparing topical negative pressure dressings with standard dressings. The six studies were for a wide variety of wound types; only one trial included venous wounds and this was a subgroup of a diverse wound population. Five of the six studies indicated that there may be some benefit to utilizing topical negative pressure dressings, though, as the authors note, only three of the studies showed a statistically significant improvement in wound healing, and two showed a trend that was not found to be statistically significant. All the studies could be questioned secondary to methodological flaws.[7]

In a recent study by Vuerstaek et al., the topical negative pressure dressing was evaluated for efficacy in the healing of chronic leg ulcers. This randomized controlled study compared topical negative pressure dressings to standard wound dressings in the healing of venous, combined venous/arterial, and arteriolosclerotic leg ulcers. All groups were evaluated for wound bed preparation time and then for time to complete wound healing following grafting. The authors found that topical negative pressure therapy resulted in a shorter wound healing time in all subgroups, though statistical significance could not be demonstrated in the combined venous/arterial subgroup. The authors also noted a trend for fewer ulcer recurrences in the topical negative pressure group when compared to the standard dressing group, though no statistical significance was found. One limitation of this study was that all patients were treated in the inpatient setting.[13]

Further well-designed studies will be needed to more definitively define the utility of the topical negative dressing as well as to elucidate the mechanism or mechanisms by which the potential benefits are gained.

REFERENCES

1. Valencia IC, Falabella A, Krisner RS, Eaglstein WH. Chronic venous insufficiency and venous leg ulceration. *J Am Acad Dermatol.* 2001;44:401–421.
2. Andros G, Armstrong DG, Attinger C, et al. Consensus statement on negative pressure wound therapy for the management of diabetic foot wounds. *Vasc Dis Manage* 2006(Suppl, July).
3. Emmanuella J, Hamori CA, Bergman S, Roaf E, Swann NF, Anastasi GW. A prospective, randomized trial of vacuum-assisted closure versus standard therapy of chronic nonhealing wounds. *Wounds.* 2000;12(3):60–67.
4. Thomas S. World wide wounds: An introduction to the use of vacuum-assisted closure. 2001. Available at: http://www.worldwidewounds.com/2001/may/Thomas/Vacuum-Assisted-Closure.html.
5. DeFranzo AJ, Argenta LC, Marks MW, et al. The use of vacuum-assisted closure therapy for the treatment of lower extremity wounds with exposed bone. *Plast Reconstr Surg.* 2001;108(5):1184–1191.
6. Argenta LC, Morykwas M. Vacuum-assisted closure: A new method for wound control and treatment: Clinical experience. *Ann Plastic Surg.* 1997;38:563–577.
7. Shirakawa M, Isseroff RR. Topical negative pressure devices: Use for enhancement of healing chronic wounds. *Arch Dermatol.* 2005;141:1449–1453.
8. Morykwas M, Faller B, Pearce D, Argenta L. Effects of varying levels of subatmospheric pressure on the rate of granulation tissue formation in experimental wounds in swine. *Ann Plast Surg.* 2001;47(5):547–551.
9. Morykwas M, Argenta LC, Shelton-Brown EI, McGuirt W. Vacuum-assisted closure: A new method for wound control and treatment: Animal studies and basic foundation. *Ann Plast Surg.* 1997;38(6):553–562.

10. Moues CM, Vos MC, Van Den Bemd GJCM, Stijnen T, Hovius SER. Bacterial load in relation to vacuum-assisted closure wound therapy: A prospective randomized trial. *Wound Repair and Regeneration.* 2004;12(1):11–17.
11. Saxena V, Hwang CW, Huang S, Eichbaum Q, Ingber D, Orgill DP. Vacuum-assisted closure: Microdeformations of wounds and cell proliferation. *Plast Reconstr Surg.* 2004;114(5):1086–1096.
12. Kirby F. Ward S, Sanchez O, Walker E, Mellett MM, Maltz SB, Lerner TT. Novel uses of a negative-pressure wound care system. *J Trauma.* 2002;53(1):117–121.
13. Vuerstaek J, Vainas T, Wuite J, Nelemans P, Neumann MHA, Veraart JCJM. State-of-the-art treatment of chronic leg ulcers: A randomized controlled trial comparing vacuum-assisted closure (V.A.C.) with modern wound dressings. *J Vasc Surg.* 2006;44(5):1029–1038.
14. Field K. Overview of wound healing in a moist environment. *J Surg.* 1994;167(1A): 2s–6s.

10

DEEP VEIN THROMBOSIS AND PREVENTION OF POST-THROMBOTIC SYNDROME

JOSEPH A. CAPRINI

INTRODUCTION

Deep vein thrombosis (DVT) refers to the formation of a thrombus in one of the deep veins of the body, usually in the leg, resulting in leg pain, tenderness, and swelling. The clinical course of DVT may be complicated by the potentially fatal conditions of pulmonary embolism (PE), recurrent DVT, and in the long term by the emergence of post-thrombotic syndrome (PTS). In its presentation, PTS is typified by a variety of symptoms and signs that can be systematically assessed, including edema, skin induration, hyperpigmentation, pain, pruritus, and paresthesia (Table 10-1). The most severe form of the syndrome will manifest as venous ulcers.[1]

As both a frequent and chronic complication of DVT, PTS interferes with patients' daily life, function, and health. The syndrome can have a profoundly adverse effect on quality of life, limiting physical activity, interfering with work and social activities, causing psychological distress, and changing patients' health perceptions.[2] Furthermore, PTS carries a high cost of medical care, with costs amounting to about 75% of the cost of a primary DVT.[3]

There are currently limited options for treating PTS, and most are aimed at the prevention or treatment of leg ulcers.[4] Indeed, the treatment of established PTS is described as frustrating for patients and physicians alike, with current therapies and management only able to improve or stabilize symptoms in around half of all patients.[4,5]

Despite extensive study of DVT and its prevention and treatment in recent years, PTS has received little attention from clinicians and research-

TABLE 10-1 Clinical Scale for Post-Thrombotic Syndrome (PTS) According to Villalta et al.[1]

PTS Symptoms	Pain
	Cramps
	Heaviness
	Pruritus
	Paresthesia
PTS Signs	Edema
	Skin induration
	Hyperpigmentation
	Ectasia
	Redness
	Pain during calf compression
PTS Score*	
0–4	No PTS
5–14	Mild PTS
≥15, or presence of ulcer	Severe PTS

* PTS score is based on the cumulative rating of the five symptoms and six signs, with each rated as 0 (absent), 1 (mild), 2 (moderate), or 3 (severe).

ers.[4] The syndrome has rarely been included as an outcome measure in DVT trials and continues to be an underdiagnosed, underappreciated, and often overlooked morbid consequence of DVT. There is poor appreciation of the burden posed by PTS and a prevailing perception that the condition is an inevitable and untreatable consequence of DVT.[4]

Prophylaxis with anticoagulant therapy is known to reduce the rate of DVT and its recurrence, and consensus guidelines on the prevention of venous thromboembolism have been formulated based on a strong body of evidence.[6,7] The prevention of DVT in the first place may therefore offer the best opportunities for optimal patient care and avoidance of PTS and venous ulcers.

RISK OF VENOUS THROMBOEMBOLISM, RECURRENCE, PTS, AND VENOUS ULCERS

Venous thromboembolism (VTE) is a major health problem. Recent estimates suggest that the total annual number of nonfatal, symptomatic VTE events in the United States exceeds 600,000, more than half of which are DVT.[8] Individual risk for DVT varies according to clinical situation, as shown in Table 10-2,[6] and potential of additional VTE risk factors including an acute infectious disease, cancer, age greater than 75 years, stroke, immobilization, and previous history of VTE.[9,10]

TABLE 10-2 Risk of Deep Vein Thrombosis (DVT) in Different Patient Populations[6]

	Probability of DVT without Prophylaxis (%)
Surgical	
Elective hip replacement	51 (48–54)
Total knee replacement	47 (42–51)
Hip fracture	44 (40–47)
General surgery	25 (24–26)
Gynecological surgery (malignancy)	22 (17–26)
Gynecological surgery (benign disease)	14 (11–17)
Medical	
General medical	12 (10–14)
Stroke	56 (51–61)
Myocardial infarction	22 (16–28)
Spinal cord injury	35 (31–39)
Medical intensive care	25 (19–32)

95% confidence intervals in parentheses.

Data from a population-based cohort study in the United States reveal that the risk for VTE recurrence after a first episode is high. As many as 5% of patients with a first VTE experience a recurrent event within 30 days, around 17% have a recurrence within 2 years, and 30% within 10 years.[11] The hazard of recurrence is highest in the first 6–12 months after the initial event, with hazard rate per 1,000 person-days ranging from 30 recurrent VTE events at 7 days, to 50 events at 30 days, and 10 at 6 months and 1 year.[11]

In a group of 335 consecutive patients who experienced symptomatic thrombosis, PTS occurred in 23% within 2 years of the VTE event, with a strong association between the development of ipsilateral recurrent DVT and risk for PTS.[12] Severe cases of PTS had manifested in 9% of patients after 5 years.[12] A recent meta-analysis has also shown that in patients in whom postoperative DVT was asymptomatic, the relative risk of developing PTS is 1.58 times (95% confidence interval [CI] 1.24–2.02) the risk in patients without evidence of DVT ($P < 0.0005$).[13] This highlights that even silent DVT predisposes the patient to the later threat of chronic PTS and risk for venous ulceration.

To date, specific patient risk factors for developing PTS have not been clearly established, but it has emerged that many patients with chronic venous ulceration have unsuspected PTS.[14,15] Several risk factors have been put forward, of which only recurrent DVT has clearly been identified as increasing the risk of PTS as much as 6-fold (Table 10-3).[4]

The annual incidence of venous ulceration is estimated to be at least 0.3%, of which around one-quarter of cases can be linked to a DVT history.[16,17] Both clinical and subclinical VTE appear to be significant pre-

TABLE 10-3 Proposed Risk Factors for Post-Thrombotic Syndrome (PTS)[4]

Proposed Risk Factor	Association with PTS
Recurrent DVT	Clearly identified risk factor, up to 6-fold increase
Asymptomatic DVT	Identified risk factor, 1.6-fold increase
Characteristics of the initial DVT; distal DVT	Distal DVT appears to be associated with PTS, although some studies suggest that the site of the initial thrombus is not predictive, or that the risk is higher in patients with proximal rather than distal DVT
Patient characteristics	Factors **predictive** of PTS (as identified in retrospective studies): • Increasing age • Female gender • Hormone therapy • Varicose veins • Abdominal surgery • Increased body mass index Factors **not predictive** of PTS (as identified in prospective studies): • Gender • Delay in initiating treatment for DVT • Risk factors for thrombosis • Family history of thrombosis • Intensity of warfarin anticoagulation

DVT = deep vein thrombosis.

disposing factors for the development of chronic venous insufficiency and venous ulcers.[18-20] Indeed, patients with a DVT history have been shown to be 2.4-fold (95% CI 1.7–3.2) more likely to develop venous stasis with its associated risk for venous ulcer formation.[19]

THROMBOPROPHYLAXIS: REDUCING THE INCIDENCE OF VTE AND LONG-TERM COMPLICATIONS

There is a compelling body of evidence that the incidence of VTE, and thereby its long-term complications, can be drastically reduced by providing effective thromboprophylaxis to patients at known risk for VTE.[6,7] Many studies of VTE prophylaxis have focused more on surgical rather than medical patients because of the very high risk of VTE following major surgical procedures. However, the absolute numbers of potentially at-risk patients reveal high numbers of medical patients who are candidates for VTE prophylaxis in accord with current guidelines on VTE prevention.[6,7,21]

Of an estimated 13.4 million U.S. residents who meet the current American College of Chest Physicians (ACCP) guideline defined risks for VTE each year, 7.6 million (57%) are medical patients.[21]

Despite internationally recognized guideline recommendations and government-backed initiatives in support of wider adoption of VTE prophylaxis in at-risk groups, the reality is that prophylaxis for VTE continues to be underutilized in clinical practice in both surgical and medical patient groups.[6,7,22-24] A recent multicenter study in Canada revealed that while most patients hospitalized for medical illness had indications for thromboprophylaxis, only 16% received prophylaxis deemed appropriate to reduce the risk of VTE.[24] In a Spanish study, it was noted that appropriate adherence to all guideline recommendations on VTE prophylaxis was observed in only 42% of patients with risk factors, with dosage of thromboprophylaxis often inappropriate according to patients' relative risk.[23]

Reasons for underprophylaxis of hospitalized patients may relate to an underestimation of VTE risk, since VTE events often arise after hospital discharge and VTE is often clinically silent, or to emergence of symptoms that are not immediately detected or detectable during patient follow-up. Inadequate levels of prophylaxis not only increase the risk of occurrence of thrombosis, but also increase the risk of PTS and, in this way, venous ulceration.[4] The key to optimal VTE prevention lies in better assessment of patient risk and appropriate intervention. Indeed, in the United States, the National Quality Forum (NQF) identifies VTE risk assessment among its 30 practices for improving patient safety, highlighting the need to evaluate all patients for VTE risk on hospital admission and stressing the need for appropriate methods to prevent VTE, a recommendation supported by the Agency for Healthcare Research and Quality (AHRQ) as well.[22]

RISK ASSESSMENT MODEL (RAM)

There is a body of literature and evidence regarding VTE risk factors and guidance on patient risk assessment to support clinicians and surgeons in identifying patients at risk of VTE in daily practice. The ACCP guidelines approach to risk assessment is to define patients as belonging to one of four broad categories of VTE risk: low, moderate, high, and highest risk.[7] However, the ACCP approach of broad risk categories and its lack of specification of what constitutes "other risk factors" does not provide an immediately accessible model for application in clinical practice, and there have therefore been attempts to develop more practical risk assessment models and tools.

For example, using a computerized alert program, it is possible to encourage greater and more successful use of prophylaxis among at-risk patients.[25] In a study conducted at a large U.S. hospital, an alert system linked to the inpatient database used eight common risk factors to determine each patient's risk status and then prompted clinicians to use thromboprophylaxis. This approach resulted in a 41% reduction in the rate of venographically confirmed VTE at 90 days compared with clinician-based judgment alone ($P = 0.001$).[25]

A risk assessment model based on detailed evaluation of a patient's exposing risk factors associated with the clinical setting and predisposing risk factors for VTE has been devised based on previous models and real clinical data[10] (Figure 10-1). The model requires that a form be completed on hospital entry to allow physicians to score thrombosis risk and select and prescribe the most appropriate prophylaxis according to whether the patient falls into a low, moderate, high, or highest risk category. Through this model, the appropriate selection of prophylaxis per patient is based on a risk factor score, and the type, duration, and intensity of prophylaxis are based on patient need.

METHODS OF THROMBOPROPHYLAXIS

A range of drugs is available for VTE prophylaxis, including warfarin, unfractionated heparin (UFH), the low molecular weight heparins (LMWHs) enoxaparin and dalteparin, and the pentasaccharide fondaparinux (Table 10-4).

WARFARIN

Data from the Global Orthopaedic Registry (GLORY) of in-hospital management and outcomes following elective total hip and knee replacement reveal that, as a contemporary choice for primary VTE prophylaxis, warfarin use is restricted almost entirely to the United States.[26] While 58% of orthopedic patients in GLORY in the United States received warfarin in the hospital, the use of this vitamin K antagonist has largely been abandoned in Europe in favor of 92% use of LMWH.[26]

Although warfarin therapy has the convenience of oral administration, its use is complicated by the difficulty in reaching and maintaining target International Normalized Ratio (INR) levels—a process which may take between 1 and 12 days.[27] Drug levels must be monitored to ensure a balance between treatment efficacy and safety, specifically concerns over bleeding, and there is a need for frequent drug monitoring to avoid the many common drug and food interactions with warfarin.[28]

Underanticoagulation with warfarin is common, and it has been shown that achieving a target INR of 2–3 is important throughout the period of patients' highest risk. For example, in a study of 125 patients undergoing total hip replacement, achievement of target INR over the entire 4 weeks

Joseph A. Caprini, MD, MS, FACS, RVT
Louis W. Biegler Professor of Surgery
Northwestern University
The Feinberg School of Medicine;
Professor of Biomedical Engineering
Northwestern University;
Email: j-caprini2@aol.com
Website: venousdisease.com

Venous Thromboembolism Risk Factor Assessment

Patient's Name:_____ Age: ___ Sex: ___ Wgt:___lbs

Choose All That Apply

Each Risk Factor Represents 1 Point

- Age 41–59 years
- Minor surgery planned
- History of prior major surgery
- Varicose veins
- History of inflammatory bowel disease
- Swollen legs (current)
- Obesity (BMI >30)
- Acute myocardial infarction (<l month)
- Congestive heart failure (<1 month)
- Sepsis (<1 month)
- Serious lung disease incl. pneumonia (<1 month)
- Abnormal pulmonary function (COPD)
- Medical patient currently at bed rest
- Leg plaster cast or brace
- Other risk factors_____

Each Risk Factor Represents 2 Points

- Age 60–74 years
- Major surgery (>60 minutes)
- Arthroscopic surgery (>60 minutes)
- Laparoscopic surgery (>60 minutes)
- Previous malignancy
- Central venous access
- Morbid obesity (BMI >40)

FIGURE 10-1 Risk assessment model. (Updated from Caprini et al.)[10]

Each Risk Factor Represents 3 Points

- ☐ Age 75 years and over
- ☐ Major surgery lasting 2–3 hours
- ☐ Venous stasis syndrome (BMI >50)
- ☐ History of SVT, DVT/PE
- ☐ **Family history of DVT/PE**
- ☐ Present cancer or chemotherapy
- ☐ Positive Factor V Leiden
- ☐ Positive Prothrombin 20210A
- ☐ Elevated serum homocysteine
- ☐ Positive Lupus anticoagulant
- ☐ Elevated anticardiolipin antibodies
- ☐ Heparin-induced thrombocytopenia (HIT)
- ☐ Other thrombophilia Type_____ _____

Each Risk Factor Represents 5 Points

- ☐ Elective major lower extremity arthroplasty
- ☐ Hip, pelvis, or leg fracture (<1 month)
- ☐ Stroke (<1 month)
- ☐ Multiple trauma (<1 month)
- ☐ Acute spinal cord injury (paralysis) (<1 month)
- ☐ Major surgery lasting over 3 hours

For Women Only (Each Represents 1 Point)

- ☐ Oral contraceptives or hormone replacement therapy
- ☐ Pregnancy or postpartum (<1 month)
- ☐ History of unexplained stillborn infant, recurrent spontaneous abortion (≥3), premature birth with toxemia, or growth-restricted infant

FIGURE 10-1 *Continued*

postoperation was vital to providing adequate DVT prophylaxis with warfarin.[29] This study found not only that average INR values were higher among patients who did not develop DVT, but also demonstrated that a larger proportion of the patients who developed DVT were outside the therapeutic INR range during the third (73% vs. 39%, $P < 0.001$) and fourth postoperative weeks (89% vs. 39%, $P < 0.001$).

TABLE 10-4 Summary of Methods of Thromboprophylaxis and Guideline Recommendations in Patients at Risk of Venous Thromboembolism (VTE) (for further details, see Geerts et al.[7] and Nicolaides et al.[6])

Prophylaxis Method	Patient Population	Dosing	Duration
Warfarin	Orthopedic surgery (THA, TKA, HFS)	INR target 2.5 (range 2–3)	TKA: 10 days, THA and HFS: 28–35 days
UFH	General surgery	Moderate risk: 5,000 units twice or three times daily High risk: 5,000 units three times daily	No recommendation available
	Medical conditions	At-risk: 5,000 units three times daily	No recommendation available
LMWH	General surgery	Moderate and high risk: enoxaparin 40 mg once daily, dalteparin 5,000 units once daily	5–10 days
	Orthopedic surgery (THA, TKA, HFS)	Enoxaparin 30 mg twice daily, enoxaparin 40 mg once daily when started pre-operatively (THA only), dalteparin 5,000 units once daily	TKA: 10 days, THA and HFS: 28–35 days
	Medical conditions	At-risk: enoxaparin 40 mg once daily, dalteparin 5,000 units once daily	6–14 days
Fondaparinux	Orthopedic surgery (THA, TKA, HFS)	2.5 mg once daily	TKA: 10 days, THA and HFS: 28–35 days
Mechanical methods	Low-risk patients, patients with contraindication to pharmacological prophylaxis, in addition to pharmacological prophylaxis in high-risk patients	—	—

HFS = hip fracture surgery; INR = international normalized ratio; THA = total hip arthroplasty; TKA = total knee arthroplasty.

UNFRACTIONATED HEPARIN (UFH)

Seminal studies of heparin-based VTE prophylaxis have shown the beneficial effects of low-dose unfractionated heparin (UFH) administered subcutaneously twice or three times daily in patients undergoing major surgery.[30,31] One of the first studies to demonstrate the preventive benefits of UFH was an international multicenter study comparing low-dose UFH given three times daily with no specific prophylaxis in 4,121 elective surgical patients, where rates of postoperative DVT fell from 24.6% to 7.7% ($P < 0.005$).[30] An early systematic review of studies with UFH for VTE prevention confirmed that perioperative UFH reduced the risk of DVT following general, orthopedic, and urologic surgery by 68% compared with control ($P < 0.001$), but that overall excessive bleeding or need for transfusion was increased relatively by between 50% and 66% in patients given UFH.[31]

Low-dose UFH (5,000 U) given twice daily has not been found as effective in reducing DVT rates in medical patients at risk of VTE, despite early reports that this form of prophylaxis reduced mortality.[32] The Heparin Prophylaxis Study group, which assessed the effects of low-dose UFH twice daily as VTE prophylaxis in patients with infectious disease, also found that while UFH appeared to delay time to fatal PE in this group of medical patients, it had no effect on overall rate of fatal PE.[33]

Indeed, although UFH twice daily is commonly used in clinical practice, a recent evaluation showed that there is no large, well-conducted, placebo-controlled trial to support the use of UFH twice daily for VTE prevention in hospitalized medical patients.[34] Future large-scale controlled studies in this field are unlikely given the known incidence of complications associated with UFH compared with LMWHs. While more frequent dosing of UFH (three times daily) may offer more effective VTE prophylaxis in medical patients, such a regimen would be associated with a higher risk of bleeding.[34]

Another factor playing against widespread use of UFH for VTE prophylaxis is the risk of heparin-induced thrombocytopenia (HIT). A recent meta-analysis of published study data spanning 20 years and including 2,478 patients found that HIT occurs in around 2.6% of patients given UFH thromboprophylaxis compared with 10-fold lower rates at 0.2% during use of LMWH to prevent VTE (odds ratio [OR] 0.10, 95% CI 0.03–0.33).[35]

Prophylaxis with enoxaparin, dalteparin, or fondaparinux reduces the rate of DVT very effectively and requires less frequent administration than use of UFH, making these drugs preferred options for the prevention of VTE (initial and recurrent episodes).

LOW MOLECULAR WEIGHT HEPARIN (LMWH)

A strong body of evidence supports the role of LMWH in the prevention of VTE in a wide range of patients at moderate to highest risk for thrombosis associated with surgery, medical conditions, and known VTE risk factors. A meta-analysis of randomized trials in orthopedic surgery comparing warfarin with other forms of VTE prophylaxis has shown the superior efficacy profile of LMWH relative to warfarin in preventing DVT (risk reduction [RR] 1.51, 95% CI 1.27, 1.79, $P < 0.001$).[36] On an individual trial basis, for example in 1,472 patients undergoing total hip arthroplasty (THA), pre- and postoperative dalteparin (initial dose 2,500 IU, followed by 5,000 IU once daily) for 6 ± 2 days was associated with lower rates of DVT of 10.7% and 13.1%, respectively, compared to 24% with warfarin ($P < 0.001$).[37] Extended dalteparin out-of-hospital to 35 days in 569 patients further reduced the cumulative incidence of DVT to 19.7% compared to 36.7% in the warfarin/placebo group ($P < 0.001$).[38]

Extending enoxaparin prophylaxis (40 mg once daily) for 21 days, after an initial course of 7–10 days, in 435 patients undergoing THA resulted in a VTE rate of 8.0% compared with 23.3% ($P < 0.001$) in the enoxaparin/placebo group.[39] Enoxaparin (30 mg twice daily) for 4–14 days has also been shown to reduce the incidence of VTE after total knee arthroplasty (TKA) in a study of 349 patients, with VTE rates of 25% in the enoxaparin group compared with 45% in the warfarin group ($P = 0.0001$) and no difference in major bleeding rates.[40] Prolonging enoxaparin for 3 weeks had no significant benefit in TKA patients.[39]

In a meta-analysis of 51 studies comparing LMWH with UFH for prevention of VTE in more than 48,000 general surgery patients, LMWH was at least as effective and safe as UFH in preventing VTE, with a suggested benefit in reducing overall clinical VTE events (RR 0.71, 95% CI 0.51–0.99).[41] To highlight one of the 51 studies included in this meta-analysis, the Enoxaparin and Cancer (ENOXACAN) study demonstrated that enoxaparin 40 mg once daily was at least as effective and safe as UFH three times daily in 1,115 patients undergoing elective surgery for cancer. The incidence rate of VTE was 14.7% in the enoxaparin group and 18.2% with UFH (OR 0.78, 95% CI 0.51–1.19), with no differences in bleeding complications.[42] Extending enoxaparin for another 21 days after an initial 6–10 days in the ENOXACAN–II study resulted in a VTE rate of 4.8% compared to 12.0% in the enoxaparin/placebo group ($P = 0.02$).[43]

A number of key trials have shown that LMWH reduces VTE incidence compared with placebo in medical patients at risk for VTE and associated complications, without compromising safety.[6] In addition, LMWH is at least as effective and safe as UFH three times daily in this patient population.

In the Prophylaxis in Medical Patients with Enoxaparin (MEDENOX) trial, comparing once daily subcutaneous enoxaparin 20 mg or 40 mg with placebo for 6–14 days in a cohort of 1,102 acutely ill hospitalized medical patients, enoxaparin 40 mg daily significantly reduced VTE to 5.5% compared with an incidence of 14.9% in the placebo group ($P < 0.001$).[44] This reduction in rate of VTE linked with medical illness and hospitalization was achieved with a similar rate of adverse events with LMWH or placebo.

Another large-scale study in 3,706 acutely ill medical patients—the Prospective Evaluation of Dalteparin Efficacy for Prevention of VTE in Immobilized Patients Trial (PREVENT)—also demonstrated the efficacy and safety benefits of dalteparin over placebo.[45] Medically ill patients assigned to dalteparin 5,000 IU daily for 14 days had a rate of VTE of 2.77% compared with 4.96% in the placebo group in the 90 days of follow-up, representing an absolute risk reduction of 2.19%, and a relative risk reduction of 45% ($P = 0.0015$). This reduction in VTE was achieved without a significant increase in the risk of major bleeding.[45]

A study which compared the effects of enoxaparin with UFH prophylaxis in medical patients with heart failure or severe respiratory disease was the Thromboembolism-Prevention in Cardiac or Respiratory Disease with Enoxaparin (THE-PRINCE) study.[46] This multicenter, open study randomized 665 patients to receive either enoxaparin 40 mg once daily or UFH (5,000 IU three times daily) for 10–12 days. The incidence of VTE was 10.4% with UFH and 8.4% with enoxaparin ($P = 0.015$, for equivalence). The use of enoxaparin was associated with fewer deaths, a lower rate of bleeding complications, and fewer adverse events (45.8% vs. 53.8%, $P = 0.044$) than UFH prophylaxis in this patient group.[46]

To date, LMWH has been used in an estimated 1.3 million patients worldwide for prevention of VTE, with the class recognized to have an excellent track record of no recalls due to lack of efficacy or safety concerns. Within the class, the LMWH enoxaparin is both the most widely used and most studied of the available drugs, and evidence relating to this LMWH is widely cited in current guidelines and consensus publications concerning recommendations for use of LMWH for VTE prevention.[6,7]

FONDAPARINUX

Fondaparinux is a pentasaccharide molecule and the first in a novel class of synthetic antithrombotic drugs. A meta-analysis of data from four randomized double-blind studies evaluating the role of fondaparinux (2.5 mg once daily) in VTE prevention in 7,344 patients undergoing major orthopedic surgery showed a significant reduction in the incidence of total VTE (6.8%) at day 11 compared with enoxaparin (13.7%) ($P < 0.001$). However, fondaparinux was not associated with a difference in the rate of symptomatic VTE and PE (0.6% vs. 0.4%, $P = 0.25$), and increased the rate of major bleeding (2.7% vs. 1.7%, $P = 0.008$).[47]

The PENTasaccharide in HIp-FRActure Surgery Plus (PENTHIFRA Plus) study evaluated the efficacy and safety of extending fondaparinux prophylaxis for up to 3 weeks after hip surgery in a group of 656 patients.[48] All patients received fondaparinux for 6–8 days after surgery, and subsequently one group was assigned to extended prophylaxis with fondaparinux while another received placebo for 3 weeks. Extended fondaparinux was reported to reduce the incidence of postoperative VTE by 96% from 35% to 1.4% in these high-risk patients ($P < 0.001$). Again, there was a trend toward more major bleeding in patients receiving fondaparinux (2.4% vs. 0.6%, $P = 0.06$).[48]

In 2,048 high-risk abdominal surgery patients in the PEntasaccharide GenerAl SUrgery Study (PEGASUS), which compared fondaparinux 2.5 mg and dalteparin 5,000 IU for 5–9 days, the rate of VTE observed up to day 10 was similar between treatment groups, at 4.6% in the fondaparinux group compared with 6.1% in the dalteparin group ($P = 0.144$).[49]

Fondaparinux has also been compared with placebo in hospitalized medical patients. In the ARixtra for ThromboEmbolism prevention in a Medical Indications Study (ARTEMIS), the efficacy and safety of 2.5 mg of fondaparinux once daily for 6–14 days were compared with that of placebo in a group of 849 elderly, acutely ill patients with conditions requiring hospitalization and bed rest.[50] VTE was detected in 10.5% of placebo patients, and this rate was reduced by 46.7% to 5.6% in patients given fondaparinux ($P = 0.029$) with no apparent increase in bleeding risk.[50]

The evidence to date suggests promise for this novel drug against preventable VTE, but until further data accumulate, fondaparinux should be reserved for use in selected patients.

MECHANICAL METHODS OF PROPHYLAXIS

Mechanical methods of prophylaxis have a long history in VTE prevention. A recent meta-analysis of data from 15 clinical studies, assessing the efficacy of intermittent pneumatic compression (IPC) for DVT prevention in 2,270 surgical patients, showed that IPC reduces the risk of DVT by 60% compared with no prophylaxis ($P < 0.001$).[51] When mechanical prophylaxis (pneumatic compression devices) was combined with pharmacological prophylaxis (low-dose UFH) in 2,551 patients undergoing cardiac surgery, reductions in VTE risk were greater still with a further 62% reduction in DVT over use of mechanical methods alone ($P < 0.001$).[52]

Preliminary data from the safety study of fondaparinux sodium to prevent venous thromboembolic events (APOLLO), which evaluates the combined effect of fondaparinux and IPC versus IPC alone for prevention of VTE after major abdominal surgery, suggest that this combination of mechanical and pharmacological prophylaxis can effect high rates of VTE reduction 32 days after surgery, reducing VTE incidence to 1.7%, a 70% reduction over use of IPC alone ($P = 0.004$).[53]

Although mechanical devices do not provide sufficient protection against VTE in moderate to high-risk patients on their own, they may be used to complement or replace pharmacoprophylaxis in certain patient groups in whom pharmacological prophylaxis is contraindicated. Compression devices and vena cava filters may be indicated as methods to reduce the risk of VTE in patients with known contraindications to anticoagulant therapy, in selected trauma cases where bleeding risk is high, or to complement anticoagulant prophylaxis during procedures where there is a very high risk of VTE.[6,7]

REMAINING ISSUES IN THE PREVENTION OF POST-THROMBOTIC SYNDROME

There is a robust body of evidence to support the use of prophylaxis to reduce the incidence of DVT, but, to date, little to none of the research and study of VTE prevention has included collection of data that demonstrate a direct impact of primary DVT prevention on PTS rates.

Although prophylaxis can help reduce the incidence of DVT, its use to prevent PTS and venous ulceration is also limited by the fact that measures to reduce DVT do not eliminate it completely. Furthermore, not all DVT can be predicted based on risk assessment. The risk of recurrence of DVT following an initial symptomatic DVT event can be reduced through long-term use of secondary prophylaxis, for which warfarin is generally used because of its ease of oral administration. As demonstrated in a meta-analysis, LMWH is an alternative option, being at least as effective as and perhaps favorable to warfarin in terms of VTE recurrence (OR 0.70, 95% CI 0.42–1.16) and resulting in fewer bleeding complications (OR 0.38, 95% CI 0.15–0.94).[54]

There is evidence that the use of therapeutic compression stockings reduces the incidence of PTS after symptomatic DVT.[55-57] In 325 outpatients with a first episode of DVT, graded compression stockings for at least 2 years were associated with a lower incidence of mild-to-moderate PTS at 20% compared to 47% in the control group without stockings ($P < 0.001$), as well as a lower incidence of severe PTS (11% vs. 23%, $P < 0.001$).[56] In a review on the value of graduated compression stockings, the pooled incidence of PTS of three studies was significantly reduced from 54% to 25.2% (RR 0.47, 95% CI 0.36–0.61), as well as the incidence of recurrent asymptomatic DVT (RR 0.20, 95% CI 0.06–0.64).[57] Thus, available data strongly suggest that after DVT, compression stockings can reduce the incidence of PTS and should therefore be used together with DVT prevention to reduce the burden posed by PTS.[55] In contrast, there is less evidence to suggest that the intensity of warfarin anticoagulation has any important impact on incidence of PTS after DVT.[15]

Clearly, more studies are needed to address the prevention and treatment of PTS in order that evidence-based data and results can be included in future guideline recommendations for improved patient care and management.

CONCLUSION

Until such time as effective treatments for PTS and its potential manifestation of venous ulceration have been found, prevention of the syndrome remains the key to reducing the impact of this condition on both patients and society. Therapeutic compression stockings associated with anticoagulant treatment after an incident of DVT may have a role in reducing the risks of progression to PTS, but better still is prevention of VTE whenever possible. Identifying patients at risk of VTE is possible, and great progress has been made in terms of accurate risk assessment of individual patients that in turn allows the best and most appropriate use of available methods of VTE prophylaxis. If more at-risk patients are provided with appropriate, clinically proven anticoagulant therapy during periods of high risk, the incidence of DVT, and consequently its long-term complications, can be reduced. Prevention of venous diseases with the potential to progress to PTS, particularly DVT prevention, may lead to a reduction in the incidence of venous ulcers.

DISCLOSURE/ACKNOWLEDGMENT

The author received editorial/writing support from Hester van Lier, PhD, in the preparation of this manuscript, which was funded by Sanofi-Aventis, NJ, USA. The author is fully responsible for content and editorial decisions in this manuscript.

CONFLICT OF INTEREST

The author is on the speaker's bureau and a consultant for Tyco, Sanofi-Aventis, GSK, and Eisai pharmaceuticals.

REFERENCES

1. Villalta S, Bagatella P, Picciolo A, Lensing A, Prins M, Prandoni P. Assessment of validity and reproducibility of a clinical scale for the post-thrombotic syndrome (abstract). *Haemostasis.* 1994;24(Suppl 1):158a.

2. Kahn SR, Hirsch A, Shrier I. Effect of postthrombotic syndrome on health-related quality of life after deep venous thrombosis. *Arch Intern Med.* 2002;162(10): 1144–1148.
3. Bergqvist D, Jendteg S, Johansen L, Persson U, Odegaard K. Cost of long-term complications of deep venous thrombosis of the lower extremities: An analysis of a defined patient population in Sweden. *Ann Intern Med.* 1997;126(6):454–457.
4. Kahn SR. The post-thrombotic syndrome: The forgotten morbidity of deep venous thrombosis. *J Thromb Thrombolysis.* 2006;21(1):41–48.
5. Bernardi E, Prandoni P. The post-thrombotic syndrome. *Curr Opin Pulm Med.* 2000; 6(4):335–342.
6. Nicolaides A. Cardiovascular Disease Educational and Research Trust; Cyprus Cardiovascular Disease Educational and Research Trust; European Venous Forum; International Surgical Thrombosis Forum; International Union of Angiology; Union Internationale de Phlebologie. Prevention and treatment of venous thromboembolism. International Consensus Statement (guidelines according to scientific evidence). *Int Angiol.* 2006;25(2):101–161.
7. Geerts WH, Pineo GF, Heit JA, et al. Prevention of venous thromboembolism: The Seventh ACCP Conference on Antithrombotic and Thrombolytic Therapy. *Chest.* 2004;126(3 Suppl):338S–400S.
8. Heit JA, Mohr DN, Silverstein MD, Petterson TM, O'Fallon WM, Melton LJ III. Predictors of recurrence after deep vein thrombosis and pulmonary embolism: A population-based cohort study. *Arch Intern Med.* 2000;160(6):761–768.
9. Alikhan R, Cohen AT, Combe S, et al., and the MEDENOX Study. Risk factors for venous thromboembolism in hospitalized patients with acute medical illness: Analysis of the MEDENOX Study. *Arch Intern Med.* 2004;164(9):963–968.
10. Caprini JA, Arcelus JI, Reyna JJ. Effective risk stratification of surgical and nonsurgical patients for venous thromboembolic disease. *Semin Hematol.* 2001;38(2 Suppl 5): 12–19.
11. Heit JA, Cohen AT, Anderson FA, et al., for the VTE Impact Assessment Group. Estimated annual number of incident and recurrent, non-fatal and fatal venous thromboembolism (VTE) events in the U.S. Poster 68 presented at: American Society of Hematology, 47th Annual Meeting; December 10–13, 2005; Atlanta, GA.
12. Prandoni P, Lensing AW, Cogo A, et al. The long-term clinical course of acute deep venous thrombosis. *Ann Intern Med.* 1996;125(1):1–7.
13. Wille-Jorgensen P, Jorgensen LN, Crawford M. Asymptomatic postoperative deep vein thrombosis and the development of postthrombotic syndrome. A systematic review and meta-analysis. *Thromb Haemost.* 2005;93(2):236–241.
14. MacKenzie RK, Ludlam CA, Ruckley CV, Allan PL, Burns P, Bradbury AW. The prevalence of thrombophilia in patients with chronic venous leg ulceration. *J Vasc Surg.* 2002;35(4):718–722.
15. Kahn SR, Kearon C, Julian JA, et al., and the Extended Low-Intensity Anticoagulation for Thrombo-Embolism (ELATE) Investigators. Predictors of the post-thrombotic syndrome during long-term treatment of proximal deep vein thrombosis. *J Thromb Haemost.* 2005;3(4):718–723.
16. Nelzen O, Bergqvist D, Lindhagen A, Hallbook T. Chronic leg ulcers: An underestimated problem in primary health care among elderly patients. *J Epidemiol Community Health.* 1991a;45(3):184–187.
17. Nelzen O, Bergqvist D, Lindhagen A. Leg ulcer etiology—A cross sectional population study. *J Vasc Surg.* 1991b;14(4):557–564.
18. Scott TE, LaMorte WW, Gorin DR, Menzoian JO. Risk factors for chronic venous insufficiency: A dual case-control study. *J Vasc Surg.* 1995;22(5):622–628.

19. Mohr DN, Silverstein MD, Heit JA, Petterson TM, O'Fallon WM, Melton LJ. The venous stasis syndrome after deep venous thrombosis or pulmonary embolism: A population-based study. *Mayo Clin Proc.* 2000;75(12):1249-1256.
20. Berard A, Abenhaim L, Platt R, Kahn SR, Steinmetz O. Risk factors for the first-time development of venous ulcers of the lower limbs: The influence of heredity and physical activity. *Angiology.* 2002;53(6):647-657.
21. Anderson FA, Zayaruzny M, Heit JA, Cohen AT. Estimated annual number of U.S. acute-care hospital inpatients meeting ACCP criteria for venous thromboembolism (VTE) prophylaxis. *Blood.* 2005;106:Abstract 903.
22. Agency for Healthcare Research and Quality. Thirty safe practices for better health care. March 2005. Available at: http://www.ahrq.gov/qual/30safe.pdf. Accessed September 2006.
23. Vallano A, Arnau JM, Miralda GP, Perez-Bartoli J. Use of venous thromboprophylaxis and adherence to guideline recommendations: A cross-sectional study. *Thromb J.* 2004;2(1):3.
24. Kahn SR, Panju A, Geerts W, et al., for the CURVE Study Investigators. Multicenter evaluation of the use of venous thromboembolism prophylaxis in acutely ill medical patients in Canada. *Thromb Res.* 2006 Mar 1; [Epub ahead of print].
25. Kucher N, Koo S, Quiroz R, et al. Electronic alerts to prevent venous thromboembolism among hospitalized patients. *N Engl J Med.* 2005;352(10):969-977.
26. Friedman R, Gallus A, Cushner F, FitzGerald G, Anderson F Jr, for the GLORY Investigators. Compliance with ACCP guidelines for prevention of venous thromboembolism: Multinational findings from the Global Orthopaedic Registry (GLORY). ASH Congress, San Diego, CA, USA; December 8, 2003. *Blood.* 2003;102(11):165a (abstract #574).
27. Lieberman JR, Sung R, Dorey F, Thomas BJ, Kilgus DJ, Finerman GA. Low-dose warfarin prophylaxis to prevent symptomatic pulmonary embolism after total knee arthroplasty. *J Arthroplasty.* 1997;12(2):180-184.
28. Holbrook AM, Pereira JA, Labiris R, et al. Systematic overview of warfarin and its drug and food interactions. *Arch Intern Med.* 2005;165(10):1095-1106.
29. Caprini JA, Arcelus JI, Motykie G, Kudrna JC, Mokhtee D, Reyna JJ. The influence of oral anticoagulation therapy on deep vein thrombosis rates four weeks after total hip replacement. *J Vasc Surg.* 1999;30:813-820.
30. Prevention of fatal postoperative pulmonary embolism by low doses of heparin. An international multicentre trial. *Lancet.* 1975;2(7924):45-51.
31. Collins R, Scrimgeour A, Yusuf S, Peto R. Reduction in fatal pulmonary embolism and venous thrombosis by perioperative administration of subcutaneous heparin. Overview of results of randomized trials in general, orthopedic, and urologic surgery. *N Engl J Med.* 1988;318(18):1162-1173.
32. Halkin H, Goldberg J, Modan M, Modan B. Reduction of mortality in general medical in-patients by low-dose heparin prophylaxis. *Ann Intern Med.* 1982;96(5):561-565.
33. Gardlund B. Randomised, controlled trial of low-dose heparin for prevention of fatal pulmonary embolism in patients with infectious diseases. The Heparin Prophylaxis Study Group. *Lancet.* 1996;347(9012):1357-1361.
34. Yalamanchili K, Sukhija R, Sinha N, Aronow WS, Maguire GP, Lehrman SG. Efficacy of unfractionated heparin for thromboembolism prophylaxis in medical patients. *Am J Ther.* 2005;12(4):293-299.
35. Martel N, Lee J, Wells PS. Risk for heparin-induced thrombocytopenia with unfractionated and low-molecular-weight heparin thromboprophylaxis: A meta-analysis. *Blood.* 2005;106(8):2710-2715.
36. Mismetti P, Laporte S, Zufferey P, Epinat M, Decousus H, Cucherat M. Prevention of venous thromboembolism in orthopedic surgery with vitamin K antagonists: A meta-analysis. *J Thromb Haemost.* 2004;2(7):1058-1070.

37. Hull RD, Pineo GF, Francis C, et al. Low-molecular-weight heparin prophylaxis using dalteparin in close proximity to surgery vs. warfarin in hip arthroplasty patients: A double-blind, randomized comparison. *Arch Intern Med.* 2000;160(14):2199–2207.
38. Hull RD, Pineo GF, Francis C, et al. Low-molecular-weight heparin prophylaxis using dalteparin extended out-of-hospital vs. in-hospital warfarin/out-of-hospital placebo in hip arthroplasty patients: A double-blind, randomized comparison. *Arch Intern Med.* 2000;160(14):2208–2215.
39. Comp PC, Spiro TE, Friedman RJ, et al. Prolonged enoxaparin therapy to prevent venous thromboembolism after primary hip or knee replacement. Enoxaparin Clinical Trial Group. *J Bone Joint Surg Am.* 2001;83–A(3):336–345.
40. Fitzgerald RH Jr, Spiro TE, Trowbridge AA, et al. Prevention of venous thromboembolic disease following primary total knee arthroplasty. A randomized, multicenter, open-label, parallel-group comparison of enoxaparin and warfarin. *J Bone Joint Surg Am.* 2001;83–A(6):900–906.
41. Mismetti P, Laporte S, Darmon JY, Buchmuller A, Decousus H. Meta-analysis of low molecular weight heparin in the prevention of venous thromboembolism in general surgery. *Br J Surg.* 2001;88(7):913–930.
42. ENOXACAN Study Group. Efficacy and safety of enoxaparin versus unfractionated heparin for prevention of deep vein thrombosis in elective cancer surgery: A double-blind randomized multicentre trial with venographic assessment. *Br J Surg.* 1997;84(8):1099–1103.
43. Bergqvist D, Agnelli G, Cohen AT, et al. Duration of prophylaxis against venous thromboembolism with enoxaparin after surgery for cancer. *N Engl J Med.* 2002;346(13):975–980.
44. Samama MM, Cohen AT, Darmon JY, et al. A comparison of enoxaparin with placebo for the prevention of venous thromboembolism in acutely ill medical patients. Prophylaxis in Medical Patients with Enoxaparin Study Group. *N Engl J Med.* 1999;341(11):793–800.
45. Leizorovicz A, Cohen AT, Turpie AG, Olsson CG, Vaitkus PT, Goldhaber SZ, and the PREVENT Medical Thromboprophylaxis Study Group. Randomized, placebo-controlled trial of dalteparin for the prevention of venous thromboembolism in acutely ill medical patients. *Circulation.* 2004;17;110(7):874–879.
46. Kleber FX, Witt C, Vogel G, Koppenhagen K, Schomaker U, Flosbach CW, and THE-PRINCE Study Group. Randomized comparison of enoxaparin with unfractionated heparin for the prevention of venous thromboembolism in medical patients with heart failure or severe respiratory disease. *Am Heart J.* 2003;145(4):614–621.
47. Turpie AG, Bauer KA, Eriksson BI, Lassen MR. Fondaparinux vs. enoxaparin for the prevention of venous thromboembolism in major orthopedic surgery: A meta-analysis of four randomized double-blind studies. *Arch Intern Med.* 2002;162(16):1833–1840.
48. Eriksson BI, Lassen MR, and the PENTasaccharide in HIp-FRActure Surgery Plus Investigators. Duration of prophylaxis against venous thromboembolism with fondaparinux after hip fracture surgery: A multicenter, randomized, placebo-controlled, double-blind study. *Arch Intern Med.* 2003;9;163(11):1337–1342.
49. Agnelli G, Bergqvist D, Cohen AT, Gallus AS, Gent M, and the PEGASUS Investigators. Randomized clinical trial of postoperative fondaparinux versus perioperative dalteparin for prevention of venous thromboembolism in high-risk abdominal surgery. *Br J Surg.* 2005;92(10):1212–1220.
50. Cohen AT, Davidson BL, Gallus AS, et al., and the ARTEMIS Investigators. Efficacy and safety of fondaparinux for the prevention of venous thromboembolism in older acute medical patients: Randomised placebo controlled trial. *BMJ.* 2006;332(7537):325–329.

51. Urbankova J, Quiroz R, Kucher N, Goldhaber SZ. Intermittent pneumatic compression and deep vein thrombosis prevention. A meta-analysis in postoperative patients. *Thromb Haemost.* 2005;94(6):1181–1185.
52. Ramos R, Salem BI, De Pawlikowski MP, Coordes C, Eisenberg S, Leidenfrost R. The efficacy of pneumatic compression stockings in the prevention of pulmonary embolism after cardiac surgery. *Chest.* 1996;109(1):82–85.
53. Turpie AG, Bauer K, Caprini J, Comp P, Gent M, Muntz J. Fondaparinux combined with intermittent pneumatic compression (IPC) versus IPC alone in the prevention of VTE after major abdominal surgery: Results of the APOLLO study. *J Thromb Haemost.* 2005;3(1):P1046 (Abstract).
54. van der Heijden JF, Hutten BA, Buller HR, Prins MH. Vitamin K antagonists or low-molecular-weight heparin for the long term treatment of symptomatic venous thromboembolism. *Cochrane Database Syst Rev.* 2002;(1):CD002001.
55. Kolbach DN, Sandbrink MW, Hamulyak K, Neumann HA, Prins MH. Non-pharmaceutical measures for prevention of post-thrombotic syndrome. *Cochrane Database Syst Rev.* 2004;(1):CD004174.
56. Brandjes DP, Buller HR, Heijboer H, et al. Randomised trial of effect of compression stockings in patients with symptomatic proximal-vein thrombosis. *Lancet.* 1997;349(9054):759–762.
57. Kakkos SK, Daskalopoulou SS, Daskalopoulos ME, Nicolaides AN, Geroulakos G. Review on the value of graduated elastic compression stockings after deep vein thrombosis. *Thromb Haemost.* 2006;96(4):441–445.

SECTION III

THERAPUTIC PROCEDURES FOR CHRONIC VENOUS INSUFFICIENCY

11

Results Comparing Compression Alone Versus Compression and Surgery in Treating Venous Ulceration

Manjit S. Gohel, Jamie R. Barwell, Mark R. Whyman, and Keith R. Poskitt

Chronic venous ulceration represents the worst extreme of a spectrum of venous disorders and is a common problem in the Western world. Multilayer compression bandaging providing 40 mmHg of pressure at the ankle and 17–20 mmHg at the calf is the mainstay of treatment for open ulcers. Surgically correctable superficial venous reflux is commonly seen in this population, although the benefit of operative intervention has been widely debated, particularly because many patients with venous ulcers are often elderly with extensive comorbidity and therefore unattractive surgical candidates. The ESCHAR venous ulcer study was a large, randomized controlled study aiming to investigate the influence of superficial venous surgery in addition to compression bandaging.

This chapter aims to summarize the clinical, anatomic, and hemodynamic findings from the ESCHAR study and other clinical trials comparing compression alone with compression plus surgery in the treatment of venous ulcers.

INTRODUCTION

EXTENT OF THE PROBLEM

Over 70% of chronic lower limb ulcers are thought to have a predominantly venous cause,[1,2] and the management of this group is estimated to cost U.S. health providers over $1 billion each year.[3] This chronic and often recurrent condition is associated with a significant impact on patient quality of life.[4] The unglamorous nature of chronic leg ulceration has meant that patient care has been supervised by nursing teams; community general practitioners; and plastic surgery, dermatologic, and vascular specialists, leading to inconsistencies in clinical practice. The true incidence and prevalence of chronic venous ulcers may be impossible to determine, although studies from Europe and the United States have estimated a prevalence of 0.5% to 1% in the adult population increasing to over 3% in those aged over 80 years.[3,5] For each patient with an open ulcer, a further 3 to 4 are thought to be at high risk of developing ulceration.[6] In an aging population and with increasing awareness of vascular disorders, it is reasonable to assume that the incidence of chronic venous disease will continue to increase.

THE RATIONALE FOR SURGERY

The dogma that venous ulcers are due to deep venous disease has been largely disproved in recent years. Duplex studies have demonstrated that incompetence in superficial veins (greater or lesser saphenous) is present in 60% to 80% of legs with chronic ulcers.[7–9] This is occasionally in combination with deep venous reflux, although isolated incompetence in deep or perforating veins is uncommon.[8,9] Chronic venous hypertension, usually secondary to venous reflux, is generally accepted to play a major role in the pathogenesis of venous ulceration.[10] Compression therapy and limb elevation are proven measures to counter this. However, compression bandages are uncomfortable and difficult to don, resulting in poor compliance and limited long-term efficacy. Surgical correction of venous incompetence may provide a more durable approach to the problem, and a number of strategies have been forwarded. Procedures to correct deep venous reflux are feasible but associated with high complication rates and questionable clinical benefit.[11] The surgical correction of superficial venous reflux has been suggested by a number of nonrandomized studies[12,13] but has lacked the endorsement of a large, randomized trial until recent years.

THE ESCHAR VENOUS ULCER STUDY

The ESCHAR study[14] aimed to investigate the Effect of Surgery and Compression on Healing And Recurrence in patients with chronic venous ulcers and was set in three vascular departments in the United Kingdom serving a target population of approximately 800,000.

STUDY DESIGN

Consecutive patients with recently healed or open leg ulcers (CEAP grades C5–6), superficial venous incompetence as assessed by color venous duplex, and normal ankle brachial pressure index (ABPI >0.85) were targeted. Those consenting to inclusion were randomized to compression therapy alone or compression plus superficial venous surgery. The study design is summarized in Figure 11-1. Patients with normal deep veins were described as having *isolated superficial reflux*. Legs with incompetence in some but not all deep veins had *superficial and segmental deep reflux*, and those with reflux throughout the deep system were termed as having *superficial and total deep reflux*. Patients with open ulcers were treated with multilayer compression bandaging aiming to achieve 40 mmHg of pressure at the ankle and 17–20 mmHg at the calf. Healed legs were fitted with class II elastic stockings providing 20 mmHg of compression at the ankle and 10–12 mmHg at the calf.

Those randomized to surgery were offered additional saphenofemoral and/or saphenopopliteal junction disconnection, stripping of the greater saphenous vein to the knee, and varicosity avulsions. Specific perforator surgery was not offered. Patients unfit for general anesthesia were offered saphenofemoral and/or saphenopopliteal junction disconnection alone under local anesthesia. To observe hemodynamic and anatomic changes in the trial population, venous refill times (VRT) were measured by digital photoplethysmography (PPG) at randomization and repeated with color duplex after 1 year (Figure 11-1). All patients were followed up in specialist nurse-led leg ulcer clinics using the Cheltenham model,[15] and analyses were performed strictly on intention to treat.

PATIENT FLOW

The study commenced in January 1999 and successfully recruited the target of 500 patients by August 2002. A total of 258 patients were randomized to compression alone and 242 to compression plus surgery. Despite consenting to recruitment, 47/242 (19%) patients refused to attend for surgery. Similar experiences have been reported by other authors and illustrate that a large group of patients with venous ulcers will not accept surgical interventions irrespective of any clinical benefit. A quarter of patients had local anesthetic procedures, and the median time from randomization to operation was 7 weeks. The 3-year mortality was 17%, highlighting that the study population were elderly with significant comorbidity. None of the deaths occurred within 30 days of surgery.

CLINICAL OUTCOMES

Impact of Surgery on Ulcer Healing

The overall 3-year healing rates were similar in the groups randomized to compression alone or compression plus surgery (89% and 93%, respectively, p = 0.73).[16] There were no differences in ulcer healing for the groups

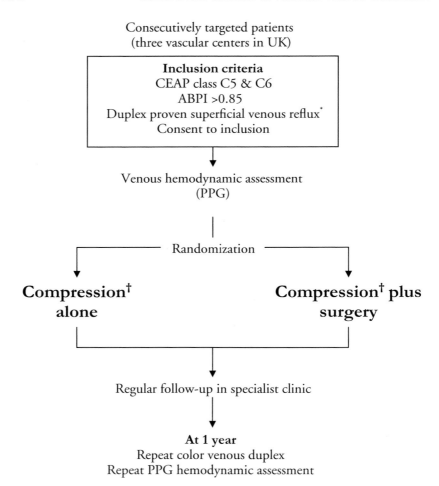

* Legs classified into those with isolated superficial reflux, superficial plus segmental deep reflux, and superficial plus total deep reflux.

† Multilayer compression for open ulcers, Class II elastic stockings for healed legs.

FIGURE 11-1   ESCHAR study design.

stratified by pattern of venous reflux (Figure 11-2). It should be noted that most patients did not receive immediate surgery, and any benefits from operative intervention were therefore delayed. A study by Bello and colleagues demonstrated that superficial venous surgery improved venous ulcer healing when compared to legs treated without compression.[17]

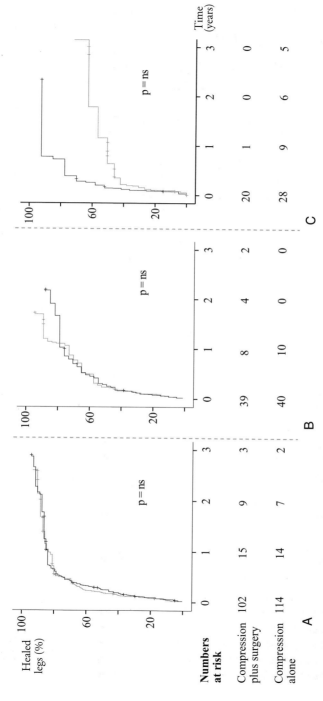

FIGURE 11-2 Ulcer healing rates in ESCHAR study. **A.** Isolated superficial reflux. **B.** Superficial with segmental deep reflux. **C.** superficial with total deep reflux.

However, other studies comparing surgery to compression have echoed the findings from ESCHAR and failed to demonstrate a healing benefit with surgery.[18,19] These findings confirm that the most effective strategy for rapid healing of venous ulcers is likely to be effective multilayered graduated compression applied by appropriately trained staff.

Impact of Surgery on Ulcer Recurrence

The ESCHAR study confirmed the findings of previous, smaller studies and demonstrated that superficial venous surgery reduces ulcer recurrence in legs with isolated superficial reflux. Three-year recurrence rates were 21% for patients randomized to surgery, compared to 51% for legs treated with compression alone (p < 0.001, Figure 11-3).[16] A significant benefit was also seen for legs with segmental deep reflux, suggesting that the presence of coexistent deep reflux should not be considered a contraindication to superficial venous surgery (Figure 11-3). For legs with total deep reflux, a significant benefit for surgery was not demonstrated, although patient numbers in this subgroup were small. Interestingly, some legs with total deep reflux did derive a hemodynamic benefit from surgery (see following section). Legs randomized to surgery experienced greater ulcer free time over 3 years compared to patients in the compression group (mean 100 vs. 85 weeks respectively, p < 0.001) and suffered significantly fewer episodes of recurrent ulceration.[16]

These results reproduced the high rates of venous ulcer recurrence reported by other authors (>50% at 3 years) despite advice to wear class II hosiery.[20] Even for legs randomized to surgery, 3-year ulcer recurrence rates were over 20%. Although it is reasonable to assume that surgery may not be successful in all cases, this does suggest that simply correcting anatomic venous incompetence does not cure the leg of chronic venous hypertension. In cases of recurrent ulceration after surgery, factors such as calf muscle pump failure, obesity, and immobility may be significant in individual patients. The use of long-term class II elastic stockings after superficial venous surgery may therefore be advisable to counter the causes of venous hypertension not addressed by saphenous surgery.

OTHER FINDINGS

Anatomic Changes with Surgery

Color venous duplex assessment in the ESCHAR study revealed that residual incompetence in greater or lesser saphenous veins was a common finding after venous surgery. In a report of 84 patients treated for greater saphenous reflux, only 22/84 (26%) were free of greater saphenous incompetence after 1 year.[21] In 30/84 (36%) patients, there was reflux in the above-knee segment of the greater saphenous vein after 1 year, including six patients in which the saphenofemoral junction was incompetent (Figure

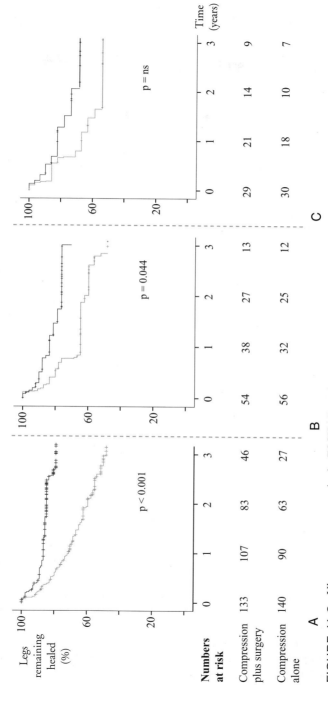

FIGURE 11-3 Ulcer recurrence rates in the ESCHAR study. **A.** Isolated superficial reflux. **B.** Superficial with segmental deep reflux. **C.** Superficial with total deep reflux.

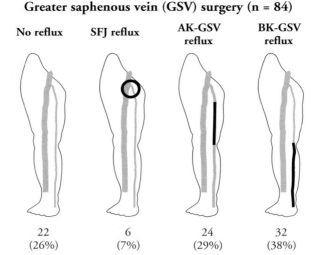
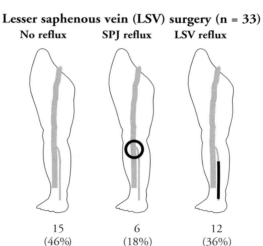

FIGURE 11-4 Residual superficial venous reflux 1 year after saphenous surgery.

11-4). Disappointing surgical results following saphenous vein surgery have been reported previously and neovascularization, incomplete surgery, or anatomic anomalies (such as bifid greater saphenous vein) may have contributed to residual reflux. Interestingly, in a subgroup analysis, the presence of residual superficial incompetence after surgery was not associated with reduced hemodynamic status, inferior ulcer healing, or higher recurrence rates (unpublished data). This observation questions the significance of residual reflux after venous surgery and suggests that repeat surgery to cure venous reflux may not confer any hemodynamic or clinical benefit. Perhaps more detailed assessment of the duration of venous incompetence and size of refluxing vein may help identify clinically significant anatomic reflux.

Previous authors have commented that deep venous reflux may be reversed following superficial venous surgery.[22] A similar effect was seen in this study, particularly for patients with segmental deep reflux, which was abolished in 45% of patients after surgery.[21] Interestingly, reversal of deep and superficial venous incompetence was seen in some patients treated with compression alone. The temporal variation of venous incompetence and influence of simple measures such as elevation, exercise, and compression remain poorly understood, and these limitations must be considered when interpreting duplex changes.

The significance of incompetent calf perforating veins in patients with venous ulcers is poorly understood and continues to be widely debated. Duplex examination revealed that incompetent perforators were present in 51% of patients prior to surgery, which reduced to 42% of legs 1 year after surgery ($p = 0.001$). This effect was seen in legs with and without deep venous reflux.[21] The reversal of calf perforator incompetence after saphenous vein surgery is well reported but widely considered to be a short-term phenomenon and less well recognized in the presence of coexistent deep reflux.

Hemodynamic Changes with Surgery

Venous refill time (VRT) reflects the time taken for a limb to return to its resting venous pressure following calf muscle pump contraction. Long VRT demonstrates normal venous function, whereas very short VRT may indicate a greater severity of venous disease. Digital photoplethysmography (PPG) is a minimally invasive procedure for the measurement of VRT, validated against invasive pressure measurements, and was used for functional limb assessment in the ESCHAR study. PPG was performed with and without a narrow below-knee tourniquet to occlude superficial veins (to simulate surgery) and repeated 1 year after recruitment. VRT >20 seconds was considered normal.

The initial median VRT in the ESCHAR population was 11 seconds (range 3–48), demonstrating significant venous dysfunction. No improvement in VRT was seen in the legs treated with compression alone. However, in the surgical group, initial VRTs of median 10 seconds (range 3–48) improved to median 15 seconds (range 4–48) 1 year after surgery ($p < 0.001$).[21] Some legs gained a hemodynamic benefit from superficial venous surgery, whereas others did not regardless of the initial pattern of venous reflux.

Hemodynamic data also demonstrated that VRT measured with a below-knee tourniquet correlated with both ulcer healing and recurrence,

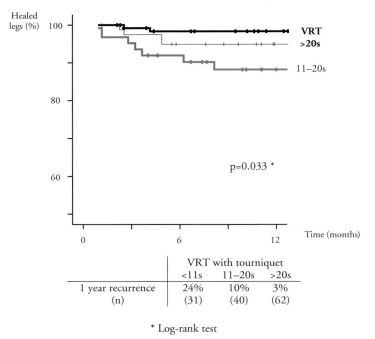

FIGURE 11-5 Kaplan-Meier curve for ulcer recurrence demonstrating the predictive value of preoperative hemodynamic assessment using PPG *with* below-knee tourniquet.

implying that PPG assessment accurately reflects actual venous dysfunction. Poor initial VRT measurements were associated with delayed ulcer healing, and using this predictive value, a prognostic scoring system for venous ulcer healing has been developed from the ESCHAR results.[23] Interestingly, a normal VRT with a tourniquet (>20 seconds) before surgery was associated with a 1-year recurrence rate of only 3% compared to 24% in legs with a poor preoperative VRT (<11 seconds) (Figure 11-5).[24] These findings strongly suggest that VRT measurement using PPG could be a useful adjunct to duplex scanning to identify good operative candidates and potentially avoid unnecessary surgery.

CRITIQUE OF THE ESCHAR STUDY

The ESCHAR study is the largest published randomized study to evaluate the role of surgery in addition to compression for treating venous ulcers. A great attribute of the study is that inclusion criteria were deliberately wide, and the study design aimed to emulate standard clinical practice to ensure that results could be widely applicable. The trial was set within an established leg ulcer service and detailed anatomic and hemodynamic assessments have augmented the clinical results and improved understanding of the overall influence of surgery.

A large number of patients refused to attend for surgery despite consenting to inclusion. Moreover, a quarter of patients were treated with local anesthetic procedures, and the median time to surgery was nearly 2 months. As the clinical outcomes were analyzed on intention to treat, it is likely that the actual benefit of superficial venous surgery may have been underestimated. It is regretful that quality of life and economic data are not available to support the clinical findings, as these are powerful drivers for changes in modern clinical practice.

RESULTS FROM OTHER STUDIES

In a small study published in 2003, Zamboni and colleagues randomized 47 legs (45 patients) to either compression alone or compression plus minimally invasive hemodynamic venous surgery performed under local anesthetic.[18] Greater saphenous stripping was not performed, and patients over 80 years with a history of deep venous thrombosis or evidence of deep reflux were excluded. The authors reported that healing rates were similar in the two groups, but the legs treated with surgery had a recurrence rate of only 9% at 3 years compared to 38% for patients in compression only.

The USABLE multicenter study aimed to randomize 1,000 patients to compression alone or compression plus surgery, but was abandoned after 18 months due to poor recruitment.[25] The early results for 76 recruited patients showed no difference in healing rates or quality of life scores.[19] The trial organizers concluded that as a large proportion of screened patients were not suitable for superficial venous surgery, the role of surgery in venous ulceration may have been overestimated. However, these observations differed from the ESCHAR study, where the proportion of screened patients suitable for surgery was far greater. Differences in the study setting and target populations may have contributed to these observations.

CONCLUSIONS

Chronic venous ulceration is widely accepted as a common, debilitating, and expensive health problem. Although most patients have superficial venous incompetence potentially amenable to surgical correction, evidence to support operative intervention has been scarce. The findings from the ESCHAR study and other research strongly suggest that superficial venous surgery reduces venous ulcer recurrence and should be considered for all patients with chronic venous ulceration. Patients deemed suitable for surgical treatment should undergo color duplex venous mapping in order to identify superficial venous incompetence potentially suitable for surgical correction. Although the advantage appears greatest for legs with isolated superficial reflux, surgery may also reduce ulcer recurrence for some patients with deep venous incompetence, and hemodynamic assessment may be a useful selection tool.

Despite any proven clinical benefit, it must be acknowledged that elderly patients with leg ulcers may be unfit for surgical intervention or unwilling to accept it. Local anesthetic surgery was performed in a quarter of patients, but long-term effectiveness is unlikely to match general anesthetic procedures. In keeping with previous studies, residual venous reflux after venous surgery was common, although this was not associated with reduced hemodynamic function or clinical outcomes.

Foam sclerotherapy may be used for the treatment of superficial venous incompetence and has the advantage of being performed using local anesthesia as an outpatient procedure. The early experience of foam sclerotherapy in chronic venous ulceration both in our unit and other units has been promising.[26] More studies are needed to evaluate the efficacy of foam sclerotherapy and other endovenous procedures, including radiofrequency ablation and endovenous laser therapy. Patient acceptance of such procedures may be greater than traditional surgery.

Selection of patients for surgery has been largely based on duplex findings, although results from the ESCHAR study suggest that measurement of venous refill time may also be beneficial. Further refinement of selection pathways may improve the identification of patients most likely to benefit from surgery.

REFERENCES

1. Baker SR, Stacey MC, Singh G, Hoskin SE, Thompson PJ. Aetiology of chronic leg ulcers. *Eur J Vasc Surg.* 1992;6:245–251.
2. Nelzen O, Bergqvist D, Lindhagen A. Leg ulcer etiology—A cross sectional population study. *J Vasc Surg.* 1991;14:557–564.
3. Lawrence PF, Gazak CE. In: Gloviczki P, Bergan JJ, eds., *Atlas of Endoscopic Perforator Vein Surgery.* London: Springer-Verlag, 1998:31–44.
4. Phillips T, Stanton B, Provan A, Lew R. A study of the impact of leg ulcers on quality of life: Financial, social and psychological implications. *J Am Acad Dermatol.* 1994;31:49–53.
5. Nelzen O, Bergqvist D, Lindhagen A. The prevalence of lower-limb ulceration has been underestimated: Results of a validated population questionnaire. *Br J Surg.* 1996;83:255–258.
6. Callam MJ. Epidemiology of varicose veins. *Br J Surg.* 1994;81:167–170.
7. Magnusson MB, Nelzen O, Risberg B, Sivertsson R. A colour doppler ultrasound study of venous reflux in patients with chronic leg ulcers. *Eur J Vasc Endovasc Surg.* 2001;21:353–360.

8. Hanrahan LM, Araki CT, Rodriguez AA, Kechejian GJ, LaMorte WW, Menzoian JO. Distribution of valvular incompetence in patients with venous stasis ulceration. *J Vasc Surg.* 1991;13:805–811.
9. Grabs AJ, Wakely MC, Nyamekye I, Ghauri AS, Poskitt KR. Colour duplex ultrasonography in the rational management of chronic venous leg ulcers. *Br J Surg.* 1996;83:1380–1382.
10. Nicolaides AN, Hussein MK, Szendro G, Christopoulos D, Vasdekis S, Clarke H. The relation of venous ulceration with ambulatory venous pressure measurements. *J Vasc Surg.* 1993;17:414–419.
11. Hardy SC, Riding G, Abidia A. Surgery for deep venous incompetence. *Cochrane Database Syst Rev.* 2004;CD001097.
12. Darke SG, Penfold C. Venous ulceration and saphenous ligation. *Eur J Vasc Surg.* 1992;6:4–9.
13. Barwell JR, Taylor M, Deacon J, et al. Surgical correction of isolated superficial venous reflux reduces long-term recurrence rate in chronic venous leg ulcers. *Eur J Vasc Endovasc Surg.* 2000;20:363–368.
14. Barwell JR, Davies CE, Deacon J, et al. Comparison of surgery and compression with compression alone in chronic venous ulceration (ESCHAR study): Randomised controlled trial. *Lancet.* 2004;363:1854–1859.
15. Ghauri AS, Taylor MC, Deacon JE, et al. Influence of a specialised leg ulcer service on management and outcome. *Br J Surg.* 2000;87:1048–1056.
16. Gohel MS, Barwell JR, Taylor M, et al. Superficial venous surgery reduces venous ulcer recurrence: 3 year results from the ESCHAR trial. *Br J Surg.* 2005;92:506.
17. Bello MSM, Hartshorne T, Bell PR, Naylor AR, London NJ. Role of superficial venous surgery in the treatment of venous ulceration. *Br J Surg.* 1999;86:755–759.
18. Zamboni P, Cisno C, Marchetti F, et al. Minimally invasive surgical management of primary venous ulcers vs. compression treatment: A randomised clinical trial. *Eur J Vasc Endovasc Surg.* 2003;25:313–318.
19. Guest M, Smith JJ, Tripuraneni G, et al. Randomized clinical trial of varicose vein surgery with compression versus compression alone for the treatment of venous ulceration. *Phlebology.* 2003;18,130–136.
20. Franks P, Oldroyd M, Dickson D, Sharp E, Moffatt C. Risk factors for leg ulcer recurrence: A randomised trial of two types of compression stocking. *Age Ageing.* 1995;240:440–494.
21. Gohel MS, Barwell JR, Earnshaw JJ, et al. Randomized clinical trial of compression plus surgery versus compression alone in chronic venous ulceration (ESCHAR study)—Haemodynamic and anatomical changes. *Br J Surg.* 2005;92:291–297.
22. Walsh JC, Bergan JJ, Beeman S, Comer TP. Femoral vein reflux abolished by greater saphenous vein stripping. *Ann Vasc Surg.* 1994;8:566–570.
23. Kulkarni SA, Gohel MS, Poskitt KR, Whyman MR. Ulcerated Leg Severity Assessment (ULSA): A new scoring tool to predict healing in venous leg ulceration. *Br J Surg.* 2007;94:189–193.
24. Gohel MS, Barwell JR, Wakely C, et al. Haemodynamic assessment is predictive of healing and recurrence in chronic venous leg ulceration. *Phlebology.* 2004;19:157–158.
25. Davies AH, Hawdon AJ, Greenhalgh RM, Thompson S. Failure of a trial evaluating the effect of surgery on healing and recurrence rates in venous ulcers? The USABLE trial: Rationale, design and methodology, and reasons for failure. *Phlebology.* 2004;19:137–142.
26. Pascarella L, Bergan JJ, Mekenas LV. Severe chronic venous insufficiency treated by foamed sclerosant. *Ann Vasc Surg.* 2006;20:83–91.

12

SUPERFICIAL SURGERY AND PERFORATOR INTERRUPTION IN THE TREATMENT OF VENOUS LEG ULCERS

WILLIAM MARSTON

INTRODUCTION

The primary goals when treating patients with leg ulcers associated with chronic venous insufficiency (CVI) are to heal the ulcers as rapidly as possible, control the patients' other symptoms manifesting from their venous disease, and minimize the potential for ulcer recurrence after healing. Because venous leg ulcers (VLUs) are a result of chronic venous hypertension, it is generally believed that interventions capable of reducing the severity of venous hypertension should impact positively on these goals of treatment.

In this chapter, data concerning the efficacy of superficial and perforator surgery on healing leg ulcers and preventing recurrence will be reviewed. The diagnostic evaluation and indications for intervention in these patients will be discussed. The options for surgical management of various anatomic types of venous insufficiency will be reviewed and contrasted with nonsurgical techniques.

PRESENTATION OF PATIENTS WITH CVI

A commonly held perception is that VLUs are always the result of deep venous reflux, and this is not amenable to surgical or interventional reconstruction. For this reason, the majority of patients treated with VLUs are never referred to a venous specialist for consideration of a corrective procedure. Several authors have defined the anatomy of reflux in patients with

CEAP clinical classes 5 and 6 CVI, and isolated saphenous or saphenous and perforator reflux is not uncommon, occurring in 20% to 35% of patients in various series. Also, as outlined by Drs. Neglen, Raju, and Kistner in other chapters, many patients with deep venous insufficiency causing ulceration may be improved with surgical or endovenous procedures, reducing symptoms and the incidence of recurrent ulcers.

For these reasons, all patients with VLUs who are candidates for corrective procedures should be studied with diagnostic studies to determine the anatomy and physiology of their individual case. Referral to a venous specialist familiar with surgical and nonsurgical options will allow the optimal method of correction to be selected.

DIAGNOSTIC TESTING FOR PATIENTS WITH VENOUS LEG ULCERS

The interventional treatment of a patient with venous leg ulcers requires the use of noninvasive studies to determine the anatomic and hemodynamic characteristics of the patient's venous system. The anatomic sites of venous dysfunction and the hemodynamic importance of this dysfunction are required to allow treatment plans to be formulated and optimal results to be achieved. Selecting surgical therapy without a knowledge of which vein segments are abnormal is essentially blind surgery and cannot result in optimal results. The optimal time to perform these noninvasive studies would be after the limb edema and wound exudate are controlled. This will allow better visualization of the veins and better patient compliance with the maneuvers required for standing venous reflux examination and plethysmography. Morbidly obese patients and those with limited ability to stand without assistance are poor candidates for accurate testing and usually are poor candidates for venous intervention as well.

DUPLEX ULTRASOUND

Imaging techniques using ultrasound combined with Doppler interrogation of the venous system have been validated as sensitive methods of diagnosis of deep venous thrombosis. Important information for patients with VLUs that would be detected with this technique includes the presence or absence of venous obstruction or other changes typical of previous DVT. This information will help to determine whether the patient's venous hypertension is due to obstruction, reflux, or both (pathophysiology). The presence of outflow obstruction in the iliac veins and/or IVC can often be detected looking at flow patterns, phasicity, and respiratory variation in the common femoral vein. In addition to an examination of the deep and superficial systems, the perforator veins are carefully examined for evidence of incompetence.

Second, venous reflux in the deep and superficial venous systems is evaluated with the patient in the standing position using duplex ultrasound and either manual compression or a rapid inflation/deflation system to elicit reflux. In studies of normal volunteers and patients with CVI, a normal valve closure time of <0.5 has been defined. Systematic interrogation of the common femoral, superficial femoral, popliteal, greater saphenous, and lesser saphenous veins is conducted, allowing an anatomic map of venous reflux in the limb to be constructed.

Using this information, the clinician can determine the etiology, anatomy, and pathophysiology of CVI for the patient. Although duplex evaluation provides detailed information on the anatomy of venous disease, it cannot define the importance of anatomic abnormalities in the venous function of the limb. Clearly, one patient with gross reflux in the saphenous vein will have no resultant symptoms, while another patient may have an active ulcer from saphenous reflux alone. The clinical assessment of the severity of CVI is often subjective, so testing that allows objective measurement of the hemodynamic performance of the lower extremity venous system would greatly assist with patient assessment and treatment decisions.

PHOTOPLETHYSMOGRAPHY

Plethysmography is defined as the determination of changes in volume, and various techniques of plethysmography have been evaluated in the noninvasive examination of the venous system. Photoplethysmography (PPG) utilizes a transducer that emits infrared light from a light emitting diode into the dermis. The backscattered light is measured by an adjacent photodetector and displayed as a line tracing. The amount of backscattered light varies with the capillary red blood cell volume in the dermis. Through use of this technology and provocative limb maneuvers, an assessment of the venous system is obtained. The primary measure obtained, the refill or recovery time (VRT), represents the time required for the PPG tracing to return to 90% of baseline after cessation of calf contraction. PPG does not produce a quantitative measure, but the refill time has been found to correlate closely with ambulatory venous pressure (AVP) measurements.

Limbs affected with CVI typically have a much shorter VRT than normal limbs. As such, PPG can provide a relatively simple measure of whether or not venous insufficiency is present. However, the technique can vary depending on the site of photosensor placement and the small sample area obtained. PPG measurements have not been proven to be a strong discriminator of the severity of CVI. Nicolaides and Miles reported that normal limbs were well identified by a PPG refill time of greater than 18 seconds with their protocol. Abnormal limbs with CVI consistently had a refill time of <18 seconds. However, in the abnormal group, PPG refill time could not differentiate between degrees of CVI, with similar PPG refill times obtained in patients with AVP measurements ranging from 45 to 100 mmHg. Therefore, PPG is a poor test for assessing the results of venous corrective surgical procedures.

In summary, PPG is a reasonable measure of the presence or absence of CVI that is best used when no further information concerning the venous hemodynamic situation is desired. It would be useful to help diagnose whether the etiology of a chronic leg ulcer was related to venous disease. However, if information concerning the severity of CVI or an evaluation of the improvement after venous surgery is required, a quantitative test will be more useful.

AIR PLETHYSMOGRAPHY

Air plethysmography (APG) utilizes a technique to improve on the shortcomings of PPG and other types of plethysmography that have limited sampling areas. It employs a low-pressure air-filled cuff measuring 30 to 40 cm in length that is applied to the lower leg, allowing quantitative evaluation of volume changes of the entire lower leg from knee to ankle. Briefly, the patient lies supine initially with the leg elevated and supported at the heel, allowing the cuff to be applied to the lower leg. The cuff is inflated to a pressure of 6 mmHg to provide snug apposition to the limb without compressing the superficial veins. A baseline volume in the supine position is obtained with the patient resting. The patient then moves to a standing position supported by a walker to remove weight from the tested limb. The volume tracing gradually increases until a plateau is reached. The patient then performs one calf contraction/tiptoe maneuver followed by rest. A subsequent series of 10 tiptoe maneuvers completes the test procedure.

Figure 12-1 illustrates the data calculated from the tracings obtained. The venous volume (VV) is the difference in limb volumes obtained in the resting and standing positions. The venous filling index (VFI) is calculated by measuring 90% of the VV and dividing this volume by the time the limb requires to refill to 90% of the VV after moving to the standing position. Expressed in cc per second, VFI measures the average filling rate of the dependent leg and is slow in normal limbs. The volume of blood ejected with one tiptoe movement divided by the VV gives the ejection fraction (EF), and the limb volume remaining after 10 tiptoe movements divided by the VV gives the residual volume fraction (RVF).

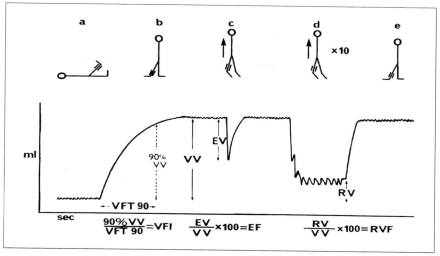

FIGURE 12-1 Diagrammatic representation of hemodynamic values obtained from air plethysmography. VV = venous volume, VFI = venous filling index, EF = ejection fraction, RVF = residual volume fraction. Reprinted with permission from Christopoulos et al. Air-plethysmography and the effect of elastic compression on venous hemodynamics of the leg. *J Vasc Surg.* 1987;5:148–159.

TABLE 12-1 Prevalence of the Sequelae of Venous Disease in Relation to VFI in 134 Limbs with Venous Disease Studied with Air Plethysmography

VFI, ml/sec	Swelling (%)	Skin Changes (%)	Ulceration (%)
<3	0	0	0
3–5	12	19	0
5–10	46	61	46
>10	76	76	58

From Christopoulos et al.[1]

In 1988, Christopoulos et al. described the use of APG for evaluation of normal limbs and those affected with CVI. A VFI <2 ml/second was associated with clinically normal limbs, and increasing levels of VFI were associated with more severe symptoms (Table 12-1).[1] The VFI is believed to provide a reasonable approximation of the global function of the lower extremity venous system in resisting reflux in the standing position.

The EF and RVF are measures of the efficacy of the calf muscle to pump blood out of the leg. The RVF was found to correlate closely with AVP throughout the range of AVP measurements, with lower RVF values representing better calf pump function (normal RVF defined as <35%).

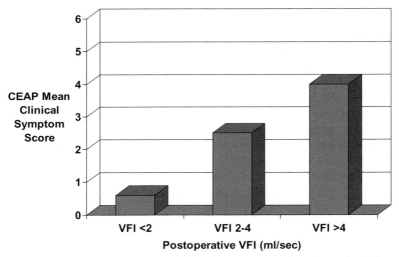

FIGURE 12-2 Comparison of mean clinical symptom score in patients who had normal postoperative VFI with mean CSS in those whose values did not normalize (abnormal >2 ml/sec). $P < 0.01$ between groups. VFI = venous filling index.

Further work with APG measurements has demonstrated that the postoperative VFI can predict the long-term symptomatic outcome for patients after venous surgical procedures. Ninety-four percent of patients in whom the VFI corrected to <2 ml/second after surgery were asymptomatic at a mean follow-up time of 44 months (Figure 12-2).[2] The VFI was also found to be predictive of the incidence of ulcer recurrence. Limbs with a VFI <4 ml/second had significantly fewer episodes of ulcer recurrence (22% at 3 years) compared to those with a VFI >4 ml/second (47% at 3 years).[3]

In summary, APG, by sampling a large portion of the calf area, provides a better measure than PPG of the global venous function of the limb. It provides a quantitative analysis that appears to be useful in the follow-up of patients with VLUs, particularly those undergoing venous intervention.

INDICATIONS FOR INTERVENTION

Intuitively, most vascular specialists believe that patients with true VLUs will experience an improved long-term course if their underlying venous hypertension can be corrected. However, there is little evidence that venous intervention can be recommended to accelerate healing of the ulcer. Patients with active ulcers can expect ulcer healing in 10–12 weeks on average using various high-compression bandaging systems.[4] Unfortunately, patients with larger ulcers and those of long duration heal more

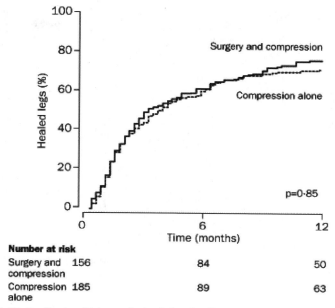

FIGURE 12-3 Kaplan-Meier analysis of ulcer healing for legs treated with compression alone compared to saphenous surgery and compression.

slowly in most cases. It is not clear whether intervention with correction of CVI will accelerate healing in these cases, but it is reasonable to perform corrective procedures if possible prior to ulcer healing. The incidence of ulcer recurrence after healing has been reported as 50% or greater at 5 years, and is somewhat dependent on compliance with the use of compression stockings. Most patients who develop one episode of ulcer recurrence will experience multiple episodes.

For patients with active or healed VLUs, Barwell et al. performed a randomized study, termed the ESCHAR study, comparing the efficacy of saphenous stripping plus compression compared to compression alone for healing and prevention of venous leg ulcers.[5] In this study, 500 patients with isolated superficial reflux (60%) or combined superficial and deep venous disease (40%) were enrolled. Demographic factors were similar in the two groups. Ulcer healing was no different, with 65% healed in each group by life table analysis at 24 weeks after randomization (Figure 12-3). In the group with healed ulcers, mean follow-up time was 14 months. Significantly fewer patients in the surgery group experienced recurrent ulceration (15%) compared to the compression alone group (34%) (Figure 12-4). Adverse events were infrequent (<5% in each group), primarily composed of compression bandage injuries in the conservative group and wound infection (n = 5) in the surgical group.

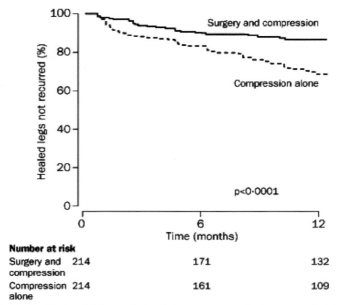

FIGURE 12-4 Kaplan-Meier analysis of ulcer recurrence for legs treated with compression alone compared to saphenous surgery and compression.

Based on the results of this trial, it is reasonable to recommend venous intervention to reduce the incidence of ulcer recurrence whenever superficial venous reflux is a prominent component of the abnormal venous function. This will comprise 30% to 50% of cases seen at most VLU centers. But intervention cannot be recommended for most patients in order to accelerate wound healing.

Anatomically, any combination of superficial, perforator, and/or deep venous disease may result in severe CVI. Marston et al. reported that 29% of limbs with CVI and leg ulceration displayed superficial or superficial and perforator disease on standing reflux examination (Table 12-2).[4] Small saphenous reflux may also be sufficient to cause leg ulceration with no other abnormalities, typically resulting in ulceration near the lateral malleolus. The contribution of incompetent perforators to global venous insufficiency remains controversial and will be discussed in detail later in this chapter, but it is clear that some leg ulcers are associated with large incompetent perforators that should be ligated.

A significant percentage of patients with VLUs are found to have abnormal venous function in multiple systems. Over 27% displayed both deep and superficial reflux in a study of 138 limbs with leg ulceration (Table 12-2).[4] It is not always clear whether these patients will experience an improvement in the severity of symptoms if their superficial or superficial and

TABLE 12-2 Anatomic Distribution of Venous Reflux in 138 Limbs with CEAP Clinical Class 6 CVI

Anatomic Site	Incidence
Deep alone	43.5%
Deep and superficial	21.0%
Deep, perforator, and superficial	6.5%
Superficial alone	18.1%
Superficial and perforator	10.9%

perforator abnormalities are corrected. This issue will be discussed in detail later in this chapter.

Hemodynamic evaluation using APG is very useful in the management of patients with multisystem venous insufficiency. Patients with deep and superficial reflux may be initially treated with superficial stripping or ablation, and the APG can be repeated to determine the degree of improvement without addressing the deep venous reflux. As noted previously, postoperative normalization of the VFI is associated with fewer episodes of ulcer recurrence and fewer symptoms at late follow-up. Therefore, a patient who has improvement in the VFI with correction of only one anatomic component of his/her venous reflux can be followed conservatively, while a patient with a persistent hemodynamic abnormality with a poor VFI can be considered for further intervention.

CONVENTIONAL SURGICAL PROCEDURES FOR CORRECTION OF CVI IN PATIENTS WITH VLU

SUPERFICIAL VENOUS REFLUX: GREAT SAPHENOUS VEIN

Traditional surgical techniques for removal of the great saphenous vein have typically employed ligation of the vein at the saphenofemoral junction and removal of the vein between the groin and knee or groin and ankle using a stripping technique. The goal of high ligation is to identify and divide all venous branches communicating with the saphenofemoral junction (SFJ) to minimize the potential for recurrent reflux pathways resulting in recurrent symptoms. Bergan and others have consistently advocated wide exposure of venous tributaries to their branch points during high ligation. Unfortunately, it appears that many patients developing recurrent venous insufficiency do so because of neovascular generation of new venous communications re-establishing the saphenofemoral junction, or dilation of pre-existing venous tributaries. At this point, it is theorized that the surgical procedure itself is the primary stimulus for neovascularization, and there is hope that endovenous techniques may prove to be associated with a lower incidence of recurrent venous insufficiency after intervention. If surgical treatment is chosen, it remains prudent to perform a complete branch ligation as previously described.

Numerous methods have been described for removal of the saphenous veins after high ligation. The current trend is toward minimizing the invasiveness of surgical intervention, and numerous alternatives to surgical stripping have been introduced. However, it should be noted that stripping procedures themselves have undergone a significant evolution. Using minimal incisions, tumescent local anesthesia, ultrasound guidance, and careful dissection, the GSV can be removed through two small incisions with relatively little bruising or postoperative discomfort in the majority of cases. Key advances have included the use of detailed preoperative venous mapping to plan surgery and the use of intraoperative ultrasound to locate the saphenofemoral junction and precisely place incisions. In the absence of significant obesity, the SFJ ligation can be performed through a 2–3-cm modified transverse incision. Using ultrasound guidance, the GSV can be located just below the knee and encircled through an incision no larger than 1 cm. We typically use tumescent anesthesia and temporary packing along the saphenous tract to minimize bruising. Patients walk upon leaving the ambulatory surgery center and routinely return to work and other routine activities within several days. In many patients, the more significant morbidity comes from the associated stab or powered phlebectomies performed concurrent to saphenous stripping to remove prominent varicosities.

The following issues in superficial venous surgery will be discussed in detail later in this chapter:

- Saphenofemoral ligation alone or ligation and saphenous stripping
- Saphenous stripping to the ankle or to the knee
- Small saphenous reflux
- Need for concomitant varicosity ablation
- Concurrent deep and superficial insufficiency

SAPHENOFEMORAL LIGATION ALONE OR LIGATION AND SAPHENOUS STRIPPING

It has long been debated whether high ligation alone or with varicosity ablation is sufficient for the treatment of superficial venous reflux. Proponents argue that preservation of the saphenous vein is preferable to allow later use as a conduit and that the rate of recurrent symptoms without stripping is acceptable. Hammarsten et al. reported a random allocation of 42 patients to high ligation with varicosity avulsion or high ligation with saphenous stripping and varicosity avulsion.[6] In both groups, incompetent perforators were ligated whenever present. At a mean follow-up of 52 months, there was no difference in the rate of recurrent symptomatic VV between the two treatment groups (12% in those with stripping and 11% in those without).

Proponents of high ligation with GSV stripping have noted the increased incidence of recurrent reflux in the saphenous vein after high ligation alone and maintain that optimal results require routine saphenous stripping. Investigators looking at the residual GSV after high ligation without stripping have identified frequent residual reflux in the GSV. McMullin et al. reported residual reflux in 24 of 52 cases (46%) after SFJ ligation and found that those with persistent reflux did not correct venous refilling times measured by PPG.[7] The Gloucestershire Vascular Group reported 5-year follow-up of 110 limbs randomized to high ligation alone compared to ligation plus stripping.[8] Although patient satisfaction was similar in the two groups, significantly fewer patients in the stripping group required reoperation for recurrent saphenous reflux with symptomatic varicosities (6%) compared to the high ligation only group (20.8%). Regardless of the type of intervention, VV recurrence appears to depend on the length of follow-up. Studies following patients longer than 10 years report that many patients develop recurrent varicosities of some severity. In Fischer's definitive report of the long-term follow-up of limbs treated with high ligation and stripping, 47% of patients followed for an average of 34 years developed clinically evident varicosity recurrence.[9] Although early recurrence was less frequent in some studies in limbs treated with stripping, this benefit is probably lost over time. The primary benefit of stripping appears to be a reduction in the number of limbs requiring reoperation. It has been suggested that this relates to the improved hemodynamic outcome in limbs treated with stripping, resulting in a lower incidence of persistent pain and swelling that would require reoperation.

Balancing the improved hemodynamic result with saphenous stripping has been the occurrence of increased complications with this procedure. While there is an increased incidence of bruising and hematoma in the thigh with stripping, this can be minimized by the use of tumescent anesthesia and other techniques, as noted previously. The most significant complication attributed to saphenous stripping involves injury to the saphenous nerve. Fully described later in this chapter, there is a significant incidence of saphenous nerve deficit after stripping, which is reduced when stripping stops at the level of the knee. Overall, few patients appear to have long-term deficits from this injury.

In summary, high ligation and varicosity ablation is likely to be an acceptable procedure for less severe classes of CVI where the reduced recovery time and reduced potential for bruising and nerve injuries are more important. But in patients with VLUs, the primary aim is to correct the hemodynamic abnormality resulting in symptoms. For many patients, the need for repeat procedures to maintain control of saphenous reflux is undesirable. The high incidence of residual saphenous reflux and significant frequency of reoperation when stripping is not performed argue in favor of high ligation and stripping as the preferred surgical operation for VLUs due to GSV reflux.

SAPHENOUS STRIPPING TO THE ANKLE OR TO THE KNEE

If GSV stripping is chosen as a component of a procedure for the treatment of CVI, the surgeon must determine the length of vein to remove. In most cases, the vein refluxes at both the saphenofemoral junction and throughout its course. Traditionally, many surgeons have elected to strip the entire vein from groin to ankle. The saphenous vein is easily identified at the ankle, and retrograde passage of the stripper is generally unobstructed. However, the saphenous nerve is in close proximity to the vein beginning just below the knee in many patients and may be susceptible to injury during stripping procedures. For these reasons, some surgeons have recommended limiting stripping at a point just below the knee.

To determine whether a limited GSV stripping to the knee would be sufficient to yield improvement in venous hemodynamics, Nishibe et al. studied 110 limbs before and after removal of the above-knee segment of GSV using duplex ultrasound APG.[10] They found that venous hemodynamics as measured by APG were markedly improved after limited GSV stripping. The majority of patients experienced correction of the abnormal preoperative VFI (4.0 ± 0.35 ml/sec) to the normal range (1.4 ± 0.15, $p < 0.001$). The incidence of apparent saphenous nerve injury on assessment at 2–3 weeks after surgery was 4.5%, with most of these patients reporting numbness, mild to moderate pain, or sensitivity to touch in the affected areas.

Holme and colleagues conducted a prospective randomized study in which 163 patients were randomized to high ligation and stripping of the GSV to the ankle (Group A) compared to high ligation and stripping to the knee (Group B).[11] Three months after surgery, 94% of patients in Group A reported good or excellent relief of symptoms compared to 97% of patients in Group B ($p = NS$). Evidence of saphenous injury was identified in 39% of limbs in Group A compared to 7% in Group B ($p < 0.001$). In a subsequent report, the same authors reported long-term follow-up of the same patient cohort. Three years after randomization, 29% of limbs in Group A were reported to display symptoms of permanent saphenous nerve injury compared to only 5% in Group B ($p < 0.01$). At 5 years of follow-up, recurrent varices were seen in 10% of patients in each group.

In a detailed study of the incidence and clinical impact of saphenous nerve injury, Morrison and Dalsing evaluated 127 limbs treated with saphenous stripping to the ankle at a mean follow-up of 4.5 years.[12] Overall, 40% of patients reported symptoms of saphenous nerve injury at some point after operation. At last follow-up, 17% reported the symptoms were persistent, but only 2.3% reported that the symptoms negatively affected their quality of life.

Although the symptoms of saphenous nerve injury may rarely be severe, minor complaints are frequent. It appears that the hemodynamic results of stripping to the knee are similar to total saphenous stripping in most cases. Therefore, most authors have recommended stripping to the knee as the treatment of choice for axial GSV reflux.

Small Saphenous Reflux

Often overlooked is the possibility that ulceration may be due solely to reflux in the SSV. In a report of 20 limbs with isolated lateral perimalleolar ulcers, Bass and colleagues found isolated SSV reflux at the saphenopopliteal junction in 15 limbs (75%).[13] After SSV ligation at the junction, all ulcers healed within 12 weeks. In general, SSVs with sufficient reflux to cause severe CVI are large, dilated veins with numerous varicose tributaries. Hemodynamic evaluation with plethysmography is often useful to determine whether isolated SSV reflux is hemodynamically significant. Patients with SSV reflux are unlikely to result in leg ulceration without an abnormal venous filling index.

Prior to intervention for patients with SSV incompetence, the anatomy of the vein must be carefully determined. The variable course of the SSV has been well documented, and the vein may terminate at numerous points into the popliteal vein, the femoral vein, the vein of Giacomini, or elsewhere. The presence of a large persistent superficial vein in the lateral thigh that does not join the deep system should alert the clinician to the possibility of a variant of Klippel-Trenaunay syndrome with hypoplastic deep veins.

Preoperative mapping of the SSV with ultrasound and marking of the saphenopopliteal junction (SPJ) will assist with operative planning. Surgeons performing SPJ ligation frequently are using intraoperative ultrasound to guide surgical decision making. The SSV may give off several branches just distal to the SPJ, and it is believed to be beneficial to ligate these branches to minimize residual collateral reflux. Similar to GSV surgery, controversy has existed concerning the need for stripping of a portion of the SSV. Although no randomized trials have compared saphenopopliteal ligation to ligation with SSV stripping, most authors have recommended stripping a portion of the SSV. Stripping should generally not involve the lower third of the calf given the increased risk of injury to the sural nerve.

Need for Concomitant Varicosity Ablation

Correction of saphenous and perforator insufficiency will improve hemodynamics with reliable reduction in VFI. For patients with VLU, there usually is no medical need to perform concomitant varicosity ablation. Some clinicians recommend removal of prominent varicose veins in the area of the VLU, believing that this will reduce perforator reflux into the periwound area. However, there is no direct evidence that this will improve the results of intervention, and most venous specialists recommend specific treatment of incompetent perforators using a procedure designed primarily to eliminate perforator reflux as described in the following text.

Results of Conventional Saphenous Surgery

The hemodynamic improvement in patients undergoing saphenous surgery has been well documented. Using APG, correction of saphenous reflux has been demonstrated to result in marked improvement in venous filling index, ejection fraction, and residual volume fraction.[2] Patient satisfaction with the procedure is generally high, but not universally so. Mackay et al. reported on 155 patients who were treated with high ligation and GSV stripping assessed by a questionnaire.[14] Nearly two-thirds of patients reported a perceived postoperative complication within the first 2 weeks after surgery, most relating to bruising, pain, and numbness. Six months after surgery, 80% of patients were satisfied with the outcome, with the most common reason for dissatisfaction being residual varicosities.

For patients with VLU, the ESCHAR study described previously found a significant benefit for saphenous stripping in reducing the incidence of ulcer recurrence, but no benefit was demonstrated in accelerating the time of ulcer healing.[5]

Chronic venous disease has been reported to negatively affect patient quality of life as assessed by a variety of outcomes measures. Using various methodology, investigators have reported that saphenous vein surgery significantly improves quality of life both initially following surgery and at mid-term follow-up several years later.

In summary, saphenous stripping procedures have a proven ability to correct venous hemodynamic dysfunction due to abnormal reflux, resulting in reduced ulcer recurrence and improved quality of life in the majority of patients. Complications of surgery are most often minor and self-limited, but an occasional patient may develop nagging discomfort from saphenous neuralgia, and a rare possibility of DVT cannot be ignored.

Alternatives to Conventional Saphenous Surgery

Numerous alternatives to conventional saphenous surgery have been promoted to effectively eliminate saphenous reflux without the need for surgical incisions or saphenous removal. They include

- Hemodynamic correction of varicose veins (CHIVA)
- External banding to restore saphenous competence
- Endovenous ablation
  - Radiofrequency
  - Laser
- Sclerotherapy
  - Ultrasound guided
  - Foam

When we consider the optimal management of patients with VLUs, the primary anatomic goal is to abolish axial reflux and prevent its recurrence. In these patients, recanalization or reopening of the previously treated saphenous vein usually results in recurrence of the preintervention symptoms, including pain, swelling, worsening skin changes, and possibly ulceration. Most patients who have suffered leg ulcers related to CVI, if given a choice, will choose an intervention that is most likely to minimize the risk of ulcer recurrence. None of the alternatives to saphenous stripping have been studied in a randomized trial to prove benefit in reducing venous ulcer recurrence. Many authors have assumed that if the GSV or SSV remains closed with no reflux throughout the length from groin to knee, the patient should experience a benefit similar to high ligation and stripping procedures.

Saphenous banding procedures to attempt to re-establish competence of the sentinel valve at the SFJ using either external banding materials or endovenous radiofrequency have met with variable results. Despite encouraging reports of success in some studies, others have generated disappointing results with high recurrence rates, such that these techniques are not widely used currently.

Sclerotherapy using ultrasound guidance has been described as a means for occluding the GSV or SSV, thereby correcting reflux. Initial attempts employed liquid sclerosants, and though many saphenous veins were successfully treated, recanalization rates were high in many studies (18.8%–23.8%). More recently, foam sclerotherapy has been studied for occlusion of the GSV, with several studies finding improved results compared to liquid sclerotherapy. In a randomized study, 88 patients were treated with either sclerosing foam or sclerosing liquid via direct puncture of the GSV under duplex guidance.[15] Three weeks after treatment, repeat examination with duplex ultrasound revealed that only 40% of patients treated in the liquid sclerotherapy group had eliminated reflux throughout the GSV compared to 84% in the foam sclerotherapy group. From these studies, it appears that the recurrence rate for liquid sclerotherapy is unacceptably high for treatment of the GSV. Using a foam sclerosant will significantly increase success rates, but long-term success may still be inferior to high ligation and stripping. Randomized studies including quality of life evaluations would further determine whether sclerotherapy has a role in the treatment of saphenous reflux for patients with VLU.

Of the listed alternatives to saphenous ligation and stripping procedures, endovenous ablation has been the most widely studied, including randomized comparisons to stripping. The goal of radiofrequency ablation (RFA) or endovenous laser ablation (EVLA) is to ablate the saphenous vein percutaneously eliminating saphenous reflux, thereby producing the same hemodynamic benefit as high ligation and saphenous stripping with no incisions, fewer complications, and faster recovery to full activity. Early studies with both techniques have demonstrated initial saphenous closure rates of over 90%. Long-term reports on the incidence of saphenous recanalization are now emerging, with acceptable 3–5-year results. In the EVOLVeS study, RFA was compared to ligation and stripping in 86 limbs, including quality of life measures and follow-up ultrasound examinations at routine intervals. Initial success rates at elimination of saphenous reflux were 100% in the stripping group and 95% in the RFA group.[16] Time to return to normal activities and return to work were significantly less in the RFA group. Quality of life surveys revealed a significantly better global score and a significantly better pain score for RFA 1 week postprocedure, but these differences progressively decreased over time. At 2 years of follow-up, two patients in the RFA group had developed recanalization of an initially closed saphenous vein (4%), but global quality of life scores still favored RFA. One patient in the RFA group and four treated with ligation and stripping were found to have evidence of neovascularization on ultrasound examination. Recurrent VV occurred in 14% of RFA limbs and 21% of stripped limbs (p = NS).

No studies have compared the outcome of endovenous ablation to saphenous stripping for patients with VLUs, and little information is available on the incidence of ulcer recurrence after saphenous ablation using any of the previously described methods. Most data on endovenous ablation have been reported on procedures performed in CEAP class 2 and 3 patients. One study reported the hemodynamic results of patients in CEAP classes 3–6 treated with radiofrequency or laser saphenous ablation. It included 29 patients in CEAP classes 5 and 6, and found that ablation resulted in reliable correction in venous hypertension.[17] Seventy-eight percent of limbs demonstrated a normal VFI postablation, and 95% had a VFI <4 ml/second. Based on previous studies of post-stripping VFI, these patients would be expected to have a low incidence of recurrent ulceration unless saphenous recanalization were to develop.

Combined Deep and Superficial Venous Insufficiency

Treatment of patients with leg ulceration and isolated superficial venous insufficiency is usually recommended given the reproducible improvement with correction of saphenous reflux and the low-risk procedures available for this patient group. In patients with CEAP classes 5 and 6 disease, superficial insufficiency is often identified in combination with deep disease. As noted previously, 27% of these limbs studied were reported as having combined reflux. In this situation, the clinician must determine whether benefit is likely from treatment of superficial reflux alone, or if the patient is more likely to require deep venous reconstruction.

Several authors have reported that when superficial reflux in the GSV or SSV is present, the more proximal deep vein segment will occasionally reflux solely due to the superficial vein incompetence. Correction of the superficial reflux reliably results in resolution of the deep vein segment reflux. In this situation GSV incompetence would be seen along with reflux in the common femoral vein, but the femoral and popliteal veins would be competent. With SSV reflux, popliteal reflux would be noted cranial to the SPJ, but not caudal to the junction. With these patterns of reflux, superficial ablative procedures are recommended using the same criteria used for superficial incompetence alone.

Treatment of limbs demonstrating true deep venous insufficiency, defined as reflux in the femoral and popliteal veins, combined with superficial reflux, is controversial. Walsh and Sales reported resolution of deep venous reflux after GSV stripping in over 90% of cases, but Scriven reported that deep venous reflux usually did not correct. Puggioni et al. reported a study of 38 limbs with combined deep and superficial reflux studied with duplex US before and after saphenous stripping.[18] Deep venous reflux was corrected in one-third of patients, and femoral vein reflux corrected more frequently when only segmental reflux was present in that vein rather than axial reflux throughout the deep venous system. The authors noted that the majority of limbs reported by Walsh and Sales demonstrated segmental reflux that may be more likely to correct with superficial surgery than axial reflux.

Padberg and colleagues reported a hemodynamic follow-up of 11 limbs with deep and superficial disease treated with superficial stripping in all cases and perforator ligation in some cases.[19] Although only 27% of limbs studied postoperatively were found to have correction of deep venous reflux, marked hemodynamic improvement was demonstrated. The venous filling index decreased from 12 ml/second preoperatively to 2.7 ml/second postoperatively, and clinical symptom scores decreased from 10 to 1.4. These data suggest that, though superficial and perforator surgery in the presence of deep venous incompetence may not correct deep reflux, it will usually result in significant hemodynamic and symptomatic improvement.

In summary, it is reasonable to consider superficial ablative intervention in patients with leg ulceration due to combined deep and superficial insufficiency. Those with proximal or segmental reflux are more likely to benefit than patients with full axial reflux in the entire deep system. Hemodynamic evaluation with plethysmography before and after saphenous stripping or ablation can determine the degree of improvement and predict long-term symptomatic improvement.

The Significance of Perforator Reflux in CVI

The incidence of perforator incompetence increases as the clinical severity of CVI worsens. The majority of limbs in CEAP clinical classes 5 and 6 have been reported to contain perforators with incompetence on duplex imaging. For this reason, some clinicians believe that incompetent perforators should be corrected whenever they are diagnosed. Unfortunately, it is difficult to clearly determine the hemodynamic significance of incompetent perforators because they are usually seen in limbs that also display superficial and/or deep system incompetence. There clearly are cases in which incompetent perforators are seen in a limb previously treated with saphenous stripping with persistent leg ulceration. In these patients, perforator interruption is necessary. But it is unclear whether perforators should routinely undergo ligation in patients with VLU at the time of saphenous ablation.

CONVENTIONAL SURGICAL LIGATION OF INCOMPETENT PERFORATOR VEINS (IPVs)

When the surgeon believes that IPVs are associated with clinical symptoms, elimination of perforator reflux can be performed using a variety of techniques. Open surgical ligation, mini-incision ligation, subfascial endoscopic ligation, and percutaneous ablation can all be considered. Until the last decade, only open surgical perforator ligation was performed, usually using the Linton procedure. As originally described by Linton, the procedure involves a medial lower limb incision placed over the site of the clinically significant IPVs. Dissection proceeds down to the level of the fascia where the perforators are located and ligated with suture ligatures (Figure 12-5). The use of skin flaps was advocated to help reduce the potential for skin breakdown at the incision site postoperatively.

Though the Linton procedure was effective at eliminating perforator reflux, it has been associated with a high incidence of complications, mostly occurring at the incision site in the area of hyperpigmented, scarred skin typical of advanced CVI. In a report of 37 limbs treated with the Linton procedure, Stuart et al. reported that calf wound complications occurred in seven patients (19%), and the average hospital time was 9 days.[20] Recur-

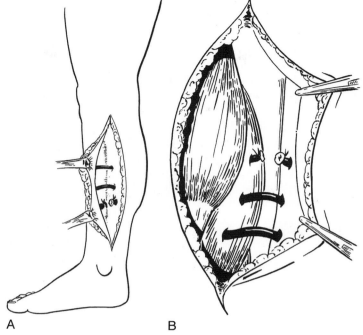

FIGURE 12-5 Incision with flap dissection for Linton procedure.

rent ulceration was reported in 7% to 22% of treated limbs at varying lengths of follow-up after the Linton procedure.

For these reasons, alternate methods were developed to ligate IPVs while eliminating the need for surgical incisions in the area of poor skin expected to be at risk for compromised wound healing. The most widely performed alternative to the Linton procedure is the use of endoscopy to facilitate subfascial perforator ligation (SEPS) through a remote incision in more healthy tissue just below the knee. The primary benefits of this technique have been reported to include more rapid recovery and fewer perioperative complications with equivalent hemodynamic results in comparison to the Linton procedure. In a retrospective comparison of SEPS (n = 27) to the Linton procedure (n = 29), Sato et al. reported similar rates of ulcer healing and recurrence, as well as a similar improvement in venous dysfunction scores in the two groups. In the Linton group, 45% of patients developed incision-related complications, compared to only 7% in the SEPS group ($p < 0.005$).[21] In a prospective comparison of the Linton procedure to SEPS, Pierik et al. randomized 39 patients to open or endoscopic perforator ligation.[22] In the open group, 53% of patients developed postoperative wound infection compared to 0% in the SEPS group (p < 0.001). Ulcer healing rates and recurrence rates were similar in the two groups.

It appears clear that when perforator ligation is deemed necessary, the SEPS procedure is preferred to open surgical ligation. Other alternate options have been reported for treatment of refluxing perforators. Perforator ligation has been reported using a mini-incisional technique, minimizing wound complications. Results have been reasonably good, but experience is limited. Initial reports of the use of endoluminal techniques using radiofrequency or laser heat sources have suggested that percutaneous ablation of perforator veins is feasible. Using ultrasound guidance, the incompetent perforator is imaged and accessed with a micropuncture needle and guidewire. A sheath can then be inserted into the perforator down to the fascial level. Through the sheath, an ablative radiofrequency or laser catheter can be placed, allowing energy to be delivered to the perforator vein at and superficial to the fascia. Some perforators are difficult to access due to tortuosity and branching, but early reports have described success in 80% to 90% of cases attempted. Larger prospective studies are needed to determine the efficacy of these less invasive methods.

A more fundamental question concerns the indications for perforator ligation. This remains controversial with proponents arguing that perforators are frequently present in limbs with VLUs and should be ligated whenever present. Skeptics argue that perforators are usually present in combination with superficial and/or deep venous incompetence, and the relative contribution of the incompetent perforator to venous insufficiency is less important. Iafrati et al. reported on the treatment of 51 limbs with perforator reflux and leg ulcers using SEPS.[23] Venous disability scores improved significantly after the procedure, and 74% of limb ulcers healed within 6 months. The recurrent ulceration rate was low at 13%. Excellent results were obtained, but 35 of the 51 limbs were treated concomitantly with saphenous or varicose vein removal. Of note, SEPS performed without saphenous surgery was associated with delayed ulcer healing.

Tawes et al. reported a large retrospective multicenter experience using SEPS in over 800 limbs with CVI.[24] The majority of patients (532) had active or previous leg ulceration. Concomitant GSV removal was performed in 55% of cases. Reported results were excellent with 92% of limb ulcers healing at 4–14 weeks after SEPS. Recurrent ulceration occurred in only 4% at a mean follow-up of 15 months. From this review, the authors concluded that until definitive level I evidence is available, SEPS is advocated as optimal therapy for patients with CVI and incompetent perforator veins.

Mendes et al. studied a common subset of patients with IPVs, those with concomitant saphenous reflux and IPVs.[25] Twenty-four limbs were studied before and after surgery with duplex ultrasound and APG. In all limbs, saphenous stripping was performed, with powered phlebectomy added in patients with prominent varicosities. No SEPS or other specific treatment for the IPVs was performed. After surgery, 71% of the limbs no longer contained IPVs. Hemodynamic improvement on APG occurred in all limbs, with the VFI improving from 6.0 ± 2.9 preoperatively to 2.2 ± 1.3 after surgery ($p < 0.001$). They concluded that either the varicosity ablation performed an extrafascial perforator ligation by removing the outflow tract for the IPVs, or the IPVs were of relatively little hemodynamic importance in comparison to saphenous reflux in this patient group.

It is not clear whether IPVs found in limbs coexisting with deep venous reflux should be ligated, particularly in the absence of corrective surgery for the deep venous system. In the North American SEPS Registry report, there was an increased incidence of leg ulcer recurrence in patients with deep venous insufficiency after SEPS. Other authors, including Tawes et al., still recommend SEPS in these patients, believing that venous hypertension will improve after perforator ligation despite the presence of continued deep venous reflux. No prospective randomized studies have been performed to further evaluate these important questions.

It is obvious that the treatment of limbs found to contain IPVs remains controversial in many situations. Perhaps the primary problem in this debate is the lack of a comprehensive definition of perforator incompetence based on their potential to cause venous hemodynamic dysfunction. We currently treat all IPVs similarly, and there is significant variation in the criteria for perforator incompetence in reported studies. This situation would be similar to considering an arterial stenosis of 30% similarly to a 90% stenosis. Delis and colleagues previously suggested that all perforators demonstrating outward flow are not equal, proposing that the volume of outward flow in 1 second after compression release (based on perforator size and velocity of reflux) may be used to define classes of perforator reflux.[26] They proposed that the early hemodynamic function of the IPV determines its clinical impact on the leg, rather than the duration of reflux. The maximum diameter of IPVs may also be important, such that a 5-mm IPV with a high velocity of reflux would be expected to cause a significantly greater hemodynamic impact than a 3-mm perforator that refluxes in the outward direction, but only at low velocity (Figure 12-6). Further research on diagnosis and management of IPVs is required to allow optimal treatment of IPVs.

CONCLUSION

In patients with severe CVI resulting in leg ulceration, the primary goal is elimination of abnormal venous reflux resulting in venous hypertension. Rational treatment of this diverse group of patients requires detailed

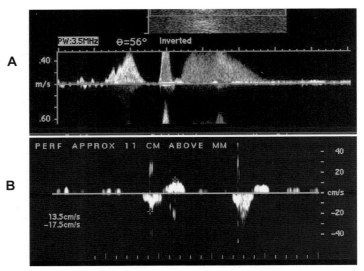

FIGURE 12-6 Varying ultrasound characteristics of "incompetent perforators." A. Large 5.2 mm diameter perforator with high velocity reflux (40–50 cm/second) lasting greater than 2 seconds after calf release. B. Smaller 3.4 mm diameter perforator demonstrating maximum reflux velocity of 13.5 cm/second lasting 0.7 seconds after calf release.

anatomic and hemodynamic assessment with duplex ultrasound and plethysmography. Postprocedure reassessment can reveal the results of therapy and direct further management. Standard surgical techniques for correction of superficial and perforator incompetence have been found to reduce the incidence of ulcer recurrence, but are being replaced by less invasive methods that appear in early and mid-term studies to have comparable symptomatic and hemodynamic results. Further long-term study will be required to evaluate the efficacy of these techniques at accelerating healing and preventing recurrence of problem VLUs.

REFERENCES

1. Christopoulos D, Nicolaides AN, Szendro G. Venous reflux: Quantitation and correlation with the clinical severity of chronic venous disease. *Br J Surg.* 1988;75: 352.
2. Owens LV, Farber MA, Young ML. The value of air plethysmography in predicting clinical outcome after surgical treatment of chronic venous insufficiency. *J Vasc Surg.* 2000;32;961–968.
3. McDaniel HB, Marston WA, Farber MA, et al. Recurrence of chronic venous ulcers on the basis of clinical, etiologic, anatomic, and pathophysiologic criteria and air plethysmography. *J Vasc Surg.* 2002;35:723–728.

4. Marston WA, Carlin RE, Passman MA, et al. Healing rates and cost efficacy of outpatient compression treatment for leg ulcers associated with venous insufficiency. *J Vasc Surg.* 1999;30:491–498.
5. Barwell JR, Davies CE, Deacon J, et al. Comparison of surgery and compression with compression alone in chronic venous ulceration (ESCHAR study): Randomized controlled trial. *Lancet.* 2004;363:1854–1859.
6. Hammarsten J, Pederson P, Cederlund CG, Campanello M. Long saphenous vein saving surgery for varicose veins. A long-term follow-up. *Eur J Vasc Surg.* 1990;4:361–364.
7. McMullin GM, Coleridge Smith PD, Scurr JH. Objective assessment of high ligation without stripping the long saphenous vein. *Br J Surg.* 1991;78:1139–1142.
8. Dwerryhouse S, Davies B, Harradine K, Earnshaw JJ. Stripping the long saphenous vein reduces the rate of reoperation for recurrent varicose veins: Five-year results of a randomized trial. *J Vasc Surg.* 1999;29:589–592.
9. Fischer R, Linde N, Duff C, Jeanneret C, Chandler JG, Seeber P. Late recurrent saphenofemoral junction reflux after ligation and stripping of the greater saphenous vein. *J Vasc Surg.* 2001;34:236–240.
10. Nishibe T, Nishibe M, Kudo F, Flores J, Miyazaki K, Yasuda K. Stripping operation with preservation of the calf saphenous veins for primary varicose veins: Hemodynamic evaluation. *Cardiovasc Surg.* 2003;11:341–345.
11. Holme JB, Skajaa K, Holme K. Incidence of lesions of the saphenous nerve after partial or complete stripping of the long saphenous vein. *Acta Chir Scand.* 1990;156:145–148.
12. Morrison C, Dalsing MC. Signs and symptoms of saphenous nerve injury after greater saphenous vein stripping: Prevalence, severity, and relevance for modern practice. *J Vasc Surg.* 2003;38:886–890.
13. Bass A, Chayen D, Weinmann EE, Ziss M. Lateral venous ulcer and short saphenous vein insufficiency. *J Vasc Surg.* 1997;25:654–657.
14. Mackay DC, Summerton DJ, Walker AJ. The early morbidity of varicose vein surgery. *J R Nav Med Serv.* 1995;81:42–46.
15. Hamel-Desnos C, Desnos P, Wollmann JC, Ouvry P, Mako S, Allaert FA. Evaluation of the efficacy of Polidocanol in the form of foam compared with liquid form in sclerotherapy of the greater saphenous vein: Initial results. *Dermatol Surg.* 2003;29:1170–1175.
16. Lurie F, Creton D, Eklof B, et al. Prospective randomized study of endovenous radiofrequency obliteration (closure procedure) versus ligation and stripping in a selected population (EVOLVeS Study). *J Vasc Surg.* 2003;38:207–214.
17. Marston WA, Owens LV, Davies S, Mendes RC, Farber MA, Keagy BA. Endovenous saphenous ablation corrects the hemodynamic abnormality in patients with CEAP clinical class 3–6 CVI due to superficial reflux. *Vasc Endovasc Surg.* 2006;40:125–130.
18. Puggioni A, Lurie F, Kistner RL, Eklof B. How often is deep venous reflux eliminated after saphenous vein ablation. *J Vasc Surg.* 2003;38:517–521.
19. Padberg FT Jr, Pappas PJ, Araki CT, Thompson PN, Hobson RW 2nd. Hemodynamic and clinical improvement after superficial vein ablation in primary combined venous insufficiency with ulceration. *J Vasc Surg.* 1996;24:711–718.
20. Stuart WP, Asam DJ, Bradbury AW, Ruckley CV. Subfascial endoscopic perforator surgery is associated with significantly less morbidity and shorter hospital stay than open operation (Linton's procedure). *Br J Surg.* 1997;84:1364–1365.
21. Sato DT, Goff CD, Gregory RT, et al. Subfascial perforator vein ablation: Comparison of open versus endoscopic techniques. *J Endovasc Surg.* 1999;6:147–154.
22. Pierik EGJM, van Urk H, Hop WCJ, Wittens CHA. Endoscopic versus open subfascial division of incompetent perforating veins in the treatment of venous leg ulceration: A randomized trial. *J Vasc Surg.* 1997;26:1049–1054.

23. Iafrati MD, Pare GJ, O'Donnell TF, Estes J. Is the nihilistic approach to surgical reduction of superficial and perforator vein incompetence for venous ulcer justified? *J Vasc Surg.* 2002;36:1167–1174.
24. Tawes RL, Barron ML, Coello AA, Joyce DH, Kolvenbach R. Optimal therapy for advanced chronic venous insufficiency. *J Vasc Surg.* 2003;37:545–551.
25. Mendes RR, Marston WA, Farber MA, Keagy BA. Treatment of superficial and perforator venous incompetence without deep venous insufficiency: Is routine perforator ligation necessary? *J Vasc Surg.* 2003;38:891–895.
26. Delis KT, Husmann M, Kalodiki E, Wolfe JH, Nicolaides AN. In situ hemodynamics of perforating veins in chronic venous insufficiency. *J Vasc Surg.* 2001;33:773–782.

13

Endovascular Techniques for Superficial Vein Ablation in Treatment of Venous Ulcers

Michael J. Stirling and
Cynthia K. Shortell

Superficial venous reflux is a common clinical problem with a variety of manifestations ranging from telangiectasias to ulceration. In 1995 the CEAP system for classification of lower extremity venous disease was introduced[1] (Figure 13-1). The disease process is treated by a wide range of specialists with varying degrees of success. The number of patients afflicted is significant because symptomatic varicose veins are present in 15% to 40% of men and 25% to 32% of women.[2,3] In a retrospective study of patients in Olmstead County, Minnesota, the findings of venous ulceration in patients with clinically diagnosed venous stasis was 263 of 1,131 patients.[4] This was calculated as an incidence of 18 per 100,000 person years.[4] Sybrandy et al.[9] estimated the incidence of venous ulceration as 1% to 2% in the Western population.

The four main risk factors for developing varicose veins that have been identified are (1) family history, (2) female hormones, (3) gravitational hydrostatic forces (exacerbated during pregnancy), and (4) hydrostatic muscular compartment force.[5] In essence, chronic venous insufficiency (CVI) and subsequent ulceration (CEAP 5, 6) are the result of hydrostatic and hydrodynamic forces acting on a susceptible vascular system.

The majority (60%–70%) of varicose veins are as a result of saphenofemoral junction (SFJ) and great saphenous vein (GSV) reflux. A further 10% to 20% are from saphenopoliteal junction (SPJ) and small saphenous

CEAP classification

Class	Clinical Signs
0	No physical signs of venous disease
1	Telangiectasia, reticular veins, malleolar flare
2	Varicosities
3	Edema without skin changes
4	Skin changes (pigmentation, venous eczema, lipodermatosclerosis)
5	Skin changes with healed ulceration
6	Skin changes with active ulceration

FIGURE 13-1   CEAP classification.[1]

vein (SSV) incompetence. The remaining result from perforator incompetence.[3] The challenge for physicians has been effective treatment of these patients and prevention of recurrent lesions.

In recent years, physicians have witnessed the development of minimally invasive techniques to treat superficial and perforator reflux. These techniques have not been studied extensively with respect to their effect on venous ulceration, but preliminary results favor their utility in this setting.

HISTORICAL BACKGROUND

Historically, the definitive treatment of venous disease has been based on surgical intervention. Failure of conservative treatment resulted in offering patients high ligation and stripping of the GSV or SSV in conjunction with removal of accessory varicosities. This technique was associated with a 5-year recurrence rate of 20%[1] and a reintervention rate of 25%.[5] Open surgical technique of perforator vein ligation (Linton procedure) for the treatment of venous ulcers had even more disappointing results, with an ulcer recurrence rate of 40%.[8] A prospective, randomized trial comparing open versus endoscopic perforator ligation demonstrated a wound complication rate of 53%.[9] Sybrandy et al.[9] also had good success rates with subfascial endoscopic division of incompetent perforating veins (SEPS) when compared with open ligation. Their experience demonstrated ulcer healing in 17 of 19 patients (89%) with a mean follow-up of 46.1 months.[9] Iafrati et al.[8] reversed the previously pessimistic recommendations of their senior author with successful utilization of the SEPS procedure. They found an ulcer recurrence rate of <13%.[8]

The incidence of superficial venous reflux in patients who have skin changes and venous ulceration ranges from 17% to 53%.[6] This has led to treatment of the superficial venous system in an effort to treat venous ulceration that has failed conservative management. As stated by Cho and Gloviczki, "The main goal of surgical treatment of chronic venous insufficiency is to achieve healing of venous ulcers and to prevent new or recurrent ulcers."[7]

ENDOVASCULAR OPTIONS

There are presently three endovascular techniques commonly utilized in the treatment of superficial venous reflux by practitioners in the United States. Originally described in 1939, foam sclerotherapy[10] was modified in the 1990s and has experienced an increase in popularity. In 1999, the FDA approved radiofrequency ablation (RFA; VNUS Closure, VNUS Medical Technologies, San Jose, California) as a technique to obliterate refluxing superficial veins.[11] This was followed in 2002 by the development and subsequent FDA approval of endovenous laser treatment (EVLT).[12] This is now commonly referred to as endovenous laser ablation (EVLA) because EVLT is a registered trademark (Diomed Inc, Andover, Massachusetts).[3]

MECHANISM OF ACTION

FOAM SCLEROTHERAPY

Sclerotherapy involves the deposit of a foreign substance (i.e., sclerosant) into a vessel lumen, resulting in thrombosis and fibrosis of the vessel. The sclerosants cause endothelial damage, which is the initiating event in this process. Endothelial layer disruption and the resulting exposure of the subendothelial collagen lead to platelet aggregation and the activation of the intrinsic pathway of coagulation.[5] This, in turn, prompts endofibrosis, leading ultimately to ablation of the vein. When the sclerosant is administered as foam, an air block is created and blood is displaced. This, when combined with a large surface area of the sclerosant (in foam form), increases endothelial exposure to the sclerosant and increased treatment efficacy.[12] Sodium tetradecyl sulfate and poly-iodinated iodine are currently the strongest sclerosing agents available.[11]

RFA

The RFA technique subjects the vein wall to high-frequency alternating current by direct contact of the catheter prongs with the endothelium of the vein (Figures 13-2 and 13-3). This leads to the loss of vessel wall

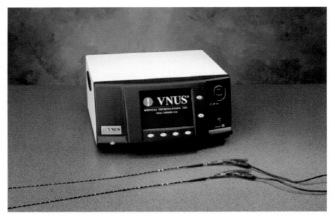

FIGURE 13-2 RFA generator.

FIGURE 13-3 RFA probes.

architecture, disintegration, and carbonization of the vessel.[13] Tissue destruction is precise, and very little thrombus is created. The requirement of direct contact of the catheter prongs with the vein wall may limit the application of RFA to patients with veins smaller than 12 mm.[16]

EVLA

The application of endoluminal laser energy (Figure 13-4) to the treatment of superficial venous reflux disease began gaining popularity in the 1990s.[14] EVLA delivers laser energy directly into the vein lumen, but the

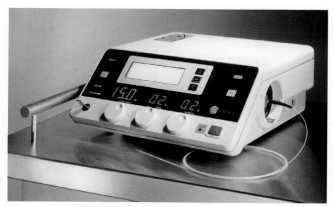

FIGURE 13-4   EVLT generator.

mechanism of action by which this brings about destruction of the vein wall is debated. The laser energy causes the blood inside the vein to boil, and it may be the diffusion of the superheated steam bubbles to the vein wall that actually destroys the vein architecture.[15] Other authors have proposed the heat generated by the steam bubbles is not sufficient, in and of itself, to destroy the vein wall. Instead, vein wall destruction requires direct contact with the laser fiber.[13] Regardless of the mechanism, the final common pathway is the same as with RFA; the heating of the vein wall results in collagen contraction and destruction of endothelium. The vein wall undergoes inflammatory reaction with the result being fibrosis of the vein.[15]

PATIENT SELECTION

Each patient considered for intervention should be subject to a thorough history and physical exam. Pertinent information includes history of superficial or deep venous thrombosis (DVT), prior treatments attempted (both surgical and nonsurgical in nature), and overall health of the patient.[16] Personal or family history of thrombosis should prompt consideration of investigation for coagulation disorder (discussed at length elsewhere in the text). Treatment of a vessel with an acute clot is contraindicated.

Physical exam should include detailed evaluation of both the arterial and venous systems. Signs of chronic venous insufficiency and stasis along with the size, location, and distribution of varicosities, if present, should be noted. Duplex evaluation of deep, superficial, and perforating systems is essential prior to planning any intervention.

Only after this in-depth patient evaluation is complete should preoperative planning be commenced. The findings of the evaluation determine which veins to treat and the order in which this will be attempted.[16] It is common to treat the GSV first because many cases of reflux in the SSV and/or deep system may resolve following treatment of severe GSV reflux. Labropoulos et al.[6] found evidence of ulceration with reflux present through the entire length of the GSV in 9 of 11 cases of ulceration. In the absence of GSV incompetence, reflux of the SSV had to be associated with reflux of the Giacomini vein or the medial and lateral gastrocnemius veins in order for ulceration to be present.[6] In these cases, treatment of the SSV will take precedence. Associated accessory veins may be treated at the same sitting or in a delayed fashion if necessary. Many of these will resolve spontaneously, so a repeat evaluation may be the most prudent strategy.

OBLITERATIVE TECHNIQUES

FOAM SCLEROTHERAPY

Injectable sclerosants have been utilized in the treatment of varicosities for over a century, with the first reported treatment occurring in 1840.[17] Foam sclerotherapy involves mixing air with approved sclerosants, resulting in the creation of a microfoam. The target vein is then identified with ultrasound and the foamed sclerosant injected. Two methods of foam creation are currently in widespread use. The Tessari and Monfreux methods may result in some inconsistently prepared foam; however, both are popular and effective in their application.[11]

First described in 1997, the Monfreux method creates the foam in a glass syringe. The sclerosant is drawn into a glass syringe, followed by air. This results in a foam with large bubbles. Variations in the foam occur secondary to the concentration and type of sclerosant, the syringe itself, and the rate of withdrawal of the syringe piston.[11]

The Tessari method has evolved into the most popular of the foaming techniques.[18] Two syringes, one containing the sclerosant and the other air, are attached to a three-way stopcock, and then the sclerosant is passed between the two syringes.[11] The optimal number of passes appears to be 20. The higher the concentration of sclerosant, the more stable the foam.[18]

Prior to the injection of the foam, the treated extremity should be elevated (to empty the superficial venous system).[17] The saphenofemoral junction (SFJ) should be compressed to prevent foam extending into the deep venous system and also to prevent microemboli from proximal flow. After the injection of the foam, the lower extremity should be elevated for a further 10 minutes (J. J. Bergan, personal communication, October 2006). The final step of the procedure is the placement of compression dressings.[12]

RFA

A single manufacturer provides the generator and catheter required for RFA.[17] The patient is placed in the reverse Trendelenburg position. As with foam sclerotherapy, ultrasound is utilized to identify the target vessel and micropuncture set used for access. The GSV is usually accessed at the knee, while the SSV is accessed at the ankle. Once access has been obtained, the patient is switched to the Trendelenburg position, and the catheter is advanced to 0.5–1 cm inferior to the SFJ (again with ultrasonographic guidance). The ideal position of the tip of the catheter is immediately inferior to the junction with the superficial epigastric vein.[12] The administration of tumescent anesthesia (dilute mix of 1% lidocaine, epinephrine, and sodium bicarbonate) in the saphenous sheath separates the vein from surrounding tissue (and the skin). The benefits of tumescent anesthesia are 3-fold: (1) analgesia, (2) protection against thermal injury of the skin (acts as a heat sink), and (3) compression of the vein, which improves contact between the endothelium and the catheter prongs.[16]

Two sizes of electrodes are available for treatment of vessels up to 12 mm. A 6Fr catheter treats vessels up to 8 mm, and an 8Fr catheter treats vessels from 8 to 12 mm. Adjunctive strategies such as increasing tumescent volume, Esmarch bandage application, extreme Trendelenburg position, and tumescent administration superior to the SFJ have been utilized for the successful treatment of vessels greater than 12 mm in diameter.[16]

Probe temperature is set to 85°C and, once this temperature has been achieved, the probe is withdrawn at a rate of 2.5–3.0 cm/minute.[19] In an attempt to reduce treatment time, some physicians have found an increase in probe temperature will allow faster withdrawal rates. If the probe is set to 90°C, it can be withdrawn at 4 cm/minute with no loss of treatment efficacy.[12] Temperature, impedance, and generator output should be monitored throughout the treatment process, with appropriate adjustments to catheter withdrawal rates being dictated by readings.

The patient is returned to a horizontal position after the catheter has been withdrawn completely. Ultrasound is used to confirm the patency of the common femoral and superficial epigastric veins, as well as the successful closure of the GSV. If there is persistent and significant flow in the vein, the treatment may be repeated. If flow is minimal and the vessel walls are thickened, treatment can be terminated because a thrombus plug will most likely form.[12] If a second pass with the catheter is unsuccessful, proximal ligation of the GSV through a small incision may be performed.[20]

Patients are to utilize compression therapy on the treated extremity for 1 week and are encouraged to ambulate immediately and regularly. A postprocedure ultrasound is performed at 72 hours to confirm successful intervention and ensure there has been no extension of clot into the common femoral vein.[16]

EVLA

A benefit of EVLA is that it involves short exposure time with maximal ablation of target tissue.[16] As with RFA, the patient is placed in the reverse Trendelenburg position, and ultrasound is used to facilitate micropuncture access of the GSV at the level of the knee. A 5Fr sheath is then placed in the vein.[15] Again, as with RFA, the patient is converted to the Trendelenburg position. At this point, a 600-µm fiber is positioned 1 cm inferior to the SFJ. Tumescent anesthesia is administered in the same manner as with RFA. Device position is then confirmed using ultrasound and visualization through the skin of the red aiming beam.[11,15] Laser energy is delivered at a wavelength of 810 nm. Alternatively, wavelengths of 940, 980, and 1,320 nm have also been used with success.[11] The device is then turned on and withdrawn relatively rapidly. Manual compression of the vein may aid with obliteration of the lumen.[12] Again, as with RFA, patients wear full-length compression stockings for 1 week, are encouraged to ambulate, and get a follow-up ultrasound at 72 hours.

OUTCOMES

FOAM SCLEROTHERAPY

An Italian study over a 3-year period included 453 patients treated with foam sclerotherapy. Using the Monfreux and Tessari methods of foam preparation, they reported immediate success rates of 88.1% and 93.3%, respectively.[21] A prospective, multicenter study was conducted in France which randomized 88 patients to treatment with foam and liquid sclerosants. In the foam group, GSV reflux was absent in 84% at 3 weeks, and at 6 months, there were only two patients with evidence of recanalization. At 12 months, there were no further failures.[22] Pascarella et al.[30] compared foam sclerotherapy with compression for treatment of CEAP 4, 5, and 6 limbs. When they used ulcer healing alone as a marker of successful treatment, they found better efficacy with foam sclerotherapy versus compression (70% and 50%, respectively), and 62.5% of ulcers that failed to heal with compression healed within 2 weeks of foam sclerotherapy.[30] This group's experience will be expanded upon elsewhere in the text.

RFA

The RFA experience is being recorded as a multicenter registry. A review of 890 patients (1,078 limbs) was attempted at 1 week; 6 months; and 1, 2, 3, and 4 years. The number of limbs examined at each time interval was 858, 446, 384, 210, 114, and 98, respectively. The vein occlusion rates were 91%, 88.8%, 86.2%, 84.2%, and 88.8%, respectively.[23] A retrospective review of a single center's experience demonstrated complete GSV obliteration in 99% of 332 limbs evaluated 72 hours after treatment.[20] A prospective, randomized comparison of RFA and ligation with stripping (EVOLVeS) demonstrated RFA to be at least equivalent to surgery for the obliteration of GSV reflux at 2 years. Furthermore, quality-of-life scores were higher in the RFA group at 2 years.[24]

EVLA

EVLA has also shown excellent results in early evaluation of GSV obliteration rates. A reintervention rate of 0% in EVLA was found in a retrospective review of 92 consecutive patients (130 limbs) at the Mayo Clinic (vs. 17% in RFA) with an immediate GSV occlusion of 100%.[25] These results exceeded those of a larger trial that included 499 limbs in 423 patients and documented an immediate GSV occlusion rate of 98.2% (490 of 499) and a 2-year success rate of 93.4% (113 of 121 followed at 2 years).[15] An earlier report from the same author demonstrated a 6-month success rate of 99%.[26] These results have been approached by other authors with smaller sample sizes, including a study of 29 patients with a 1-month success rate of 97%.[27]

Marston et al.[31] reviewed the efficacy of RFA and EVLA in CEAP 3, 4, 5, and 6 patients with similar improvements in clinical symptom score (approximately 10% in both groups) and GSV ablation rates. They did not report objective results for ulcer healing, however.[31]

COMPLICATIONS

FOAM SCLEROTHERAPY

The majority of complications arising from foam sclerotherapy are minor and resolve spontaneously. Clot extension beyond the SFJ is rare.[28] A prospective trial of large-volume sclerotherapy found no symptomatic limbs to have DVT.[17] Perivenous injection will result in local ulceration, and intra-arterial injection will result in arterial occlusion. Although severe, these complications are rare.[13]

Much debate has focused on the potential consequences of microembolization of foam particles. In a series of 869 patients, Bergan had four ocular complications, which have been attributed to microembolization in patients with a patent foramen ovale (J. J. Bergan, personal communication, October 2006). None of these complications occurred after the technique was modified to include elevation of the injected limb.

Side effects have been reviewed in canine models and include transient pulmonary hypertension and decreased systolic blood pressure. Both of these phenomena have resulted in no significant adverse events.[18] Consequently, dosing regimens have remained <80 ml per treatment.[18]

RFA

RFA complications include DVT, pulmonary embolus (PE), focal paresthesias, phlebitis, skin burns, and infection. Merchant et al. reported paresthesia rates of 12.1% at 1 week, 6.7% at 1 month, and 2.0% at 4 years.[23] Subgroup analysis revealed these rates to be lower in the presence of tumescent anesthesia and at the higher volume centers. The Mayo Clinic retrospectively compared RFA with EVLA and found the additional treatment rate for RFA to be 17% at the time of original procedure (in contrast to EVLA, which was virtually nil).[25]

Local thrombus at the SFJ and DVT occurred in 0.5% of patients. One of these patients developed a PE.[23] In sharp contrast, Hingorani et al. reported a DVT rate of 16%.[29] They reported no episodes of PE.[29]

Merchant et al. experienced a skin-burn rate of 0.5% with the utilization of tumescent anesthesia and 1.7% in its absence.[23] Phlebitis was 3.3% at 1 week and 0.2% at 6 months.[20]

The Mayo Clinic review observed a higher incidence of immediate complications in the EVLA group (16.8%) compared with the RFA group (7.6%).[25] These complications ranged from urinary retention to excessive pain.

EVLA

The most common complaint after EVLA therapy is the sensation of "pulling" along the course of the ablated vein, reported in 90% of patients. This resolved in all patients within 10 days.[15] Thrombophlebitis and pigmentation change were the next most common complaints, but they also required no further intervention. Min et al. reported bruising in 24% of 504 patients at 1 week (all resolved at 1 month) and superficial thrombophlebitis in 5%.[15] They had no events of skin burn, paresthesias, DVT, or PE.[15]

The Mayo Clinic experience demonstrated extension of thrombus into the femoral vein in 2.3% of patients.[25] These patients were asymptomatic, and thrombus was discovered at follow-up ultrasound. One patient required inferior vena cava filter; the others had resolution of clot by 12 weeks with anticoagulation treatment.[25]

DISCUSSION

The advent of endovenous therapies for superficial venous reflux is rendering older surgical therapies obsolete. We currently lack prospective, randomized trials comparing the endovenous therapies discussed in this chapter. Comparing the pros and cons of each treatment will leave the individual practitioner to fit the treatment modality to patients.

The role of endovenous therapy in the treatment of venous ulceration is evolving. There is evidence that surgical stripping has aided in the treatment of venous ulceration, but the morbidity of the surgery may be too much for many patients to tolerate in addition to the pre-existing ulcerations.

The three endovenous ablative techniques have demonstrated short- and long-term success rates equivalent to or surpassing those of surgical vein stripping while reducing recovery time and complication rates.[16] As more evidence is accumulated in CEAP 5 and 6 patients, we expect to see support for these techniques in the treatment of venous ulceration. The most efficacious (and cheapest) of these methods may be foam sclerotherapy with lower extremity elevation. This will combine ablation of the more proximal superficial system with direct ablation of the vessels at the site of venous ulceration without additional insult to the regional integumentary system.

REFERENCES

1. Porter EP, Moneta GL. Reporting standards in venous disease: An update. International Consensus Committee on Chronic Venous Disease. *J Vasc Surg.* 1995;21:635.
2. Callam MJ. Epidemiology of varicose veins. *Br J Surg.* 1994;81:167–173.
3. Bough MJ. Endovenous laser therapy for varicose veins: Current evidence and recommendations. In Wyatt MG, Watkinson AF, eds. *Endovascular Therapies: Current Evidence.* Shrewsbury, UK: tfm Publishing Ltd., 2006:197–210.
4. Heit JA, Rooke TW, Silverstein MD, et al. Trends in the incidence of venous stasis syndrome and venous ulcer: A 25 year population-based study. *J Vasc Surg.* 2001;33:1022–1027.
5. Bergan JJ. Excision of varicose veins. In: Ernst CB, Stanley JC, eds. *Current Therapy in Vascular Surgery,* 4th ed. St Louis, MO: Mosby, 2001;838–840.
6. Labropoulos N, Leon M, Nicolaides AN, Giannoukas AD, Volteas N, Chan P. Superficial venous insufficiency: Correlation of anatomic extent of reflux with clinical symptoms and signs. *J Vasc Surg.* 1994;20:953–958.
7. Cho J-S, Gloviczki P. Surgical management of lower extremity venous insufficiency. In: Hobson AW, Wilson SE, Veith FJ, eds. *Vascular Surgery: Principles and Practice.* New York: Marcel Dekker, 2004;991–1002.
8. Iafrati MD, Pare GJ, O'Donnel TF, Estes J. Is the nihilistic approach to surgical reduction of superficial and perforator vein incompetence for venous ulcer justified? *J Vasc Surg.* 2002;36:1167–1173.
9. Sybrandy JEM, van Gent WB, Pierik EGJM, Wittens CHA. Endoscopic versus open subfascial division of incompetent perforating veins in the treatment of venous leg ulceration: Long-term follow-up. *J Vasc Surg.* 2001;33:1028–1033.
10. Barrett JM, Allen B, Ockelford A, et al. Microfoam ultrasound-guided sclerotherapy of varicose veins in 100 legs. *Dermatol Surg.* 2004;30:6–12.
11. Sadick NS. Advances in the treatment of varicose veins: Ambulatory phlebectomy, foam sclerotherapy, laser and radiofrequency closure. *Dermatol Clin.* 2005;23:443–455.

12. Teruya TH, Ballard JL. New approaches for the treatment of varicose veins. *Surg Clinics NA.* 2004;84:1397–1417.
13. Coleridge Smith PD. Sclerotherapy in the treatment of varicose veins. In: Negus D, Coleridge Smith PD, Bergan JJ, eds. *Leg Ulcers. Diagnosis and Management*, 3rd ed. New York: Oxford University Press, 2005;163–171.
14. Bone C. Trateamiento endoluminal de las varices con laser de diodo: Studio preliminary. *Rev Patol Vasc.* 1999;5:35–46.
15. Min RJ, Khilnani N, Zimmet SE. Endovenous laser treatment of saphenous vein reflux: Long-term results. *JVIR.* 2003;14:991–996.
16. Stirling MJ, Shortell CK. Endovascular treatment of varicose veins. *Sem Vasc Surg.* 2006;19:109–115.
17. Morrison N, Rogers C, Neuhardt D, Melfy K. Large-volume, ultrasound-guided Polidocanol foam sclerotherapy: A prospective study of toxicity and complications. Abstract presented at: UIP World Congress Chapter Meeting; August 27–31, 2003; San Diego, CA.
18. Hsu T. Foam sclerotherapy: A new era. *Arch Dermatol.* 2003;139:1494–1496.
19. Manfrini S, Gasbarro V, Daniellson G, et al. Endovenous management of saphenous vein reflux. *JVS.* 2000;32:330–342.
20. Shortell C, et al. Radiofrequency ablation for superficial venous reflux: Improved outcomes in a high volume university setting. Presented at: SVS Annual Meeting; June 12, 2005; Chicago, IL.
21. Frullini A, Cavezzi A. Sclerosing foam in treatment of varicose veins and telangiectases: History and analysis of safety and complications. *Dermatol Surg.* 2002;28:11–15.
22. Cabrera J, Cabrera Jr J, Garcia-Olmedo A. Treatment of varicose long saphenous veins with sclerosant in microfoam form: Long-term outcomes. *Phlebology.* 2000;15:19–23.
23. Merchant RF, Pichot O, Myers KA. Four-year follow-up on endovascular radiofrequency obliteration of great saphenous reflux. *Dermatol Surg.* 2005;31:129–134.
24. Lurie F, Creton D, Eklof B, et al. Prospective randomized study of endovenous radiofrequency obliteration (Closure) versus ligation and vein stripping (EVOLVeS): Two-year follow-up. *Eur J Vasc Endovasc Surg.* 2005;29:67–73.
25. Puggioni A, Kalra M, Carmo M, et al. Endovenous laser therapy and radiofrequency ablation of the great saphenous vein: Analysis of early efficacy and complications. *J Vasc Surg.* 2005;42:488–493.
26. Min RJ, Zimmet SE, Isaacs MN, Forrestal MD. Endovenous laser treatment of the incompetent greater saphenous vein. *J Vasc Interv Radiol.* 2001;12:1167–1171.
27. Proebstle TM, Lehr HA, Kargle A, et al. Endovenous treatment of the greater saphenous vein with a 940-nm diode laser: Thrombotic occlusion after endoluminal thermal damage by laser-generated steam bubbles. *J Vasc Surg.* 2002;45:729–736.
28. Bergan JJ, Pascarella L. Severe chronic venous insufficiency: Primary treatment with sclerofoam. *Sem Vasc Surg.* 2005;49–56.
29. Hingorani AP, et al. Deep venous thrombosis after radiofrequency ablation of greater saphenous vein: A word of caution. *J Vasc Surg.* 2004;40:500–504.
30. Pascarella L, Bergan JJ, Mekenas LV. Severe chronic venous insufficiency treated by foamed sclerosant. *Ann Vasc Surg.* 2006;20:83–91.
31. Marston WA, Owens LV, Davies S, Mendes RR, Farber MA, Keagy BA. Endovenous saphenous ablation corrects the hemodynamic abnormality in patients with CEAP clinical class 3–6 CVI due to superficial reflux. *Vasc Endovasc Surg.* 2006;40:125–130.

14

TREATMENT OF CHRONIC VENOUS INSUFFICIENCY WITH FOAM SCLEROTHERAPY

ALESSANDRO FRULLINI

SUMMARY

Treatment of venous insufficiency has recently been revolutionized by introduction of new, less-invasive procedures for saphenous closure as alternatives to surgery. Endovenous laser, radiofrequency, and recently steam thermal ablation have been proposed in the treatment of incompetent veins as an alternative to surgery. Sclerotherapy, the injection of drugs capable of transforming the walls of varicose veins in fibrotic cords, has been utilized for almost 50 years in the treatment of varicose veins.

The past 10 years has represented a true Renaissance of sclerotherapy. This has happened because sclerosing foam was introduced in this field.

Sclerosing foam (SF) is a mixture of gas and a liquid solution with tensioactive properties; the gas must be well tolerated or physiologic, and the bubble size should be preferably under 100 microns. The behavior of sclerosing foam is different when injected if compared to the action of a liquid solution because it forms a coherent bolus inside the vein that avoids any mixing of the drug with blood.

This paper presents a review on the properties of sclerosing foam, the current state of the art of this treatment, and future perspectives. Published data on mid- and long-term results are also presented together with different methods for foam production.

INTRODUCTION

Sclerotherapy has experienced different phases: from the initial enthusiasm of Tournay, Sigg and Fegan,[1,2] surgery has been repeatedly compared in an antagonist role against sclerotherapy in the treatment of venous insufficiency.

Despite the late advancement of echo-guided treatments, surgery was considered the gold standard for varicose vein patients until foam sclerotherapy appeared although an evidence-based study on efficacy of surgery for varicose veins has never been published.

With sclerosing foam (SF), it is possible for the first time to achieve full control of drug concentration inside the vein and time of contact between the sclerosing agent and the endothelium.[3]

Through use of a liquid sclerosing agent, the true effect on endothelium is due to the specific concentration of sclerosant (minimal effective concentration) in a vein for a desired time (minimal effective time). The combination of ideal time with ideal concentration gives optimal sclerosis. This is easily achieved in smaller veins like telangiectasias (where sclerotherapy has never presented a problem), but is difficult to obtain in large veins like the great saphenous vein.

The liquid sclerosant is diluted with the amount of blood inside the vein itself, but sclerosing foam pushes away almost all the blood, creating a virtually bloodless field where the ideal condition for sclerotherapy is achieved. We can consider that, with SF, the ideal effective time and concentration are always reached if the treatment is performed correctly.

The possibility, with a single injection, to permanently eliminate the saphenous trunk together with tributaries in one episode, with low cost, no hospitalization, and no anesthesia, with a virtually painless treatment, has completely changed the perspective of varicose vein treatment.

The idea of using air and drug in combination is quite old. In 1944, Orbach described his "air block" technique using sodium tetradecyl sulphate (STS).[4]

In 1993, Cabrera, a vascular surgeon from Spain, proposed the use of a therapeutic foam of STS or Polidocanol (POL) in the treatment of varicose veins. This represented a true revolution in the stagnant world of sclerotherapy, a true step forward in the treatment of superficial venous insufficiency.[5]

Many authors subsequently reported different methods of foam production.

In 1997 Monfreux reported a technique utilizing a glass syringe and a sterile plug to produce a weak foam.[6]

Two years later Benigni and Sadoun presented their personal technique of producing POL foam with a disposable syringe and a tap.[7]

In 1999 Mingo-Garcia reported another technique using helium and a specifically designed device.[8]

The year 2000 represented the turning point in foam sclerotherapy: Tessari presented his three-way tap technique, which was capable of preparing a very good foam with extremely reduced cost.[9] To produce Tessari's foam with STS, a three-way stopcock is needed, coupled with a small syringe filled with 1 cc of drug and a 5-ml syringe with 4–5 ml of atmospheric air. Twenty quick passages of the solution are made. After the first 10 passages, the tap on the three-way stopcock is narrowed to the maximum level possible. This will form a high-quality and consistent foam. Luer-lock® syringes are useful in avoiding unwanted detachments, but pressing the three-way on a firm surface during passage allows safe production of foam even without Luer-lock® syringes.

In the same year as Tessari's method was described, the Frullini method was presented with characteristics similar to Tessari's.[10,11] Frullini's technique generates foam in a vial of sclerosing solution, providing that the vial has a rubber cap. Foam generation is due to the Venturi effect that occurs when forceful passage of a fluid (detergent liquid) is performed in a narrow opening.

A small (facultative) connector was used in the original description to couple the syringe and vial. A minimum of five passages is generally necessary to create a good foam (similar to Tessari's). It is necessary to properly choose the size of the vial and the syringe because a small syringe cannot be coupled with a vial that is too large.

Another advance was then developed by the introduction of a mixture of CO_2/O_2 in the preparation of the extemporary foam, usually with the Tessari method. My personal method is to utilize an elastomeric pump as temporary tank for the two gases. From that, I aspirate the desired volume of gas (CO_2/O_2) with the three-way stopcock to prepare the sclerosing foam (Figure 14-1). A similar foam can also be made using a tank with premixed O_2/CO_2 in a certified ratio. This allows the production of a standardized foam.

As time passed, the pharmaceutical companies started to be interested in foam, and now a British company is developing a standardized "industrial grade" microfoam (Varisolve™, Provensis, UK), which could represent the ultimate foam for varicose vein treatment.[12]

FOAM AND SCLEROSING FOAM

Sclerotherapy is the injection of drugs capable of transforming the wall of a varicose vein into a fibrotic cord. The typical endpoint of sclerotherapy should be permanent occlusion, but this does not always occur with liquid

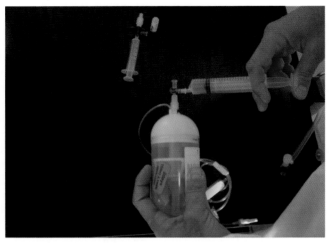

FIGURE 14-1   The production of a CO_2/O_2 extemporary sclerosing foam.

sclerosants. The main factor causing insufficient sclerotherapy is the volume of blood in which the drug is diluted.

With liquid sclerosants, the injection inside a vein segment elevates the drug concentration to a peak level, followed by a quite fast decrease of sclerosant concentration. The shape of the concentration curve is ruled by the speed of injection (mostly related to the pressure exerted on the piston and the caliber of the needle), the ratio of injected volume, the size of the vessel, and the blood flow. Sclerosis will be triggered only if a minimal effective concentration of the drug is present for a sufficient period of time.

In telangectasias, we expect a prompt rise in concentration and a relatively long plateau effect in which only the drug will be present inside the vessel. This is clearly observed when performing a spider vein injection when the needle is held inside the vein for a longer time.

In a large great saphenous vein (GSV) with a significant reflux, the peak will be reached more slowly than in the previous example and will be related to the size of the needle and the fluidity of the injected material. The maximal concentration of the liquid sclerosant in that vein segment will be related to the volume of blood with which it will be diluted. In a vein where 10 ml of blood is present, an injection of 1 ml of a liquid sclerosant like Polidocanol 3% gives a final concentration of drug in the vein of nearly 0.3%. This explains why sclerotherapy has never been a problem of drug power for telangectasias where the dilution of drug is virtually absent and why saphenous sclerosis has always been difficult where the agent is always diluted.

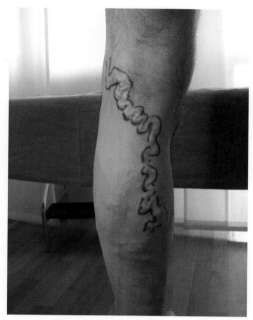

FIGURE 14-2   The large, varicose tributary vein is shown here before treatment.

The introduction of SF has completely changed this perspective: When foam is injected, it forms a coherent bolus inside the vein. Due to its properties, this bolus has controlled and uniform activity and can be controlled in site for a definite time. This will lead to optimal and, for the first time, controlled sclerosis (Figures 14-2, 14-3 and 14-4).

Foam is a nonequilibrium dispersion of gas bubbles in a relatively small volume of liquid which contains surface-active macromolecules (surfactants). These preferentially adsorb at the gas/liquid interfaces and are responsible both for the tendency of a liquid to convert in a foam and for the stability of the produced dispersion.

The sclerosing foam is a mixture of gas and a liquid solution with tensioactive properties; the gas must be well tolerated or physiologic, and the bubble size should be preferably under 100 microns. The behavior of sclerosing foam is different when injected, if compared to the action of a liquid solution.[11]

The most common mistake with foam is to consider it as a single entity. In fact, according to the method chosen, it is possible to produce very different foams, with different characteristics, complication rates, and therapeutic indications.

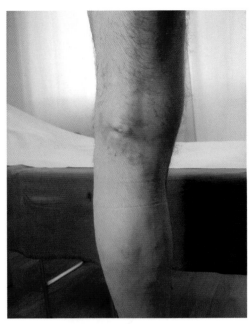

FIGURE 14-3 This is the same case as in Figure 14-2, 2 weeks after one treatment with foam sclerotherapy.

We can classify foams by a bubble's diameter (froth, foam, minifoam, or microfoam) or by the relative quantity of liquid (the shape is the result of the competition between surface tension and interfascial forces) as wet foam (nearly spherical bubbles; wetness or the volume fraction of liquid is over 5%) or dry foam (polyhedral bubbles; the volume fraction of liquid is below 5%). The wet foam is more fragile, and if a small needle is used when injecting such foam, most of the structure is lost at the tip of the needle.

Wet foam has the maximum stability because, when the bubble is polyhedral, as in dry foams, there is a greater competition between surface tension and interfacial forces. Uniform diameters also mean more stability because smaller bubbles empty into larger ones, according to Laplace's law, because for smaller diameters there will be a higher internal pressure.

Another way to classify foam is to consider the standard of production: It could be low or medium grade for extemporary foam, but is maximal only for the industrial high standard foam.

Even when it seems very stable, foam is always in evolution. Three factors will introduce disorder: drainage (draining of liquid from foam), disproportionation (change in bubble size distribution), and coalescence

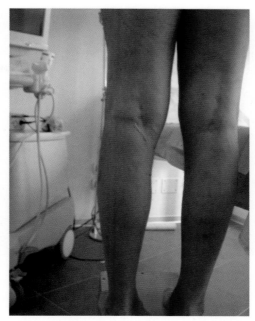

FIGURE 14-4 This is the same case as in Figures 14-2 and 14-3, 2 years after foam sclerotherapy. No further treatments were done.

(fusion of bubbles). These three factors, and the time factor, will lead to complete dissolution. This dissolution can be slowed in several ways (e.g., freezing), but for medical use, it is not necessary because therapeutic properties are more important than lasting time.

We must therefore stress the role of time between the preparation and injection of foam, particularly for low standard foam with low drug concentration. An ideal injection of air-based SF should be made no more than 15 seconds after generation with the Tessari method.

Another important aspect of foam is its response to forces or rheology: Foam can exhibit features of different basic states of matter, acting as a solid, a liquid, or a gas.

In fact, SF exhibits striking mechanical properties because it elastically resists to pushing if done gently (as inside a vein) or reacts as a liquid if pushed forcefully in a syringe (syringeability). Moreover, bubbles' disruption ultimately converts the foam in liquid and gas.

Sclerosing foam shows peculiar properties: adhesiveness and compactness (with the possibility to manipulate the foam after injection and displacing effect on blood), syringeability (or possibility to be injected with a small needle without losing its characteristics), greater volume for the same quantity of liquid agent (possibility to treat longer vein segment), long

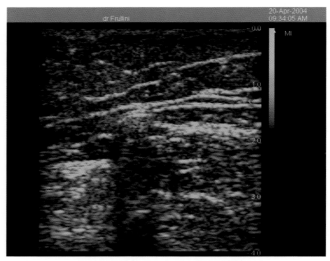

FIGURE 14-5 This ultrasound scan photo shows the echovisibility of the GSV with sclerosant foam, showing the large acoustic shadow after foam injection.

duration (long enough for therapeutical action), enhanced spasm generation (less risk of blood collection inside the sclerosed vein), echovisibility (Figure 14-5), enhancement of sclerosing power with reduced drug dose and concentration, and selectivity of action on endothelium (lesser risk in case of extravasation).

Again, it must be stressed that each type of foam has different properties and characteristics.

CLINICAL APPLICATIONS OF FOAM SCLEROSANTS

Foam can be used in every kind of classical sclerotherapy. Moreover, new and exciting indications have been tried with success.

In our experience, inguinal neovascularization and saphenous trunks are very good indications, and mid-term results are encouraging.

The first series on saphenous trunks treated with Cabrera's microfoam was published in 1997 and reported on 261 great saphenous veins treated with echoguided injection of standardized microfoam.[12]

In 1999 the first paper in English literature on foam sclerotherapy was published,[13] and in 2000 Frullini and Cavezzi presented their results on 167 large veins treated with echosclerosis, utilizing foam with very good results.[14]

The treatment with STS or POL foam is generally performed in saphenous trunk with echo-color-Doppler guide.

Sodium tetradecylsulfate 3% or Polidocanol 3% are usually chosen to prepare a thick foam with Tessari's or Frullini's method. According to the Tegernsee Consensus paper,[15] up to 10 ml of this foam can be used, but 5 ml is enough to treat most GSVs less than 10 mm if a short cannula is used, and the leg is elevated in order to speed up the flow in the deep system and reduce the caliber of superficial veins.

A maximum of 10 ml POL or STS 3% of extemporary air foam is recommended to keep the treatment safe. Industrial grade foam can be used safely in larger volumes, but the coalescence rate of extemporary air foam is not yet proven to be controlled enough to perform a safe injection exceeding 10 ml of foam. On the contrary, CO_2-based extemporary foam seems to have higher safety even when larger volumes are injected.[16]

In 2000, Tessari proposed the use of a ureteral catheter to inject SF in the GSV.[17] A balloon catheter for foam sclerotherapy has been recently produced. The balloon is used to temporarily occlude the junction during treatment in order to prevent foam passage in the deep system.[17]

In fact, this injection must be considered as something very different from simple needle or short cannula injection; the presence of a balloon makes this an injection against pressure, as the blood cannot flow freely from the GSV to the femoral vein and a gradient of pressure will direct all the blood to thigh perforators and in distal tributaries. This could be an explanation for the relatively larger incidence of deep vein thrombosis (DVT) using the balloon catheter to deliver SF.

An injection with more pressure could theoretically achieve better results in terms of long-time occlusion rate, as a deeper penetration of sclerosant in the vein wall could produce a better sclerosis.

I have only limited experience with such catheters with good results, but the time for treatment is almost five times that of injection of SF with a plain needle, and the cost is much higher. The role of these catheters may be very important for larger veins (>14 mm) where the caliber of the vessel does not allow easy treatment.

Telangectasia is, in my opinion, the worst indication for sclerosing foam, but this treatment is widely utilized. I generally prefer to use liquid sclerosants (POL) for standard telangectasias. POL foam (or very seldom STS) is reserved for resistant cases or for telangectatic matting that, on the contrary, is a very good indication for SF made with POL foam 0.25% to 0.5%.

Tessari's foam is difficult to push in these small vessels, but reducing the liquid-to-air ratio to 1 : 2 or 1 : 1 ameliorates quality and durability of foam, enhancing safety and results. Moreover, glass syringes may create a safety problem due to poor sterilization.

Complications of foam sclerotherapy are similar to those of liquid sclerosants with few exceptions.

The use of air in production of SF was reported to cause a stroke in one patient with a patent foramen ovale (PFO). This condition is present in almost 25% of the normal population and is usually linked to a history of migraine; therefore, a specific question about migraine must always be directed to the patient who is to have foam sclerotherapy. If I have a patient with a history of migraine with aura, I generally use a different treatment.

A report on a reversible but long-lasting neurological deficit has been published. In this case, a very low standard foam (POL 0.5%) in too large a volume (20 ml) was used in the treatment of a GSV of medium size.[19] No data were given on the time elapsed from the preparation of the SF and the injection. There is probably no other report of such a complication when SF is properly used.

It must be stressed again that concentration of drug and volume are directly linked (i.e., a larger volume of foam has better quality if a higher concentration of drug is used). With regard to this, I can reach up to 10 ml of foam in a single session only if a 3% POL is used. STS 3% is never used in more than 5 ml as in plastic syringes because it has less stability than POL in higher concentrations.[20]

Some reports have stressed a reduction of neurologic side effects using CO_2 alone or in combination with O_2 in a ratio of 70% to 30% in the preparation of SF with the Tessari method.

The role of these different factors has not yet been analyzed. Strejcek demonstrated the presence of circulating endothelial cells (CEC) after sclerotherapy with liquid. The number of CECs after foam sclerotherapy is significantly greater. What this means is not known. We have not studied the role of circulating cells or mediators released from the area treated with sclerosants.[21]

Among the major complications, a partial deep vein thrombosis (DVT) has been reported after sclerotherapy for varices.[22] However, in every case, the DVT was asymptomatic and without sequelae at 3 months. Unfortunately, varicose vein patients treated with classical sclerotherapy or standard surgery are rarely studied with duplex ultrasound after the treatment as are the patients treated with echoguided injections. Therefore, a comparison with SF patients cannot be made.

Lowering the concentration of the sclerosing drug used for foam generation and avoiding large volume injections in short saphenous vein (SSV) (<2.5 ml) will be extremely important to lessen the chance of a DVT. This has to be considered a rare and generally limited complication, but the need for routine color Doppler control after treatment has to be stressed. It is extremely important to clearly identify before treatment those cases in which a gastrocnemius vein empties into the terminal part of the SSV. In those cases, an increased risk of sclerosis of the deep vein is present.

Minor complications have often been reported with foam, as is also true with liquid sclerosants. The unique exception is with the treatment of telangectasias, where pigmentation and skin necrosis have been reported even by experienced doctors. This is due to the higher sclerosing power of foam.

PHILOSOPHICAL CONCLUSIONS

The treatment of varicose veins is changing. Many new options are now offered to patients and doctors: surgery, thermal ablation (endovenous laser ablation, radiofrequency closure, and steam obliteration),[23] and endovenous chemical ablation (foam sclerotherapy). I have several observations on the theoretical premises of thermal ablation (what happens in a very thin vein, what amount of smoke is produced inside the vein, and how much gets to the right heart, etc.), but I would like to say something on the whole philosophy of these new methods.

We need a treatment that is capable of achieving good results in a safe and less invasive manner than surgery. The technique should be capable of treating tributaries at the same time with the same method as the saphenous vein. The method should be cheap, easy to learn, and repeatable in case of recurrence.

Thermal techniques are only a different way of treating the saphenous trunk using a method that is something very similar to surgical intervention: The patients are in an (semi)operating room, a sterile field is prepared, and anesthesia is given. Then we gain venous access and place a catheter in the GSV. Phlebectomies will be done on tributaries in most cases. The only difference is that, in the stripping operation, the same fiber (stripper) is retrieved, and a junctional ligature is made. Moreover, gentle invaginating stripping under local anesthesia seems to me less invasive, when properly done, than laser ablation or radiofrequency closure.

Considering that almost 50% of GSV insufficiency is due to terminal valve incompetence alone, a proper indication for treatments that do not include sapheno-femoral junction ligation will be found in only a few cases.

If a different treatment of saphenous insufficiency is to be found, at this time, only foam sclerotherapy has the ideal prerequisites: It is cheap, time saving (10 minutes for the whole procedure), and easy for a trained doctor to perform; plus, tributaries are treated at the same time as the saphenous vein, and a return to daily activity is resumed after a 15-minute walk. No anesthesia is required at all, and post-treatment pain is absolutely absent. The volume of air or physiologic gas injected with foam is probably far less then the uncontrolled volume of smoke produced by the laser in the venous system.

Moreover, venous insufficiency is not simply a defect that must be eliminated. It is a true disease with a natural evolution. Every treatment is capable, when correctly performed, to bring the leg back to a prior condition, but no treatment will stop the natural evolution of the illness. The simplicity of foam sclerotherapy makes the treatment easily acceptable to patients, even if more sessions are needed in subsequent years to stabilize the condition. Neither surgery nor thermal ablation can claim the same benefits.

The ultimate study with a comparison of different methods is still lacking, and the few studies available do not compare homogeneous cases or properly done treatments. It must be stressed that alternative treatments to surgery must not unnecessarily complicate the work of doctors and the lives of patients in a search for a fashionable procedure with indefinite advantages over standard treatment. If foam sclerotherapy will demonstrate in large Good Clinical Practice (GMP) studies that it is capable of achieving results similar to surgery, a consistent step toward the control of venous insufficiency will have been made.

At this stage, it is extremely important that foam sclerotherapy is used in a wise manner respecting the guidelines of the Tegernsee Consensus conference: The 10-ml volume limitation with air must be respected or, when a higher volume is needed, CO_2/O_2 sclerosing foam must be used.

My hope is that national health systems and insurance companies ultimately understand that venous insufficiency is not just a cosmetic condition but a true disease, with consistent morbidity and mortality and that, at least in my opinion, foam sclerotherapy (also called endovenous chemical ablation) is going to become the ultimate treatment for venous insufficiency, replacing every other treatment for most cases.

REFERENCES

1. Tournay R. *Terapia sclerosante delle varici.* Milan: Raffaello Cortina Editore, 1984.
2. Fegan G. Continous compression technique of injection varicose veins. *Lancet.* 1963;20:108–112.
3. Frullini A. Advanced sclerotherapy: The sclerosing foam. In: Goldman MP, Weiss RA, eds. *Advanced Techniques in Dermatologic Surgery.* London: Taylor & Francis, 2006:317–330.
4. Orbach EJ. Sclerotherapy of varicose veins: Utilization of intravenous air block. *Am J Surg.* 1944:362–366.
5. Cabrera Garrido JR, Cabrera Garcia-Olmedo JR, Garcia-Olmedo Dominguez MA. Elargissement des limites de la schlérothérapie: Noveaux produits sclérosants. *Phlébologie.* 1997;50(2):181–188.
6. Monfreux A. Traitement sclérosant des troncs saphènies et leurs collatérales de gros calibre par la méthode MUS. *Phlébologie.* 1997;50(3):351–353.

7. Benigni JP, Sadoun S, Thirion V, Sica M, Demagny A, Chahim M. Télangiectasies et varices réticulaires. Traitement par la mousse d'Aetoxisclérol a 0,25%. Présentation d'une étude pilote. Phlébologie. 1999;52(3):283–290.
8. Mingo Garcia J. Esclerosis venosa con espuma: Foam Medical System. *Revista Espanola de Medicina y Cirugia Cosmética.* 1999;7:29–31.
9. Tessari L. Nouvelle technique d'obtention de la sclero-mousse. *Phlébologie.* 2000; 53(1):129.
10. Frullini A. New technique in producing a sclerosing foam in a disposable syringe: The Frullini method. *Derm Surg.* 2000;26:705–706.
11. Frullini A, Cavezzi A. Sclerosing foam in the treatment of varicose veins and teleangiectases: History and analysis of safety and complications. *Dermatol Surg.* 2002; 28:11–15.
12. Cabrera J, Cabrera J Jr, Garcia-Olmedo MA. Treatment of varicose long saphenous veins with sclerosant in microfoam form: Long term outcomes. *Phlebology.* 2000;15: 19–23.
13. Cavezzi A, Frullini A. The role of sclerosing foam in ultrasound guided sclerotherapy of the saphenous veins and of recurrent varicose veins: Our personal experience. *Australian and New Zealand J Phlebol.* 1999;3:49–50.
14. Frullini A, Cavezzi A. Echosclérose par mousse de tétradécylsulfate de sodium et de Polidocanol: Deux années d'espérience. *Phlébologie.* 2000;53–54:431–435.
15. Breu FX, Guggenbichler S. European consensus meeting on foam sclerotherapy. *Dermatol Surg.* 2004;30(5):709–717.
16. Morrison N, Neuhardt D. *Comparison of the Safety of Air-Based vs. CO_2 Based Foam Sclerotherapy.* Abstract ACP 20th Annual Congress. Ponte Vedra, FL, USA. November 9–12, 2006:104.
17. Gonzalez R. *Zeh Transcatheter ECHO Therapy: A Novel Approach to the Treatment of Varicose Veins 2 Years Follow-Up.* Abstract 8th International Phlebological Symposium, "Sclerotherapy 2006," Bologna, Italy, October 20–21, 2006.
18. Frullini A, Cavezzi A, Tessari L. Scleroterapia delle varici degli arti inferiori mediante schiuma sclerosante di Fibro-vein® con il metodo Tessari: Esperienza preliminare. *Acta Phlebologica.* 2000;1:43–48.
19. Forlee MV, Grouden M, Moore DJ, Shanik J. Stroke after varicose vein foam injection sclerotherapy. *J Vasc Surg.* 2006;43:162–164.
20. Rao J, Goldman MP. Stability of foam in sclerotherapy: Differences between sodium tetradecyl sulfate and Polidocanol and the type of connector used in the double-syringe system technique. *Dermatol Surg.* 2005;31:19–22.
21. Strejcek J. The evaluation of the efficacy of sclerosing potential between liquid sclerosing agents (LSA) and sclerosing foam (SF) by counting of circulating endothelial cells (CEC) in peripheral blood (PB). Abstract 8th International Phlebological Symposium, "Sclerotherapy 2006," Bologna, Italy, October 20–21, 2006.
22. Tessari L, Cavezzi A, Frullini A. Preliminary experience with a new sclerosing foam in the treatment of varicose veins. *Dermatol Surg.* 2001;27:58–60;1:55–72.
23. Milleret R, Mehier H, Foray J, Humbert S, Rauber N. Hyperheated steam: A new thermal technique for varicose veins obliteration. Abstract ACP 20th Annual Congress. Ponte Vedra, FL, USA. November 9–12, 2006:108.

15

FOAM TREATMENT OF VENOUS LEG ULCERS: THE INITIAL EXPERIENCE

JUAN CABRERA AND PEDRO REDONDO

INTRODUCTION

Venous leg ulceration is a widespread and debilitating chronic condition. It is most commonly observed in the elderly but can also affect young adults. It is characterized by slow healing, pain, and frequent ulcer recurrence. The pain is characteristically worse at night, limits mobility and leisure activities, and disturbs sleep. Bandages and dressings are often malodorous, contributing to the depression and social isolation that can be experienced by sufferers. Most patients describe the pain, which is greater with larger ulcer size,[1] as the symptom with most impact on their quality of life. Chronic venous leg ulcers also represent a considerable economic burden and consume 1% to 2% of health care budgets, comparable with expenditure on diabetes or stroke.

Venous hypertension (VH) is the underlying cause of venous ulceration (VU), although the pathogenic steps involved are not fully understood. The pathophysiology of venous ulcers is centered on a hemodynamic dysfunction that may involve superficial, perforator, or deep veins together or separately and is frequently compounded by immobility. The following mechanisms have been implicated: dysfunction of valves in superficial and/or perforator veins due to congenital or acquired incompetence; dysfunction of valves in the deep system due to congenital absence or thrombotic damage; deep venous outflow obstruction; and muscle dysfunction leading to calf muscle pump failure.[2] A number of authors have assessed the distribution of valvular incompetence in patients with venous ulcers by means of duplex ultrasound, the state-of-the-art technique in venous diagnosis.[3] Using this approach, a study of 887 limbs with venous ulcers showed the cause to be isolated superficial venous reflux in 46%, deep venous incompetence in 5%, and perforators in 2% of cases; multisystem reflux or obstruction was found in 44%, and no reflux was detected in 3% of ulcers.[4] Therefore, the successful treatment of incompetent superficial and perforating veins could be expected to eliminate VH and thereby permit ulcer healing, reducing the risk of recurrence by approximately 50%. Unfortunately, it has not been possible to take advantage of this advance, since an adequate treatment has not been available. Among other reasons, the explanation is largely that surgery has been the main therapeutic approach to incompetent veins, and elderly patients are reluctant to undergo the operation, while surgeons are reluctant to perform it in the presence of an ulcer due to the risk of infection. However, access to effective treatment of ulcerated limbs has recently improved with the advent of less invasive treatments, including radiofrequency ablation, laser ablation, and, least invasive of all, microfoam sclerotherapy. This chapter describes our initial experience using ultrasound-guided microfoam sclerotherapy in the management of ulcer healing.

TREATMENT OPTIONS

The primary aim of chronic VU treatment is to reverse the effects of VH. Treatments used to heal open VUs include standard compression therapy with elastic stockings or bandages, skin grafting, and direct vein treatment by venous surgery or sclerotherapy. Compression therapy is recognized as the mainstay of ulcer treatment and prevention, using multilayer compression bandaging for treatment and compression hosiery for prevention.[6,7]

Simple superficial venous surgery by stripping or ligature theoretically removes the underlying venous incompetence in legs with isolated superficial venous reflux. Surgery to correct venous reflux in the deep veins is complex and of unproven value.[8-10] Over the past 2 decades, however, the emergence of minimally invasive surgical techniques has led to debate about the most appropriate surgical therapy for severe chronic venous insufficiency and venous ulcers. Subfascial endoscopic perforator surgery (SEPS) has become the surgical technique of choice for perforator ablation.[11] Nevertheless, the efficacy of perforator ablation remains highly controversial, regardless of the technique used. Postoperative duplex imaging studies showed that ablative superficial venous surgery reversed around half of the cases of segmental reflux in deep veins[12] and of reflux in calf perforators.[13] Surgical correction of superficial reflux may abolish incompetence in some calf perforators and offer protection against recurrence of perforator incompetence.[14]

Percutaneous ultrasound-guided coil embolization does not appear to be as effective as SEPS in the treatment of incompetent perforator veins.[15]

FROM CLASSIC SCLEROTHERAPY TO MICROFOAM SCLEROTHERAPY

Liquid sclerotherapy represented a useful complementary therapeutic model in the treatment of VU. In 1990, Queral et al. demonstrated it to be an efficacious adjuvant to unna's boot in the treatment of VU.[16] Sclerotherapy with ligation of incompetent veins was found to be a safe and effective treatment of VU, with the combined procedure achieving an improvement in venous hemodynamics and subjective symptoms comparable to that obtained by high ligation and stripping.[17] In 2000, Guex[18] reported his experience with direct perforator treatment under ultrasound guidance. The sclerosing agent used was Sotradecol (3%) or Polidocanol (3%) for veins >4 mm and a more dilute solution for veins <4 mm. He estimated that 90% of perforators could be eliminated after 1–3 injection sessions. However, liquid sclerotherapy has many limitations.[19] A given dose of liquid sclerosing agent cannot provide a similar sclerosing action in all cases, since its intravenous concentration varies according to the hemodynamic conditions of the injected vessel and is always unknown. Moreover, contact with the substance is lost after its injection, and little can be done to direct it or modify its action. This inability to control liquid agents can sometimes cause undesirable effects in the deep venous system. Thus, the major drawbacks of liquid sclerosants include a progressive dilution in the blood, increasing deactivation, irregular distribution on the venous endothelium, absence of control over the duration of sclerosant-endothelium contact, and inadequate volumes at therapeutic dosage. Moreover, they cannot be manipulated, perceived on duplex, or seen on ultrasonography after injection. In contrast, microfoam sclerotherapy offers the following advantages: mechanical action displacing the blood, greater sclerosing ability and volumes than the liquid form, manageability or remote control (microfoam has a high degree of internal cohesion that allows it to be aspirated and re-injected), selectivity of action, stability of microbubbles, rapid elimination of the gas, and real-time perceptibility on duplex and visibility on ultrasound,[20–22] Further benefits of microfoam sclerotherapy include the absence of anesthesia and hospitalization, with no loss of working time by the patient. Indeed, sclerotherapy appears to be a safer and less costly procedure compared with surgery.

TECHNIQUE OF MICROFOAM SCLEROTHERAPY

Liquid sclerotherapy had been demonstrated to be a valuable adjunct to other therapies but has not been widely accepted as primary therapy for large vein incompetence, which must be treated in most venous ulcers. Microfoam sclerotherapy pioneered by the principal author has been shown to be highly effective and safe in the management of incompetent veins of all sizes in the superficial venous system, and he has now applied this technique to the most severe end of the spectrum of venous disease: leg ulcers.

Microfoam sclerotherapy is performed in a treatment room without anesthesia for the nonsurgical elimination of superficial reflux and incompetent perforator veins in ulcerated limbs. Duplex ultrasound is used to identify incompetent vessels and sources of reflux that cause the VH. After detailed anatomical mapping, patients are placed in the supine position. Polidocanol microfoam (15–30 ml) is injected under ultrasound guidance into the main incompetent vein to fill the vein up to the source of reflux.[23] The access route is variable according to the size of the vein. Cannulation with an Abbocath 20G cannula is performed in great saphenous veins (GSVs) or direct needle puncture (23–25G) in smaller veins. Duplex ultrasound is used during the procedure to confirm the correct location of the needle, the spread of the microfoam, and the onset of vasospasm. The ultrasound transducer (probe) is placed at the saphenofemoral junction to show the arrival of the injected microfoam and to occlude the GSV. The microfoam is then directed distally, replacing the probe with a finger to block the proximal saphenous vein. Perforator veins are selectively treated by direct ultrasound-guided injection, placing the needle in a superficial vein that is connected to but 2–3 cm distant from the perforator vein. Injection of Polidocanol microfoam into this area allows control over the volume, avoiding extension of its action into the deep venous system (DVS). A volume of 1–4 ml is usually adequate, and digital or transducer compression is performed to halt any spread of the microfoam toward the DVS. Another method to protect the DVS is to inject the microfoam with the limb in elevation (safety angle). This position creates a gradient whereby the DVS has a higher blood pressure, preventing drainage of microfoam into the DVS. As a safety measure during injection of leg trunk veins and perforator veins, advantage is taken of the blocking of intramuscular veins by calf muscle contraction. Thus, the foot is dorsally flexed by using the operator's hand, or cooperative patients voluntarily sustain contraction of the medial gastrocnemius muscle. These "closed-door" procedures enhance the safety of the procedure by preventing the microfoam from draining into the DVS. After the injection, the patient should remain supine for 15–20 minutes before standing up and before application of compression in order to avoid dislocation of the microfoam column. Treatment is completed by placement of pads and application of a compression stocking (Struva 23-mmHg stocking; MediBayreuth, Barcelona, Spain), which is worn for 7–15 days after the injection.

RESULTS

During a 115-month period, 116 consecutive patients with VU were treated by ultrasound-guided injection of Polidocanol microfoam (UIPM).[23] Their mean age was 57 years (range, 25–85 years). Sclerosis of incompetent trunk veins was followed at 2–4 weeks by further injections to sclerose tributaries and residual incompetent perforators under ultrasound guidance. The process was continued until all identifiable incompetence was eliminated. The number of sessions per patient ranged from 1 to 17 (mean = 3.6 sessions). Perforators were present in the majority of patients, usually close to the ulcer. Microfoam was injected into the superficial venous system close to the perforators. Two weeks later, the treated perforator could be visualized on ultrasound but was not compressible and no flow could be perceived by Doppler examination. Treatment with UIPM achieved complete healing in 86% of patients at 6 months, with a median time to healing of 8 weeks; seven patients were never cured and one patient was lost to follow-up (Figures 15-1 through 15-4). We observed a lower success rate in ulcers with higher chronicity and in patients with incompetent deep vein system. However, a worse prognosis was associated with perforator vein versus saphenous vein incompetence. The worst outcomes were obtained when all three segments were affected (64% of this group completely healed at 6 months) (Table 15-1).

In comparison with control patients in a previously reported series,[24] we obtained a significantly better outcome at 6 months in all subgroups except for patients with ulcers that had been open for >12 years (n = 24). However, these comparisons probably underestimate the real effectiveness of UIPM, since they always referred to the percentage of complete healing in the whole active treatment series[24] and not to the success rate in specific subgroups with poorer prognosis (Figure 15-5).

In the multivariate Cox model, presence of deep vein incompetence and high chronicity of the ulcer were identified as independent predictors of failure to achieve complete healing, although the independent contribution of the latter variable was small (hazard ratio = 1.02 for each additional year the ulcer had been open).

There were recurrences in 10 patients. The 24-month recurrence rate was 6.3% overall and was below 8.5% across all subgroups except for older patients, patients with deep vein incompetence, and those with ulcers of largest size or highest chronicity.

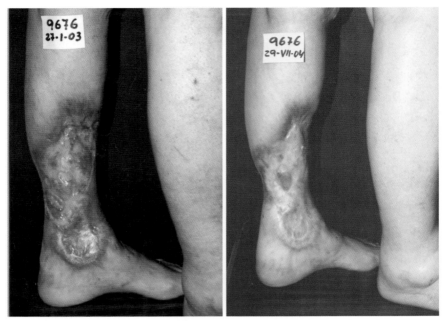

FIGURE 15-1 Extensive ulcer with chronicity of 15 years. Stable long-term outcome.

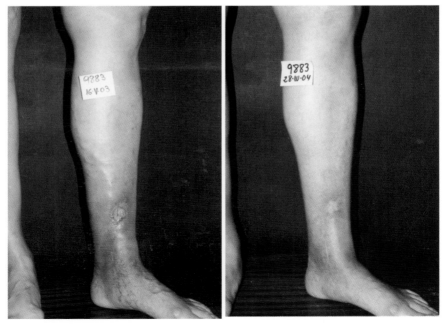

FIGURE 15-2 Varicose ulcer closure after sclerosing microfoam treatment of proximal saphenous vein.

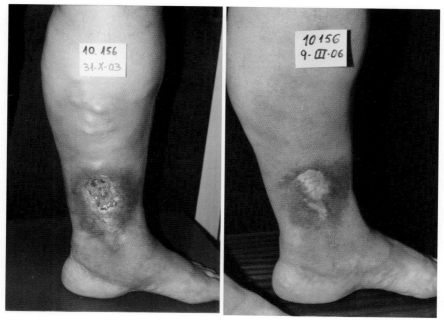

FIGURE 15-3 Stable long-term closure after microfoam sclerotherapy of varicose saphenous vein and perforator.

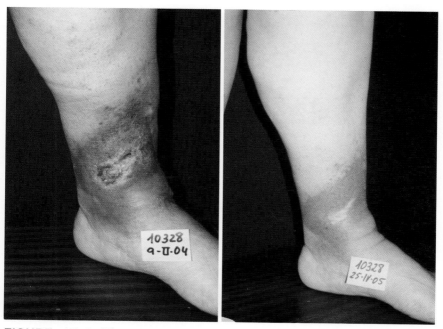

FIGURE 15-4 Ulcer closure and significant improvement in periulcerous trophic disorders after sclerosing microfoam treatment.

TABLE 15-1 Percentage Completely Healed by 6 Months (Kaplan-Meier Estimates)

	Percentage	p (vs. Reference Series*)
Overall	86%	<0.001
Gender		
Men (n = 39)	81%	<0.001
Women (n = 77)	89%	<0.001
Age		
<50 years (n = 35)	90%	<0.001
50–65 years (n = 44)	88%	<0.001
>65 years (n = 37)	80%	0.001
Ulcer chronicity (months)		
≤6 months (n = 44)	81%	<0.001
>6 to 12 months (n = 16)	83%	0.002
>12 to 24 months (n = 14)	83%	0.001
>24 to 72 months (n = 18)	86%	0.015
>72 months (n = 24)	66%	0.19
Ulcer area (cm^2)		
≤2 cm^2 (n = 55)	71%	0.001
>2 to <6 cm^2 (n = 24)	88%	<0.001
>6 cm^2 (n = 72)	81%	<0.001
Localization		
Exclusively saphenous (n = 34)	90%	<0.001
Exclusively perforator vein (n = 8)	85%	0.14
Exclusively saphenous and perforator (n = 41)	94%	<0.001
Exclusively saphenous and deep vein (n = 11)	85%	0.03
Deep vein ± perforator (n = 13)	79%	0.04
All three segments (n = 11)	64%	0.33
Previous surgery		
No (n = 96)	88%	<0.001
Yes (n = 20)	77%	0.06

* Chi-square test using as expected events the results in control patients of the study by Falanga et al. (Rapid healing of venous ulcers and lack of clinical rejection with an allogeneic cultured human skin equivalent. Human Skin Equivalent Investigators Group. *Arch Dermatol.* 1998;134:293–300), who received only compression therapy and showed complete healing of 49% at 6 months.

No major complications were noted. No patient had thrombosis of the femoral-popliteal DVS, pulmonary embolism, or neurological lesions. Deep vein thrombosis (DVT) has been a concern with this technique. Before utilization of "closed-door" procedures, five patients developed DVT in the calf after varicose vein treatment and were managed with LMW heparin, which achieved recanalization in all patients. There have been no further cases of calf DVT since the introduction of these safety maneuvers. Pigmentation was seen in 20% of the patients but was sponta-

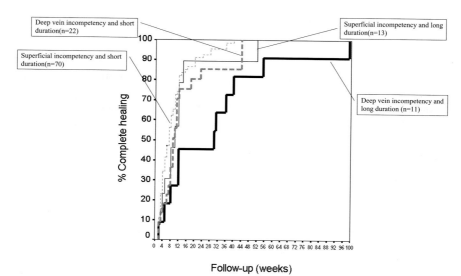

FIGURE 15-5 Cumulative incidence of complete healing according to ulcer chronicity and to localization of venous incompetence. Kaplan-Meier estimates.

neously resolved in 90% of them after 6 months. Transient visual disturbances (<5 minutes) were observed in two patients.

SPECIFIC CHARACTERISTICS OF MICROFOAM

Sclerosing foam is generally defined as a mixture of gas and liquid sclerosing solution (detergent type) with tensioactive properties.[25] Besides being composed of different specific ingredients, foams can differ in their internal cohesion, which is related to the size of the air bubbles. Microfoam is composed of bubbles smaller than 250 µm.[26,27]

The durability of foam is related to the bubble size, the tensioactivity of the liquid solution, and the conditions under which the foam is formed and kept. Foam should be sufficiently durable to avoid its separation into gas and liquid components during injection but sufficiently ephemeral to break down once injected. The gas must be physiologically tolerated at therapeutic doses.

Sclerosant microfoam is composed of microbubbles of oxygen (O_2) and carbon dioxide (CO_2). CO_2 is a nontoxic and highly soluble physiological gas, and large amounts can be administered. When both gases are mixed into the surfactant liquid sclerosant, microbubbles of small diameter can be obtained with sufficient stability to be injected into the vessels.[28,29] The area of liquid on their surface is enormously increased in inverse proportion to the diameter of the bubbles.[29] Alongside the high blood solubility and pulmonary diffusibility of the gas used, this increased surface area also facilitates its metabolism.[29] Polidocanol in microfoam form displaces the blood from the lesion, permitting homogeneous contact between liquid sclerosant and endothelium and facilitating endothelial destruction. Since the 1950s, the intravenous injection of 50 to 100 ml of CO_2 has been used as a contrast for radiological and pericardial diagnosis and in echocardiography.[30] Several researchers[31,32] reported the injection of CO_2 into the right side of the heart without complications, even after administering up to 450 ml in one procedure, and described its use as a contrast in aorto-arteriography. CO_2 is sufficiently fluid to allow fine catheters to be used. It is nontoxic and large amounts can be administered. Other variously obtained conventional foams are formed with atmospheric air that is rich in nitrogen, whose low solubility coefficient and low diffusibility in body fluids hamper its elimination and therefore limit the total volume that can be injected.

There are as yet no clinical data on the optimal volume of foam per injection. There are two distinct approaches: low concentration/high volume and high concentration/low volume. In the former, concentrations of approximately 0.5% Polidocanol are injected at volumes that are often in excess of 15 ml per treatment visit. In the latter approach, 5% Polidocanol is injected at lower volumes of approximately 6 to 8 ml per treatment session using the Tessari/double-syringe technique and a volume of around 4 ml per session using the Monfreux technique, with a maximum of 3 ml injected into small saphenous veins. Use of microfoam allows larger volumes to be injected. The concentration and volume of microfoam can be adjusted according to the disease treated. In the case of home-made foams, however, the injectable gas volume is limited by the low solubility of nitrogen, and only the concentration can be modified. In large vascular malformations, our group has used up to 100 ml of sclerosing microfoam without respiratory or neurologic complications. The ability to use such large volumes may rely on the use of CO_2 instead of room air.[33]

SAFETY OF MICROFOAM

The efficacy and safety of microfoam sclerotherapy is reproducible only if certain preconditions are met (i.e., a perfect mastery of conventional liquid sclerotherapy and an understanding of the technique for handling and injecting microfoam).

Sclerotherapy is an effective treatment method with a low incidence of complications (e.g., allergic reaction, skin necrosis, excessive sclerosing reaction (thrombophlebitis), pigmentation, matting, nerve damage, scintillating scotomas, orthostatic collapse, and thromboembolism). For patients with presence or history of migraine, there may be an increased incidence of transient visual disturbance. Contraindications for microfoam sclerotherapy are the same as for classic liquid sclerotherapy (i.e., allergy to the sclerosant, local or systemic infection, immobility, and acute superficial or deep vein thrombosis). We have observed scotoma with or without migraine (in all cases mimicking the patient's previous migraine aura) in 1% of cases, but there were no cases of neurological deficit. With the use of liquid foams from diluted liquid sclerosing solutions, there is probably a higher incidence of visual disturbance and migraine in patients predisposed to these conditions.[34] When larger doses of foam are used, there can be episodes of dry cough, chest discomfort, and transient ischemic attacks.[35] Larger amounts of foam or its administration in closer proximity to deep veins may more easily provoke DVT.[36] It has been suggested that patients with known symptomatic patent foramen ovale should be treated with special care, and the use of restricted amounts of foam is recommended, although no data on this issue have yet been published for the Provensis microfoam.

It was recently demonstrated, in an *in vivo* experiment in rat cremaster muscle, that the perfusion of two patented types of Polidocanol microfoam (Varisolve®, the name of a product yet to be commercialized by Provensis, Inc., West Conshohocken, PA, USA) through the arteriolar system does not disturb normal circulatory dynamics. In contrast, the arteriolar system was blocked, and the circulation was detained by a massive gas embolism in the same experiment when a different type of sclerosing microfoam was used, obtained by the two-syringe method and using room air.[37] Another study has described paradoxical cerebral gas cerebral embolism via patent foramen ovale after "home-made" foam treatment of a patient with varicose veins.[38] The controlled bubble size and specific gases in the patented sclerosing microfoam distinguish it from other home-made foams that use atmospheric air and lack these key elements for a safe and effective treatment.

COMMENTS

Currently, one of the most worrying aspects of venous ulcer disease is the high rate of recurrence. At 3 months after healing, recurrent rates range from 26% to 33%[39] in patients who are compliant with compression hosiery and up to 56% in patients who are not. VH is widely accepted to be the underlying cause of VU, and skin closure occurs rapidly when the VH is improved. Ablation of superficial and perforator vein incompetence yields clinical and hemodynamic improvements in patients with chronic VU.[40] A randomized controlled trial showed that superficial venous surgery confers no added benefit over multilayer elastic compression in the healing of VU but leads to a reduction in ulcer recurrence in patients with isolated superficial venous reflux.[41] Postphlebitic recurrences are the most common and show the worst response to treatment, and some of them occur with no ultrasound findings susceptible to treatment.

Several studies have indicated that SEPS is equally as effective as open perforator ablation during early follow-up, with the additional benefit of significantly fewer wound complications, thereby establishing it as the procedure of choice for perforator vein surgery when indicated.[42,43] Although early reports have been enthusiastic, one study found that 50% of incompetent perforators within 10 cm of the sole of the foot, identified preoperatively by duplex ultrasound, were not accessible by the SEPS approach.[44] Small interconnected collaterals between the perforators and the skin frequently convey the VH to the skin area, and surgical procedures are inherently unable to close these vessels. Thus, small interconnected collaterals can often remain open after surgical closure of the perforator, probably leading to a new increase in venous pressure and eventual recurrence of the ulcer.

Microfoam sclerotherapy appears to be a promising alternative approach to VU. The use of color duplex scanning should be incorporated into the assessment of all ulcerated limbs so that patients with superficial venous reflux and incompetent perforator veins can be offered simple, effective treatment by microfoam sclerotherapy. UIPM has no anatomical limitation, including ankle ulcers, and can be performed in open ulcers. Moreover, by a "sponge" effect, the injected microfoam is uniformly distributed to the adjacent area and, besides closing the vessels with reflux, can also close all small pathways that transmit VH to the skin. In fact, our current results confirm this effect of UIPM, which accelerates ulcer closure and reduces recurrences. This microfoam sclerotherapy is a safe and easily repeatable procedure that addresses the root of the problem and does not involve surgery with its associated limitations and complications.

In an updated series of more than 310 patients treated with UIPM, the same percentage (87%) of healing was achieved in a short time period (8 weeks), with a global recurrence rate of 15% at 3 years. Recurrences were in part due to the appearance of new varicose veins and recanalization of treated perforating veins. With UIPM, retreatment of these patients is simple and equally effective and should be performed before reopening of the ulcer, paying attention to early symptoms such as irritation or eczema.

The pain produced by the ulcer is one of the main problems encountered by these patients. Adequate pain control results in better compliance with treatment, and pain relief promotes a better quality of life.[45] In our experience, more than half of patients with severe pain from their ulcer have reported a reduction in pain to moderate or minor levels immediately after microfoam treatment.

UIPM is a minimally invasive procedure, originally invented to replace surgical treatment for varicose veins. In this chapter we have illustrated its utility in treating venous ulceration, one of the most difficult chronic conditions. Reports of the outcomes obtained suggest that UIPM may become the gold standard treatment for chronic venous insufficiency with ulceration in the near future.[27]

Note: The precise procedure to prepare the microfoam is the subject of a confidential agreement with a third party who is the proprietary. It is, however, based on the procedure described in granted European and U.S. patents EP 656 203 and US 5 676 962, respectively.[28] The Polidocanol microfoam manufactured by Provensis has successfully undergone phase III clinical trials in Europe and is currently in the process of obtaining FDA approval.

REFERENCES

1. Hyland ME. Quality of life of leg ulcer patients: Questionnaire and preliminary findings. *J Wound Care.* 1994;3:249–298.
2. Gourdin FW, Smith JG. Etiology of venous ulceration. *South Med J.* 1993;86:1142–1146.
3. Baker SR, Burnand KG, Sommerville KM, Thomas ML, Wilson NM, Browse NL. Comparison of venous reflux assessed by duplex scanning and descending phlebography in chronic venous disease. *Lancet.* 1993;341:400–403.
4. Myers KA, Ziegenbein RW, Zeng GH, Matthews PG. Duplex ultrasonography scanning for chronic venous disease: Patterns of venous reflux. *J Vasc Surg.* 1995; 2:605–612.
5. Wilson NM, Rutt DL, Browse NL. Repair and replacement of deep vein valves in the treatment of venous insufficiency. *Br J Surg.* 1991;78:388–394.
6. NHS Centre for Reviews and Dissemination. *Compression Therapy for Venous Leg Ulcers.* York: NHS Centre for Reviews and Dissemination, 1997.
7. Meyer FJ, McGuinness CL, Lagattolla NRF, Eastham D, Burnand KG. Randomized clinical trial of three-layer paste and four-layer bandages for venous leg ulcers. *Br J Surg.* 2003;90:934–940.
8. Abida A, Hardy SC. *Surgery for Deep Venous Incompetence (Cochrane Review).* Oxford: Update Software, 2000.
9. Linton RR. The communicating veins of the lower leg and the operative technique for their ligation. *Ann Surg.* 1938;107:582–593.
10. DePalma RG. Linton's operation and modifications of the open technique. In: Gloviczki P, Bergan JJ, eds. *Atlas of Endoscopic Perforator Vein Surgery,* 1st ed. London: Springer-Verlag, 1998:107–113.
11. Hauer G. Endoscopic subfascial discussion of perforating veins. Preliminary report. *Vasa.* 1985;14:59–61.
12. Walsh JC, Bergan JJ, Beeman S, Comer TP. Femoral venous reflux abolished by greater saphenous vein stripping. *Ann Vasc Surg.* 1994;8:566–570.
13. Stuart WP, Adam DJ, Allan PL, Ruckley CV, Bradbury AW. Saphenous surgery does not correct perforator incompetence in the presence of deep venous reflux, *J Vasc Surg.* 1998;28:834–838.

14. Gohel MS, Barwell JR, Wakely C, et al. The influence of superficial venous surgery and compression on incompetent calf perforators in chronic venous leg ulceration. *Eur J Vasc Endovasc Surg*. 2005;29:78–82.
15. van Dijk LC, Wittens CH, Toonder IM, Lameris JS, du Bois NA, Pattynama PM. Ultrasound-guided percutaneous coil embolization of incompetent perforating veins: Not effective for treatment of venous ulcers and recurrent varicosities. *J Vasc Interv Radiol*. 1999;10:1271–1274.
16. Queral LA, Criado FJ, Lilly MP, Rudolphi D. The role of sclerotherapy as an adjunct to Unna's boot for treating venous ulcers: A prospective study. *J Vasc Surg*. 1999;11:572–575.
17. Recek C. Saphenofemoral junction ligation supplemented by postoperative sclerotherapy: A review of long-term clinical and hemodynamic results. *Vasc Endovascular Surg*. 32004;8:533–540.
18. Guex JJ. Ultrasound guided sclerotherapy (USGS) for perforating veins (PV). *Hawaii Med J*. 2000;59:261–262.
19. Weiss RA, Goldman MP. Advances in sclerotherapy. *Dermatol Clin*. 1995;13:431–445.
20. Cabrera J, Cabrera J Jr, García-Olmedo MA. Elargissement des limites de la sclerotherapie: Nouveaux produits sclerosants. *Phlebologie*. 1997;2:181–188.
21. Cabrera J, Cabrera J Jr, García-Olmedo MA. Treatment of varicose long saphenous veins with sclerosant in microfoam form: Long-term outcomes. *Phebology*. 2000;15:19–23.
22. Cabrera J, Cabrera J Jr, Garcia-Olmedo MA, Redondo P. Treatment of venous malformations with sclerosant in microfoam form. *Arch Dermatol*. 2003;139:1409–1416.
23. Cabrera J, Redondo P, Becerra A, et al. Ultrasound-guided injection of Polidocanol microfoam in the management of venous leg ulcers. *Arch Dermatol*. 2004;140:667–673.
24. Falanga V, Margolis D, Alvarez O, et al. Rapid healing of venous ulcers and lack of clinical rejection with an allogeneic cultured human skin equivalent. Human Skin Equivalent Investigators Group. *Arch Dermatol*. 1998;134:293–300.
25. Wollmann J-CGR. The history of sclerosing foams. *Dermatol Surg*. 2004;30:694–703.
26. Frullini A. Review of the use of foams. Abstract presented at: UIP World Congress Chapter Meeting; August 27–31, 2003; San Diego, CA.
27. Hsu TS, Weiss RA. Foam sclerotherapy: A new era. *Arch Dermatol*. 2003;139:1494–1496.
28. Cabrera J, Cabrera J Jr. BTG International Limited inventors; assignee injectable microfoam containing a sclerosing agent. US patent 5676962. October 16, 1997.
29. Bikerman JJ. *Foams*. New York: Springer-Verlag, 1973.
30. Paul RE, Durant TM, Oppenheimer MJ, Stauffer HM. Intravenous carbon dioxide for intracardiac gas contrast in the Roentgen diagnosis of pericardial effusion and thickening. *Am J Roentgenol Radium Ther Nucl Med*. 1957;78:224–225.
31. Meltzer RS, Serruys PW, Hugenholtz PG, Roelandt J. Intravenous carbon dioxide as an echocardiographic contrast agent. *J Clin Ultrasound*. 1981;9:127–131.
32. Bendib M, Toumi M, Boudjellab A. CO_2 angiography and enlarged CO_2 angiography in cardiology. *Ann Radiol (Paris)*. 1977;20:673–686.
33. Kerns SR, Hawkins IF Jr. Carbon dioxide digital subtraction angiography: Expanding applications and technical evolution. *AJR*. 1995;164:735–741.
34. Benigni JP. Polidocanol 400 foam injection and migraine with visual aura. American College of Phlebology 16th Annual Congress; November 7–10, 2002; Ft. Lauderdale, FL, Oakland, CA: ACP,139.
35. Morrison N, Rogers C, Neuhardt D, Melfy K. Large-volume, ultrasound-guided Polidocanol foam sclerotherapy: A prospective study of toxicity and complications. Abstract

presented at: UIP World Congress Chapter Meeting; August 27–31, 2003; San Diego, CA.
36. Cavezzi A, Frullini A, Ricci S. Treatment of varicose veins by foam sclerotherapy: Two clinical series. *Phlebology.* 2002;17:13–18.
37. Eckmann DM, Kobayashi S, Li M. Microvascular embolization following Polidocanol microfoam sclerosant administration. *Dermatol Surg.* 2005;31:636–643.
38. Forlee MV, Grouden M, Moore DJ, Shanik G. Stroke after varicose veins foam injection sclerotherapy. *J Vasc Surg.* 2006;43:162–164.
39. Mayberry JC, Moneta GL, Taylor LM, Porter JM. Fifteen-year results of ambulatory compression therapy for chronic venous ulcers. *Surgery.* 1991;109:575–581.
40. Pierik EG, van Urk H, Hop WC, Wittens CH. Endoscopic versus open subfascial division of incompetent perforating veins in the treatment of venous leg ulcerations: A randomized trial. *J Vasc Surg.* 1997;26:1049–1054.
41. Barwell JR, Davies CE, Deacon J, et al. Comparison of surgery and compression with compression alone in chronic venous ulceration (ESCHAR study): Randomised controlled trial. *Lancet.* 2004;363:1854–1859.
42. Pierik EGJM, van Urk H, Wittens CHA. Efficacy of subfascial endoscopy in eradicating perforating veins of the lower leg and its relation with venous ulcer healing. *J Vasc Surg.* 1997;26: 255–259.
43. Whiteley MS, Smith JJ, Galland RB. Tibial nerve damage during subfascial endoscopic perforator vein surgery. *Br J Surg.* 1997;84:512.
44. Stacey MC, Burnand KG, Layer GT, Pattison M. Calf pump function in patients with healed venous ulcers is not improved by surgery to the communicating veins or by elastic stockings. *Br J Surg.* 1988;75:436–439.
45. Persoon A, Heinen MM, van der Vleuten CJM, de Rooij MJ, van de Kerkhof PCM, van Achterberg T. Leg ulcers: A review of their impact on daily life. *J Clin Nurs.* 2004;13: 341–354.

16

Foam Treatment of Venous Leg Ulcers: A Continuing Experience

Van L. Cheng, Cynthia K. Shortell, and John J. Bergan

As indicated elsewhere in this volume, most forms of venous abnormalities are very common. Some are just cosmetically unsightly, but the most serious, leg ulcer, is medically important. Fortunately, this occurs in limbs with uncomplicated varicose veins in fewer than one patient in 20. When leg ulcers caused by venous insufficiency develop, surgery becomes the dominant method of treatment. This always follows compression therapy, and often such compression is effective in closing the open wounds. However, compression treatment does not correct the fundamental pathophysiologic cause of the skin breakdown.

In one of the best analyzed experiences in surgical treatment of venous leg ulcers, the results were disappointing but typical of all published reports.[1] The estimated 5-year recurrence rate of leg ulcers was 19%. A long history of venous ulcer before treatment was the most significant risk factor for ulcer recurrence, but presence of reflux after treatment also affected long-term results. Such reflux, residual after deep venous thrombosis, is not affected by conventional surgical treatments. In a meta-analysis of the chronic venous insufficiency/leg ulcer problem, the post-thrombotic state was present in 36% of cases, but deep venous reflux was present in 56% and elements of venous obstruction in 12%.[2] The Goteborg group evaluated the distribution of superficial and deep reflux in a retrospective study of 186 patients with chronic leg ulcers. Superficial reflux predominated.[3] In limbs without a history of deep venous thrombosis, 49% had superficial reflux, and an additional 35% had superficial and deep venous reflux. It is in these limbs without a history of venous thrombosis

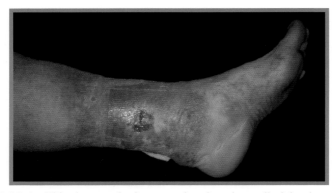

FIGURE 16-1 This photograph of a venous leg ulcer shows all of the characteristics of a typical, long-standing ankle ulceration. It is in the gaiter area, in a leg with ankle edema, hyperpigmentation, lipodermatosclerosis, healthy granulation tissue, and a corona phlebectatica or venous flare.

that the best results of all forms of treatment of the venous problem occur. Treatment of superficial venous reflux is quite simple, while treatment of deep reflux in the post-thrombotic state can be complex if not merely difficult.[4,5]

Knowing that treatment of superficial venous incompetence in limbs with venous ulcer was effective and beginning to see the results of foam sclerotherapy in limbs with varicose veins, we turned our attention to treatment of venous ulcer with sclerosant foam in 2002. Cabrera's first report on this subject appeared later, in 2004.[6] Our first report on this subject appeared a few months later in March 2005[7] (Figure 16-1).

Foam sclerotherapy has the advantage of real-time visualization as an effective contrast agent on ultrasound imaging, allowing manual displacement of the foam into selected veins. Additionally, the minimally invasive nature of the procedure permits the patient to return immediately to daily activity. Because of its efficacy, efficiency, and low cost, foam sclerotherapy has been applied to the treatment of a variety of venous conditions.[8,9] Among these is the treatment of venous ulcers. A summary of our experience with foam treatment of venous leg ulcer follows.

METHODS

Foam sclerotherapy is performed in the office setting, since neither anesthesia nor analgesics are required. The patient with the venous ulcer undergoes a standing lower extremity ultrasound examination (Sonosite 180 Plus ultrasound system with the 5–10-MHz transducer) to map the normal and pathologic anatomy, including refluxing veins and exit and

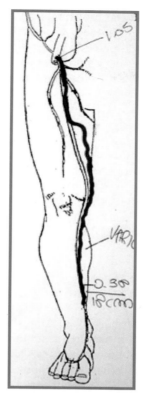

FIGURE 16-2 The venous map displays a diagram on which the refluxing veins can be indicated. This map of a leg with a medial posterior tibial ulceration shows reflux at the saphenofemoral junction into a medial accessory saphenous vein. An associated exit perforating vein 0.3 cm in diameter at the fascial level is indicated at 15 cm from the floor with the patient standing.

re-entry perforating veins[10] (Figure 16-2). Following the vein mapping, the great saphenous vein can be cannulated under ultrasound guidance. Alternatively, any varix can be accessed without imaging assistance. A 25-gauge butterfly needle is adequate for most refluxing varices and can be taped securely in place after cannulation is assured. Treatment proceeds with the patient in the supine position. The intravenous location of the foam (5–12 ml of 2–3% Polidocanol or other detergent sclerosant) is proved by ultrasound imaging. As the foam is injected, manual displacement of the foam can be accomplished using the ultrasound probe or the treating physician's hand. Prior to the development of ultrasound imaging, foam direction was accomplished by noting crepitus.[11] Foam is directed into the refluxing saphenous or accessory saphenous veins and pushed to groin level. If foam is seen to enter the deep system via a perforator, the patient is asked to flex and extend the foot forcefully to contract the gastrocnemius-soleus muscles. Resultant deep venous blood flow clears these veins of foam particles. Then the treated lower extremity is elevated for 5 to 10 minutes, during which time the foam returns to its liquid state. This avoids adverse events such as migraine, scotomata, chest pain, and dry cough.

A fitted 30–40-mmHg compression stocking is applied and reinforced with elastic bandaging. Focal pressure with gauze pads, dental rolls, or foam rubber under the bandaging but over the stocking is carefully applied to empty and keep empty those vessels that might trap blood in the post-treatment period. Careful attention to the details of focal pressure will ensure successful fibrosclerotic sealing of the affected veins and absence of blood trapping. The elastic wrap needs to include the foot and heel to prevent distal edema. The pressure dressing is left in place 72 hours or longer before it is removed. The patient is instructed to wear the compression stocking without overwrap during the day for an additional 2 weeks and to abstain from heavy lower extremity exercise.

STATISTICAL ANALYSIS

Data points for patients treated from August 2001 through August 2006 were established for gender, age, venous clinical severity score (VCSS) before/after treatment, and duration of treatment. The VCSS is based on nine clinical characteristics of chronic venous disease. They are graded 0 to 3 (absent, mild, moderate, severe) with specific criteria to avoid overlap or arbitrary scoring. Zero to 3 points were added for differences in previous conservative therapy (compression, elevation) to produce a 30-point maximum flat scale. Ulcer healing is defined as the time for a lesion to stop producing exudate.

The unpaired, two-tailed Student's t-test was performed to determine differences between two experimental groups. Values are reported as mean ±SD, and $p < 0.05$ was considered statistically significant.

RESULTS

Group I included 17 extremities that were treated only with compression. Group III included 14 extremities that had treatment with foam sclerotherapy without preliminary prolonged compression. Group II included those patients who were initially managed with compression but later crossed over to be treated with sclerosant foam (Figure 16-3). The analysis of these patients will be observational and is discussed later.

Patients in Groups I and III were evenly matched for age and gender. Six of the 17 extremities in Group I were post-thrombotic, compared with

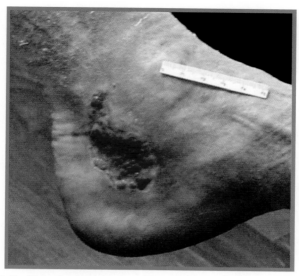

FIGURE 16-3 This photograph of a recalcitrant venous leg ulcer is typical of lesions that crossed over from Group I to Group II. Its classification is C_{6s} for unhealed symptomatic leg ulcer, E_s for post-thrombotic or secondary, $A_{S,rDr,Pr}$ for the refluxing superficial deep and perforating vein venous systems, and, finally, P_r for Pathophysiological reflux. After 6 months of unsuccessful treatment in an unna boot, the ulcer healed after 6 weeks and two sessions of foam sclerotherapy.

TABLE 16-1 Characteristics of the Treatment Groups

	Group I	Group III
Male gender	58%	57%
Age (years)	65 ± 19.4	68 ± 14
VCSS (before)	17.7	20.4
VCSS (after)	8.9	7.4
Change in VCSS	−8.9	−12.6
Duration of ulceration	9.2 weeks	4.3 weeks

five of the 14 extremities in Group III, which were post-thrombotic (Table 16-1). Group I had an average VCSS score of 17.8, which was reduced to 8.9 after treatment. This resulted in a change of 8.9. Group III had an average pretreatment VCSS score of 20.4, which dropped to 7.4 after treatment, resulting in a change of 12.6 ± 4.2 (Figures 16-4A and 16-4B). It was determined that the change in VCSS between those treated with foam versus those treated with compression was statistically significant ($p = 0.02$).

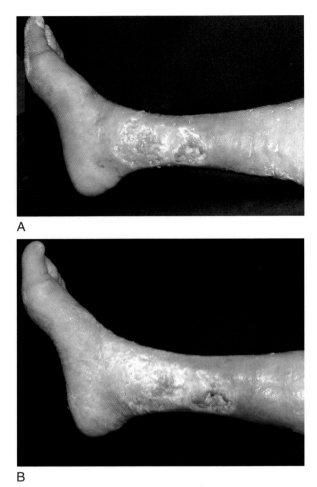

FIGURE 16-4 These photographs show the medial leg ulcers that were present for over 2 years in an otherwise healthy 92-year-old woman with an ancient history of great saphenous stripping. **A.** This photograph shows the clean ulcers on initiation of foam therapy. **B.** This photograph shows the ulcers at 14 days after injection of foam when the ulcers had begun circumferential healing. The ulcers were healed at 28 days.

Patients had ultrasound studies at 1 week and 1 month post sclerotherapy, during which time further injections were made of the residual refluxing veins as needed. At the conclusion of these treatments, the previously refluxing veins were not compressible and had no flow by ultrasound examination.

The time of healing from initiation of treatment to complete healing of the foam treatment group was, on average, 4 weeks. Compared with the duration of time from initiation of treatment to complete healing in the compression group (9 weeks), the difference was statistically significant (p = 0.027).

Group II patients initially received treatment with compression. However, when standard therapy failed to heal the ulcers, these patients received foam sclerotherapy. The mean age of this group was 64.6 years, and five of the eight had primary venous insufficiency, while three were post-thrombotic. They had an average VCSS of 16.4 on initiation of foam treatment, and this improved to a score of 7 after complete healing. This statistic included one patient who did not respond to compression or sclerosant foam. His venous clinical severity score remained at 17 despite all interventions. Later, he was subjected to resection of an aortic aneurysm and had a femoro-tibial bypass to revascularize diabetic tibial artery occlusions.

No major complications were observed. There were no temporary or permanent neurological deficits, no DVT or pulmonary emboli, and no visual abnormalities. The previously rare events of dry cough, visual disturbance (<1 minute), and migraine were eliminated after post-treatment leg elevation of the treated extremity was included in the protocol. The single adverse reaction that remained troublesome was discoloration of the skin overlying treated veins. This problem generally resolved within 6 months. Calculation of the recurrence rate will be ascertained and reported in a future paper.

DISCUSSION

It has been pointed out that venous ulceration occurs in a small proportion of the population, but this cohort commands considerable health care resources to manage the problem.[12] Venous ulcers tend to run a prolonged course, taking a number of months to achieve complete healing. Roughly half of patients attending a leg ulcer service will achieve ulcer healing within 4 months, while half will not.[13] Healing is often followed by relapse. The ulcer recurrence rate ranges from 15% to 70% with 3% to 28% recurring within 1 year. Controlling superficial reflux by ablation of superficial veins can reduce ulcer recurrence. One careful study has shown that the ulcer recurrence rate was 28% with compression alone, but this was reduced to 12% when surgical stripping was added to compression therapy.[14] Details regarding this study are found in Chapter 13.

Despite various methods of treatment, venous leg ulcer is characterized by recurrence (Table 16-2). Well-performed surgery in which attention is paid to total ablation of superficial reflux and perforator outflow has the lowest rate of recurrence. Conversely, when the ulcerated surface is covered without attention to the abnormal hemodynamics that produced the deformity, the rate of recurrence is the greatest. This experience is replicated in

TABLE 16-2 Recurrence of Leg Ulcer According to Treatment Received*

Report	Treatment	Recurrence	Follow Up
Hansson[15]	Surgery	5%	5 years
Henry[16]	Sclerotherapy	30%	5 years
Porter[17]	Compression	29%	5 years
Hansson[18]	Miscellaneous	44%	3 years
O'Donnell[19]	Perforator Ligation	55%	5 years
Reiter[20]	Skin Grafts	71%	2.5 years

* All patients treated with ancillary compression.
Modified from Coleridge Smith PD. Chapter 2. In: Negus D, Coleridge Smith PD, Bergan JJ, eds. *Leg Ulcers*, 3rd ed. London: Hodder Arnold, 2006:12.

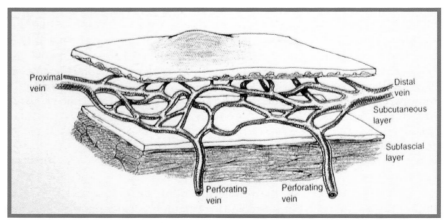

FIGURE 16-5 This diagram shows how proximal venous pressure transmitted through superficial reflux and muscular contraction pressure through incompetent perforating veins is delivered to the tangle of veins under lipodermatosclerotic skin and the venous ulcer. From DePalma RG. In: Bergan JJ, Gloviczki P, eds. *Atlas of Endoscopic Perforator Vein Surgery*. London: Springer-Verlag, 1997.

wound care centers worldwide where attention is paid to the effects of abnormal venous pressure rather than its causes.

Abnormal venous pressure is associated with a cascade of inflammatory events that seriously damage tissues such as venous valves, venous walls, subcutaneous tissues, and skin[21] (Figure 16-5). The venous system of the lower extremities in man is exposed to high pressures. A mean pressure of 95.4 ± 5.5 mmHg was recorded by Kügler in dorsal foot veins during quiet standing in a group of 20 healthy volunteers.[22] Similar values were recorded by Stick.[23] Some vessels in the calf, including the perforating veins, are subjected to even higher pressures, which are generated by calf muscle contraction. Peak intramuscular pressures in the human soleus muscle have been recorded as 181 mmHg during walking and 269 mmHg during running.[24]

Proximal venous valves may be exposed to high pressures generated within the abdomen. This intra-abdominal pressure is transmitted to groin vessels, the common femoral vein, the great saphenous vein, and the profunda femoris vein. These pressures have been reported in healthy adults to reach 107 mmHg during coughing and 171 mmHg during jumping.[25]

In the normal human lower extremity venous system with competent valves and a functioning calf muscle pump, high pressures are experienced only intermittently. However, when proximal valves are damaged or destroyed by the inflammatory effects of abdominal pressure impulses, high pressures are transmitted distally. Eventually, high pressures reach the subulcer tangle of venules where the tissue-destructive actions of the inflammatory cascade wreak havoc. High pressures transmitted through incompetent perforating veins also reach the veins under the ulcer. Summation of these forces explains the failure of ulcer treatment and the reasons for recurrence of venous ulcer after successful compression and/or surgical treatment.

De Palma's solution, and that of many surgeons, is to excise the entire ulcerated area down to and including the deep fascia. This does effectively remove the end organ of ulcer genesis, the tangle of subulcer venules. A subsequent skin graft would be predicted not only to heal the ulcer bed but also to prevent ulcer recurrence. Experience teaches that ulcer recurrence after successful ulcer excision and skin grafting is at the edges of the graft in tissue that continues to be pressurized by superficial reflux or perforator outflow. It is not surprising that ulcer recurrence occurs most frequently in limbs with previous deep vein thrombosis where the thrombotic process has destroyed deep valve function and may leave behind a degree of venous outflow obstruction that perpetuates the destructive forces of distal venous hypertension.

The rapid success of foam sclerotherapy in treatment of venous ulcers is easily explained by the fact that its effect is directly on the subulcer veins rather than indirect as in subfascial perforator interruption and vein stripping. The target venules under the ulcer are easily visualized during the administration of foam sclerotherapy. They are seen to be closed and unresponsive to transducer pressure after ulcer healing. The same effects are seen in treating severe lipodermatosclerosis and atrophie blanch.

CONCLUSIONS

Foam sclerotherapy reaches its highest pinnacle of success in treating venous leg ulcer. This is easily explained by the fact that the tangle of pressurized veins that cause and perpetuate the leg ulcer are closed by the actions of the foamed sclerosant.

REFERENCES

1. Magnusson MB, Nelzen O, Risberg B, Sivertsson R. Leg ulcer recurrence and its risk factors: A duplex ultrasound study before and after vein surgery. *Eur J Vasc Endovasc Surg.* 2006;32(4):453–461.
2. Delis KT, Ibegbuna V, Nicolaides AN, et al. Prevalence and distribution of incompetent perforating veins in chronic venous insufficiency. *J Vasc Surg.* 1998;28(5): 815–825.
3. Magnusson MB, Nelzen O, Risberg B, Sivertsson R. A colour Doppler ultrasound study of venous reflux in patients with chronic leg ulcers. *Eur J Vasc Endovasc Surg.* 2001;21(4):353–360.
4. Bergan J, Pascarella L, Mekenas L. Venous disorders: Treatment with sclerosant foam. *J Cardiovasc Surg.* 2006;47:9–18.
5. Neglen P, Hollis KC, Raju S. Combined saphenous ablation and iliac stent placement for complex severe chronic venous disease. *J Vasc Surg.* 2006;44:828–833.
6. Cabrera J, Redondo P, Becerra A, et al. Ultrasound-guided injection of Polidocanol microfoam in the management of venous leg ulcers. *Arch Dermatol.* 2004;140: 667–673.
7. Bergan JJ, Pascarella L. Severe chronic venous insufficiency: Primary treatment with sclerofoam. *Semin Vasc Surg.* 2005;18:49–56.
8. Pascarella L, Bergan JJ, Yamada C, Mekenas L. Venous angiomata: Treatment with sclerosant foam. *Ann Vasc Surg.* 2005;19:457–464.
9. Pascarella L, Bergan JJ, Mekenas LV. Severe chronic venous insufficiency treated by foamed sclerosant. *Ann Vasc Surg.* 2006;20:83–91.
10. Mekenas LV, Bergan JJ. Venous reflux examination. *J Vasc Tech.* 2001;26:139–146.
11. Flückiger P. Nicht-operative retrograde Varicenverödung mit Varsylschaum. *Schweiz Med Wochenschr.* 1956;48:1368–1370.
12. Coleridge Smith PD. Epidemiology—the extent of the problem. In: Negus D, Coleridge Smith PD, Bergan JJ, eds. *Leg Ulcers,* 3rd ed. London: Hodder Arnold, 2006.
13. Skene AI, Smith JM, Dore CJ, Charlett A, Lewis JD. Venous leg ulcers: A prognostic index to predict time to healing. *Br Med J.* 1992;;305(6862):1119–1121.
14. Barwell JR, Davies CE, Deacon J, et al. Comparison of surgery and compression with compression alone in chronic venous ulceration (ESCHAR study): Randomised controlled trial. *Lancet.* 2004;363:1854–1859.
15. Hansson LO. Venous ulcers of the lower limb. A follow-up study 5 years after surgical treatment. *Acta Chir Scand.* 1964;128:269–277.
16. Mayberry JC, Moneta GL, Taylor LM Jr, Porter JM. Fifteen-year results of ambulatory compression therapy for chronic venous ulcers. *Surgery.* 1991;109:575–581.
17. Dinn E, Henry M. Treatment of venous ulceration by injection sclerotherapy and compression hosiery: A 5 year study. *Phlebology.* 1992;7:23–26.
18. Hansson C, Andersson E, Swanbeck G. A follow-up study of leg and foot ulcer patients. *Acta Derm Venereol.* 1987;67:496–500.
19. Burnand K, Thomas ML, O'Donnell T, Browse NL. Relation between postphlebitic changes in the deep veins and results of surgical treatment of venous ulcers. *Lancet.* 1976;1:936–938.
20. Reiter H. Ulcer cruris. Cure rate after treatment by skin grafting according to Reverdin. *Acta Derm Venerol.* 1954;34:439–445.
21. Pascarella L, Penn A, Schmid-Schönbein GW. Venous hypertension and the inflammatory cascade: Major manifestations and trigger mechanisms. *Angiology.* 2005;56(Suppl 1):S3–S10.
22. Kügler C, Strunk M, Rudofsky G. Venous pressure dynamics of the healthy human leg. *J Vasc Res.* 2001;38:20–29.

23. Stick C, Jaeger H, Witzleb E. Measurements of volume changes and venous pressure in the human lower leg during walking and running. *J Appl Physiol.* 1992;72:2063–2068.
24. Ballard RE, Watenpaugh DE, Breit GA, Murthy G, Holley DC, Hargens AR. Leg intramuscular pressures during locomotion in humans. *J Appl Physiol.* 1998;84: 1976–1981.
25. Cobb WS, Burns JM, Kercher KW, Matthews BD, Norton HJ, Heniford BT. Normal intraabdominal pressure in healthy adults. *J Surg Res.* 2005;129:231–235.

17

ULTRASOUND GUIDANCE FOR ENDOVENOUS TREATMENT

LUIGI PASCARELLA, LISA MEKENAS, AND JOHN J. BERGAN

INTRODUCTION

Duplex ultrasound sonography is the best choice for evaluation of venous reflux in the lower extremities.[1] It is inexpensive, noninvasive, and generally acceptable to the patient.[2] It provides direct visualization, localization, and quantitation of venous reflux with a surprisingly high sensitivity (95%) and specificity (100%).[3] Duplex ultrasound findings have been confirmed by angioscopic observation of incompetent vein valves in advanced chronic venous insufficiency.[4] Yamaki has demonstrated that high peak reflux velocities (>30 cm/second), reflux duration greater than 3 seconds, and an enlarged valve annulus measured by duplex ultrasonography at the SFJ are all closely related to angioscopically deformed and incompetent terminal valves (Type III and Type IV of Hoshino).[4]

PREOPERATIVE ASSESSMENT

The examination should always begin with the recording of a complete medical history. A family and personal venous history are of great importance. Symptoms such as aching heaviness, limb fatigue, itching, and burning pain are typically present. History of pregnancy deliveries, limb trauma, fractures, and confining illnesses should be recorded. Any history of deep venous thrombosis and treatment must be paired with its method of diagnosis. A detailed description of previous venous treatments must be obtained. Comorbidities, allergies, and pharmacologic history must be documented.[2] The BMI is calculated from the patient's height and weight and should be recorded.

The patient should be examined in a standing position to demonstrate visual patterns of telangiectasias, reticular, and varicose veins.[5] This should be supplemented by cold light trans-illumination of the skin with a bright white light, such as the Vein Light, to identify reticular veins. A handheld Doppler device can serve to perform a screening examination, but it should be considered to be inadequate to plan specific treatment.[5]

Three levels of pathologic veins are evaluated using the techniques cited in the preceding paragraphs. Telangiectasias in the skin are visually inspected; reticular veins are trans-illuminated with the vein light; and varicosities and the saphenous veins, with ultrasound. Clinical data should be integrated into the CEAP classification.[6,7]

EQUIPMENT

The ultrasound duplex scanner should be able to detect blood flow rates as low as 6 cm/second.[2] Dedicated high-resolution vascular scanners with color and/or Power-Doppler functions as well as the continuous wave Doppler can do this. Linear transducers in the range of 4–7 megahertz (MHz) are used.[5,8] The inferior vena cava, pelvic veins, and deep veins of the limbs in obese patients may be imaged with lower frequency, 3-MHz transducers.[1] Linear hockey-stick transducers in the range of 5–12 MHz will provide detailed imaging of smaller veins and perforating veins.

After the year 2000, advances in technology allowed duplex scanners to become smaller, more transportable, and more operator friendly.[5] These miniaturized devices feature transducers designed with advanced architecture that allow a single probe to image across a greater range of depths within an application and across applications.[5] Thus, in many ways, the portable ultrasound instrument supplements the stethoscope in patient evaluation. The transducer for peripheral vascular examinations operates from 10–15 MHz and provides resolution from skin surface to 7 cm in depth.[5] The technology incorporates power Doppler sonography, tissue harmonic imaging, and direct connectivity to a personal computer.[5] The overall performance of miniaturized ultrasound devices is comparable to the more traditional and much larger ultrasound equipment that allowed establishment and growth of the vascular laboratory.[5]

VENOUS TESTING AND MAPPING

Not only is a detailed U.S. duplex study of the normal and pathologic venous anatomy essential, but a map should be created to guide therapy.

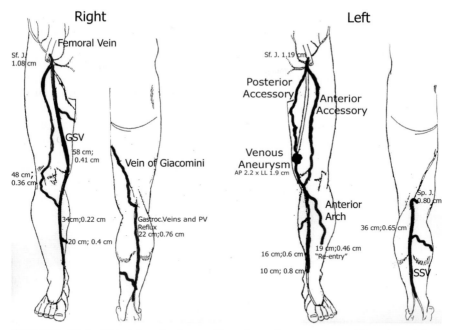

FIGURE 17-1   This data entry form outlines the saphenous veins and the relevant deep veins. Refluxing veins are added in heavy black lines. Location of perforating veins and aneurysms can be added and distance from the floor indicated. Diameters of perforating veins at the fascial level should also be noted.

A detailed verbal description of the examination and its findings is useful for record purposes, but it is the map that is most useful during therapy. A clear and illustrative graphic notation of significant vein diameters, anomalous anatomy, superficial venous aneurysms, perforating veins, and the presence and extent of reflux should always be recorded during the examination (Figure 17-1).[2,5]

As mentioned, the ultrasound examination is conducted with the patient standing.[9] This position has been found to maximally dilate leg veins and challenges vein valves. Sensitivity and specificity in detecting reflux are increased in examinations performed with the patient standing rather than when the patient is supine.[8,9] Supine examinations for reflux are unacceptable.

The veins are scanned by moving the probe vertically up and down along their course. Transverse scans are most informative, but a mental reconstruction must be created to record the venous map. Duplicated segments, sites of tributary confluence, large perforating veins, and their deep venous connections are identified as well as the very common superficial venous aneurysms. Location of abnormalities as measured in centimeters from the floor assists in preparing a therapeutic guide. Measurements from the medial malleolus are commonly recorded but are not as precise. Transverse, in, and longitudinal scans combined with continuous scanning provide a clear mapping of the venous system.

Patency of peripheral veins is usually assessed by compression of the vein with the transducer.[8] Residual, ancient thrombus, partial patency, and extrinsic compression should all be noted in the verbal description of findings. Reflux is detected by flow augmentation with compression and release maneuvers of the thigh and calf. The Valsalva maneuver is used only at the saphenofemoral junction[8] because presence of a competent proximal valve negates the value of the examination.

Automated rapid inflation and deflation cuffs are in use but are cumbersome. However, they do offer the advantage of a standardized stimulus,[10-12] which allows timing of reflux. Although reflux greater than 500 ms is considered pathologic,[9,13] this is only precisely accurate when a standard stimulus is applied.

The diameter of the saphenofemoral junction and femoral vein are recorded for use in preparation for radiofrequency VNUS closure® and endovenous laser ablation.[14-16] Important information is also gained from diameters of the GSV at mid-thigh and distal thigh. Radiofrequency ablation is commonly used to treat veins from 2 to 12 mm in diameter,[16] although diameters recorded with the patient standing do not apply when treatment is given with the patient supine.

The supragenicular, infragenicular, or immediate subgenicular great saphenous veins are often the access point for its laser or radiofrequency ablation.[16,17] Therefore, the depth of the GSV in these regions should be recorded.

Accessory veins, by definition, run parallel to the GSV in the thigh[18] (Figure 17-1); therefore, it is imperative to map their course accurately and to note their eventual communication with GSV (Figure 17-1), as well as their presence or absence of reflux with relationship to varicosities. The accessory veins are easily confused with the GSV during continuous longitudinal scanning when the saphenous vein appears to leave the saphenous compartment.[18] This error is avoided when transverse scans are done. The great saphenous vein is scanned throughout the leg and thigh so that its tributaries are identified (Figure 17-1).

The diameters of the popliteal vein and the small saphenous vein (SSV) are recorded at the junction as well as diameters of the SSV along its course in the leg. Intersaphenous veins should be identified and the extreme variability of the SSV termination carefully recorded.

The venous reflux examination also includes the mapping of exit and re-entry perforating veins (PV) wherever they are identified.[19] PV reflux is detected as outward flow duration greater than 350 ms during the release phase of the flow augmentation maneuver (distal compression has higher sensitivity in detecting PVs reflux).[1] Perforating veins should be accurately mapped in their different locations in the leg, and their position should be measured as distance (cm) from the floor.[18,20]

ULTRASOUND MONITORING DURING ENDOVENOUS THERAPY

Endothermic coagulation is caused by the generation of or application of heat to the endothelial surface of targeted veins.[16,21,22] It is been suggested that the coagulation process is related to the intravascular vaporization of blood (steam) during laser therapy with intimal denudation and collagen fiber contraction. Vein wall thickening and rapid reorganization of the vessel to form a fibrotic cord follow.[21,22] Occlusion is usually visualized within 10–20 seconds of the laser or radiofrequency energy application.[22] Both techniques have been proven to be safe and effective.[23] Percutaneous introduction of the laser or RF catheter has made surgical intervention less acceptable to patients because of post-treatment pain, number of cutaneous incisions, and subsequent postprocedural disability and recovery time.[15,16]

Before the procedure starts, the patient's limb should undergo repeat scanning for procedural planning and identification of the venous access site and its relationship to perforating veins, varicosities, and areas of tortuosity, stenosis, or dilation. In this preparatory phase, some anatomic landmarks should be clearly recognizable. They include the femoral vein, saphenofemoral junction, saphenous compartment, great saphenous vein, and the small saphenous junctional anatomy.

The supragenicular saphenous vein is usually the access point of choice for varicose vein treatment (Figure 17-2).[17] Introduction of the echogenic needle and the introducer sheath is performed percutaneously using the Seldinger technique. The intraluminal position of the sheath is ascertained by easy passage of the guidewire and aspiration of nonpulsatile venous blood. The sheathed laser fiber or a 6- or 8-F VNUS catheter is advanced to a point just distal to the entrance of the epigastric vein.[17] Position of the laser fiber is confirmed by direct visualization of the red aiming beam and that of the VNUS catheter by ultrasound[16] (Figures 17-3 and 17-4).

The catheter or sheath appears as a hyperechoic line in the GSV lumen.[14,15] Its placement must be precisely at the SFJ 1 cm distal to the epigastric vein (Figure 17-4).[16]

Administration of the low concentration anesthesia into the saphenous compartment is monitored closely by ultrasound.[17] Correct placement of anesthesia within the saphenous compartment is crucial for procedural safety and success. The vein is seen as "floating" in the anechogenic sea of the anesthetic solution (Figures 17-5A and 17-5B). It is always wise to

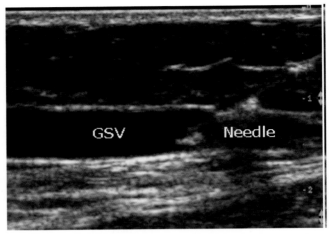

FIGURE 17-2 The great saphenous vein is cannulated using the Seldinger technique. The puncturing needle is echogenic and can be easily visualized. (Adapted from Pichot O, *Atlas of Ultrasound Images*, Copyright VNUS® Closure.)*

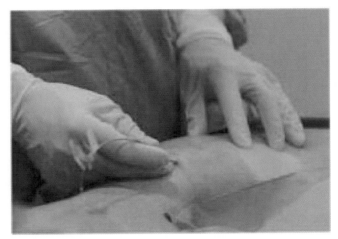

FIGURE 17-3 The laser catheter is advanced proximally toward the saphenofemoral junction. Position of the laser fiber is confirmed by direct visualization of the red aiming beam through the skin. (Adapted from Navarro L, Min RJ, Bone C. Endovenous laser: A new minimally invasive method of treatment for varicose veins: Preliminary observations using an 810 nm diode laser. *Dermatol Surg.* 2001;27(2):117.)

*http://www.vnus.com/navigation/PDFs/pichotultrasoundatlasweb.pdf

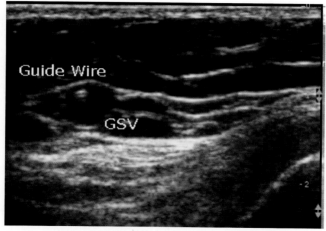

FIGURE 17-4 Position of the guidewire and radiofrequency catheter is monitored by ultrasound visualization. (Adapted from Pichot O, *Atlas of Ultrasound Images*, Copyright VNUS® Closure.)*

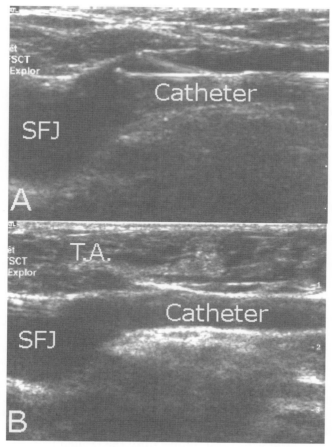

FIGURE 17-5 Administration of the tumescent anesthesia into the saphenous compartment is monitored by ultrasound. SFJ: saphenofemoral junction; T.A.: tumescent anesthesia. (Adapted from Pichot O, *Atlas of Ultrasound Images*, Copyright VNUS® Closure.)*

*http://www.vnus.com/navigation/PDFs/pichotultrasoundatlasweb.pdf

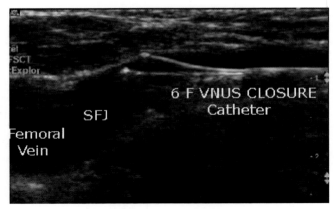

FIGURE 17-6 The ablation starts at the saphenofemoral junction and proceeds in a distal direction. It is always wise to recheck the catheter position at SFJ prior to the application of energy. (Adapted from Pichot O, *Atlas of Ultrasound Images*, Copyright VNUS® Closure.)*

recheck the catheter position at SFJ after correct placement of the limb and prior to the application of energy (Figure 17-6).[22]

Endovenous ablation starts at the saphenofemoral junction and proceeds in a distal direction.[16] Successful obliteration is confirmed by contraction of the saphenous vein to a residual diameter of <2 mm.[16] Patency of the common femoral artery and vein are confirmed by ultrasound (Figures 17-7A and 17-7B). Any thrombus is seen as a hyperechogenic core in the vessel, although this is very uncommon (Figure 17-7B).[15,24]

Early post-treatment duplex scanning should be performed. Conventionally, this is done at 1 and 7 days postprocedure. Many have eliminated the 1-day examination because deep venous thrombosis is uncommon and thrombus protruding into the femoral vein is so benign. Evidence of a protruding thrombus from the saphenous vein into the femoral vein should be looked for, however, in every postprocedural examination (Figure 17-8).[24] Evidence of a noncompressible GSV with thickened walls and absence of flow on color ultrasound analysis are signs of successful obliteration (Figure 17-9).[9]

ULTRASOUND MONITORING DURING SCLEROFOAM TREATMENT

Advent of foam sclerotherapy has added a new tool for the treatment of chronic venous insufficiency. Sclerosant agents provoke endothelial damage by several mechanisms.[25] They change either the surface tension of the plasma membrane (detergents) and/or the intravascular pH and osmolarity. The final result is a chemical fibrosis of the treated vessel.[25]

*http://www.vnus.com/navigation/PDFs/pichotnavigationatlasweb.pdf

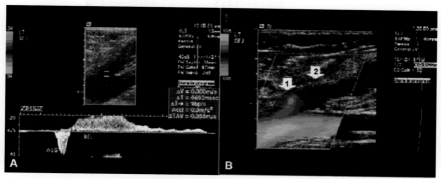

FIGURE 17-7 Duplex examinations (longitudinal views) of the great saphenous vein (GSV) at the saphenofemoral junction (SFJ). **A.** A pretreatment scan demonstrated an incompetent SFJ after augmentation. **B.** Intraoperative color duplex interrogation showed successful occlusion of the GSV with a patent, 3-mm proximal stump (arrow 1) and absence of flow within the treated segment (arrow 2). (Adapted from Puggioni A, Kalra M, Carmo M, Mozes G, Gloviczki P. Endovenous laser therapy and radiofrequency ablation of the great saphenous vein: Analysis of early efficacy and complications. *J Vasc Surg.* 2005;42(3): 488–493.)

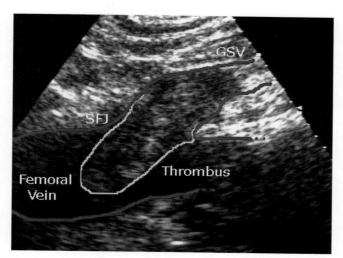

FIGURE 17-8 Early post-treatment duplex scanning should be performed. Evidence of a protruding thrombus from the saphenous vein into the femoral vein should be looked for. (Adapted from Pichot O, *Atlas of Ultrasound Images*, Copyright VNUS® Closure.)*

Sclerosing foams (SF) are mixtures of gas and a liquid solution with surfactant properties. In 1993, Cabrera re-introduced the use of SF, made of sodium tetradecyl sulfate or Polidocanol, in the treatment of varicose veins.[26] One of the intrinsic limits of liquid sclerosants in treatment of varicose veins is their dilution by blood, with reduction of their efficacy.[27] Also, they are rapidly cleared by the moving bloodstream. Sclerosing foams

*http://www.vnus.com/navigation/PDFs/pichotultrasoundatlasweb.pdf

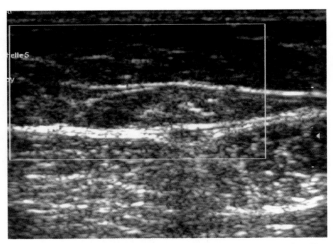

FIGURE 17-9 Evidence of a noncompressible GSV with thickened walls and absence of flow on color ultrasound analysis are signs of successful obliteration. (Adapted from Pichot O, *Atlas of Ultrasound Images*, Copyright VNUS® Closure.)*

do not mix with blood and instead remain in the vessel where they strip the endothelium.[27] Persistence of the agent in the vessel causes an increased contact time with the intimal surface. Foam preparation is remarkably simple.[27] Tessari's three-way stopcock method is the most commonly used.[27,28]

As with endovenous ablation, the treatment starts with clear ultrasound mapping. Varicose veins can be accessed by the placement of a 25-G butterfly needle, or the great saphenous or small saphenous vein can be directly cannulated with an angiocath, an echogenic Cook® needle, or a butterfly needle.[27,29,30] Nearly all descriptions of the technique explain direct ultrasound-guided access to the saphenous vein.[27,31] In contrast, many experienced operators achieve a satisfactory and rapid obliteration of the GSV and SSV by cannulating a peripheral varicosity.[30] Although the saphenous vein cannot be cannulated with a catheter by way of a varicosity because of its angle of connection, there is no such obstacle to the flow of foam.

Foam functions as an efficient ultrasound (US) contrast medium because of its air content. Its instillation is easily monitored. Its US appearance is that of a solid hyperechogenic core with an accompanying acoustic shadow (Figure 17-10).

Foam is introduced into a varix or the saphenous vein with the patient supine. As the foam reaches the SFJ as monitored by ultrasound, compression of the SFJ or the SPJ is usually done in order to preserve foam in the limb.

*http://www.vnus.com/navigation/PDFs/pichotultrasoundatlasweb.pdf

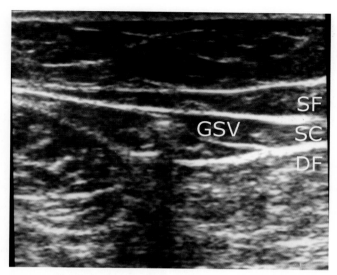

FIGURE 17-10 Foam functions as an efficient ultrasound contrast medium because of its air content. Its injection can be easily monitored. Its US appearance is that of a solid hyperechogenic core with an acoustic shadow projected in the tissue below.

Vasoconstriction and vasospasm can be induced by intermittent compression of the vein by the ultrasound transducer and by elevating the limb. This minimizes the blood content of the treated veins. Foam will be seen by ultrasound to flow distally in the elevated limb. It flows selectively through incompetent valves and is blocked by competent valves. Femoral compression and leg elevation have the effect of prolonging the action of the foamed sclerosant on the distal intima, improving the efficacy of the entire treatment. In addition, leg elevation for 10 minutes or so allows the foam to revert to its liquid state so that foam particles will not reach the right atrium or a patent foramen ovale.

The femoral, popliteal, and deep veins of the leg are scanned intermittently throughout the entire procedure. Foam particles are washed out of deep veins such as the gastrocnemius or tibial veins by flexion-extension maneuvers of the foot. Repetitive dorsiflexion of the foot completely clears the deep veins of any foam particles.

Despite much consternation, major thrombotic events have rarely been described with use of sclerofoam. In a study of over 12,000 sclerotherapy sessions by Guex, over half of which involved foam only, a single femoral vein thrombus was encountered.[32]

Thromboses of the gastrocnemius, tibial, and peroneal veins have been reported occasionally, and intra-arterial injections are even more rare.[30,33]

Ultrasound sonography has confirmed the presence of a tangled network of small varicose veins, reticular varices, and incompetent perforating

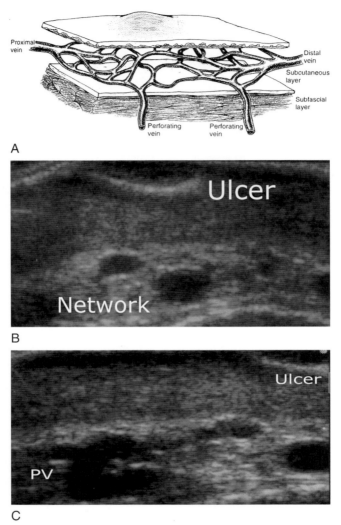

FIGURE 17-11 Ultrasound sonography has confirmed the presence of a tangled network of varicose veins of small caliber, reticular varices, and incompetent perforating veins under lipodermatosclerotic plaques and under venous ulcers.

veins under lipodermatosclerotic plaques and under venous ulcers (Figure 17-11).[30] Ultrasound monitoring should confirm the fact that these vessels are filled with foam during treatment.

Ultrasound guidance is also used in treatment of incompetent perforating veins. Direct cannulation and controlled injection of the SF under direct visualization are effective.[27] Often, superficial peripheral veins can be directly injected with the objective of obliteration of the attached perforator and its network of incompetent veins.

DISCUSSION

Compression therapy and surgery have been the cornerstone of CVI treatment for years, and they are still useful. New, minimally invasive techniques, including radiofrequency ablation of saphenous veins, EVLT, and venous ablation with sclerofoam, have been demonstrated to be safe, effective, and more acceptable to patients.[16] The contribution of ultrasound in general and duplex technology in particular has given reliability to the diagnosis of CVI and has enhanced the development of minimal treatments. Intraprocedural and postprocedural US duplex ultrasound monitoring provides the best control of the entire procedure. This provides early prevention of complications and minimizes the chance of failure.

CONCLUSIONS

Duplex ultrasound is essential in every phase of CVI patient care and has played a pivotal role in the recent revolution in venous therapy. Experience, critical thinking, uniform testing, and insight in the pathology are necessary to achieve satisfactory results.

REFERENCES

1. Labropoulos N, Leon LR Jr. Duplex evaluation of venous insufficiency. *Semin Vasc Surg.* 2005;18(1):5–9.
2. Ballard J, Bergan J, Delange M. Venous imaging for reflux using duplex ultrasonography. In: Aburahma AF, Bergan JJ, eds. *Noninvasive Vascular Diagnosis*, 1st ed. London: Springer-Verlag, 2000:339–334.
3. Depalma RG, Kowallek DL, Barcia TC, Cafferata HT. Target selection for surgical intervention in severe chronic venous insufficiency: Comparison of duplex scanning and phlebography. *J Vasc Surg.* 2000;32(5):913–920.
4. Yamaki T, Sasaki K, Nozaki M. Preoperative duplex-derived parameters and angioscopic evidence of valvular incompetence associated with superficial venous insufficiency. *J Endovasc Ther.* 2002;9(2):229–233.
5. Mekenas L, Bergan J. Venous reflux examination: Technique using miniaturized ultrasound scanning. *J Vasc Tech.* 2002;2(26):139–146.
6. Kistner RL, Eklof B, Masuda EM. Diagnosis of chronic venous disease of the lower extremities: The "CEAP" classification. *Mayo Clin Proc.* 1996;71(4):338–345.
7. Eklof B, Rutherford RB, Bergan JJ, et al. Revision of the CEAP classification for chronic venous disorders: Consensus statement. *J Vasc Surg.* 2004;40(6):1248–1252.

8. Lynch TG, Dalsing MC, Ouriel K, Ricotta JJ, Wakefield TW. Developments in diagnosis and classification of venous disorders: Non-invasive diagnosis. *Cardiovasc Surg.* 1999;7(2):160–178.
9. Labropoulos N, Tiongson J, Pryor L, et al. Definition of venous reflux in lower-extremity veins. *J Vasc Surg.* 2003;38(4):793–798.
10. Masuda EM, Kistner RL, Eklof B. Prospective study of duplex scanning for venous reflux: Comparison of Valsalva and pneumatic cuff techniques in the reverse Trendelenburg and standing positions. *J Vasc Surg.* 1994;20(5):711–720.
11. Markel A, Meissner MH, Manzo RA, Bergelin RO, Strandness DE Jr. A comparison of the cuff deflation method with Valsalva's maneuver and limb compression in detecting venous valvular reflux. *Arch Surg.* 1994;129(7):701–705.
12. Delis KT, et al. Enhancing venous outflow in the lower limb with intermittent pneumatic compression. A comparative haemodynamic analysis on the effect of foot vs. calf vs. foot and calf compression. *Eur J Vasc Endovasc Surg.* 2000;19(3):250–260.
13. Vasdekis SN, Clarke GH, Nicolaides AN. Quantification of venous reflux by means of duplex scanning. *J Vasc Surg.* 1989;10(6):670–677.
14. Pichot O, et al. Role of duplex imaging in endovenous obliteration for primary venous insufficiency. *J Endovasc Ther.* 2000;7(6):451–459.
15. Min RJ, Khilnani N, Zimmet SE. Endovenous laser treatment of saphenous vein reflux: Long-term results. *J Vasc Interv Radiol.* 2003;14(8):991–996.
16. Sadick NS. Advances in the treatment of varicose veins: Ambulatory phlebectomy, foam sclerotherapy, endovascular laser, and radiofrequency closure. *Dermatol Clin.* 2005; 23(3):443–455, vi.
17. Puggioni A, Kalra M, Carmo M, Mozes G, Gloviczki P. Endovenous laser therapy and radiofrequency ablation of the great saphenous vein: Analysis of early efficacy and complications. *J Vasc Surg.* 2005;42(3):488–493.
18. Caggiati A, Bergan JJ, Gloviczki P, Jantet G, Wendell-Smith CP, Partsch H. Nomenclature of the veins of the lower limbs: An international interdisciplinary consensus statement. *J Vasc Surg.* 2002;36(2):416–422.
19. Delis KT, et al. In situ hemodynamics of perforating veins in chronic venous insufficiency. *J Vasc Surg.* 2001;33(4):773–782.
20. Caggiati A, Bergan JJ, Gloviczki P, Eklof B, Allegra C, Partsch H. Nomenclature of the veins of the lower limb: Extensions, refinements, and clinical application. *J Vasc Surg.* 2005;41(4):719–724.
21. Weiss RA. Comparison of endovenous radiofrequency versus 810 nm diode laser occlusion of large veins in an animal model. *Dermatol Surg.* 2002;28(1):56–61.
22. Weiss RA, Weiss MA. Controlled radiofrequency endovenous occlusion using a unique radiofrequency catheter under duplex guidance to eliminate saphenous varicose vein reflux: A 2-year follow-up. *Dermatol Surg.* 2002;28(1):38–42.
23. Morrison N. Saphenous ablation: What are the choices, laser or RF energy. *Semin Vasc Surg.* 2005;18(1):15–18.
24. Pichot O, et al. Duplex ultrasound scan findings 2 years after great saphenous vein radiofrequency endovenous obliteration. *J Vasc Surg.* 2004;39(1):189–195.
25. Goldman MP. Mechanisms of action of sclerotherapy. In: Goldman MP, ed. *Sclerotherapy: Treatment of Varicose and Telangiectatic Leg Veins*, 2nd ed. St. Louis, MO: Mosby, 1995:244–279.
26. Cabrera J. Dr J. Cabrera is the creator of the patented Polidocanol microfoam. *Dermatol Surg.* 2004;30(12, Pt 2):1605; author reply, 1606.
27. Coleridge Smith P. Saphenous ablation: Sclerosant or sclerofoam? *Semin Vasc Surg.* 2005;18(1):19–24.
28. Tessari L, Cavezzi A, Frullini A. Preliminary experience with a new sclerosing foam in the treatment of varicose veins. *Dermatol Surg.* 2001;27(1):58–60.

29. Cabrera J, et al. Ultrasound-guided injection of Polidocanol microfoam in the management of venous leg ulcers. *Arch Dermatol.* 2004;140(6):667–673.
30. Bergan JJ, Pascarella L. Severe chronic venous insufficiency: Primary treatment with sclerofoam. *Semin Vasc Surg.* 2005;18(1):49–56.
31. Guex JJ. Foam sclerotherapy: An overview of use for primary venous insufficiency. *Semin Vasc Surg.* 2005;18(1):25–29.
32. Guex JJ, Allaert FA, Gillet JL, Chleir F. Immediate and midterm complications of sclerotherapy: Report of a prospective multicenter registry of 12,173 sclerotherapy sessions. *Dermatol Surg.* 2005;31(2):123–128; discussion, 128.
33. Bergan JJ, Weiss RA, Goldman MP. Extensive tissue necrosis following high-concentration sclerotherapy for varicose veins. *Dermatol Surg.* 2000;26(6):535–541; discussion, 541–542.

18

REPAIR OF VENOUS VALVES IN SEVERE CHRONIC VENOUS INSUFFICIENCY

MICHEL R. PERRIN

INTRODUCTION

According to the CEAP classification "the term chronic venous insufficiency (CVI) implies a functional abnormality of the venous system and is usually reserved for patients with more advanced disease including those with edema (C3), skin changes (C4) or venous ulcers (C5–6)."[1]

Severe CVI is harder to define. For example, the transient edema that occurs with primary varices would not qualify a patient for this designation, while permanent edema due to deep venous reflux or obstruction would.

With regard to skin disorders, the updated CEAP classification divides class C4 into subgroups C4a and C4b. We suggest that only patients with class C4b disease and higher be deemed to have severe CVI.

Because repair of superficial venous valves is performed primarily in patients without severe CVI,[2–7] only repair of deep valves will be addressed in this chapter.

Deep vein valve dysfunction may be due to a variety of causes. The most common etiology of CVI is secondary in nature; post-thrombotic syndrome (PTS) accounts for an estimated 60% to 85% of patients with severe CVI.[8–10] Post-thrombotic syndrome is the result of deep vein thrombosis and the subsequent inflammation of the valve cusps and vein wall during the process of recanalization. This leads to scarring and shortening of the cusps, which in turn results in failure of the cusps to achieve normal coaptation and the attendant reflux. Reflux secondary to the PTS is not amenable to direct surgical repair of the damaged or destroyed valve.

Primary reflux is the result of structural abnormalities in the vein wall and valve itself. Redundant, malopposed cusps and venous dilation permit valve prolapse and reflux. In primary reflux, unlike PTS, there is no evidence of previous thrombosis or inflammation near the valve. It is possible, however, to have both proximal primary reflux and distal PTS in the same leg. A rare cause of congenital reflux is the complete or partial absence of valves secondary to agenesis.

RATIONALE FOR REPAIR OF DEEP VENOUS VALVES

The goal of therapy is to correct the deep venous reflux (DVR) below the level of the inguinal ligament. Infrainguinal reflux leads to permanent venous hypertension, unmitigated by the activity of the calf venomuscular pump. It is important to note, however, that DVR frequently occurs in association with superficial and perforator reflux and that all these abnormalities must be addressed in order to correct the venous hypertension.

METHODS

Valve repair techniques for treating DVR can be classified into two groups: those that involve phlebotomy and those that do not involve phlebotomy.

TECHNIQUES WITH PHLEBOTOMY

Internal Valvuloplasty

Since the first procedure using a longitudinal phlebotomy (Figure 18-1) described by Kistner[11] in 1968, various procedures have been proposed.

Raju[12] advocated a supravalvular transverse venotomy, while Sottiurai[13] utilized a hybrid T-shaped, supravalvular incision (Figure 18-2). In 2002, Tripathi[14] proposed an internal trap-door valvuloplasty. In all cases the redundant valve cup is plicated to the vein wall using multiple interrupted or continuous 7-0 prolene reefing sutures. It has been estimated that plication of approximately 20% of the leaflet length should restore competence, although the best gauge remains visual inspection.

Venous Segment Transfer

The venous segment transfer procedure was designed for patients with a competent great saphenous or deep femoral vein valve in their proximal segment (Figure 18-3) and DVR in the femoro-popliteal axis related to post-thrombotic changes. The purpose of venous segment transfer is to

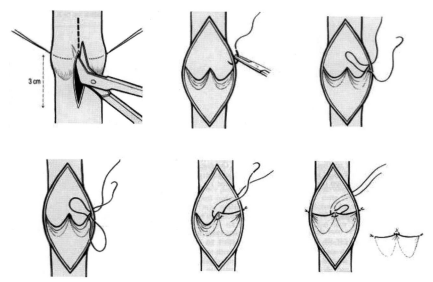

FIGURE 18-1 Internal valvuloplasty according to Kistner. Using a longitudinal phlebotomy, each valve is repaired by interrupting a series of sutures placed at the commissures. Each suture progressively shortens the leading edge of the cusp.

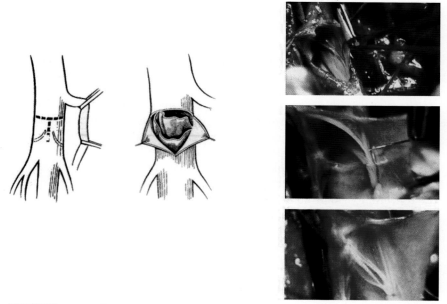

FIGURE 18-2 Internal valvuloplasty according to Sottiurai. After T-shaped phlebotomy, the valve cusp is repaired.

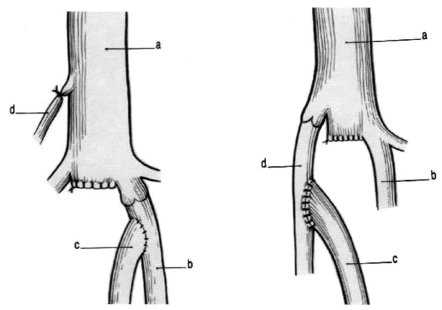

FIGURE 18-3 **A.** Venous segment transfer. The refluxive femoral vein (c) may be transposed into the great saphenous vein (d) or the profunda femoral vein (b) provided they have a competent proximal valve above the transposition. **B.** (a) common femoral vein; (b) profunda femoral vein; (c) femoral vein; (d) great saphenous vein.

transpose a competent valve-bearing venous segment into the axial deep venous system (i.e., the femoro-popliteal axis at the groin level). Several surgical variations of venous segment transfer have been employed using end-to-end or side-to-end anastomosis, depending on the competence of the various valves in the veins of the inguinal area.[15] Unfortunately, this technique can be used only in approximately 20% of patients with PTS, because the remaining 80% do not have a competent valve in the proximal segment of the great saphenous or deep femoral vein.

Vein Valve Transplantation

Figure 18-4 shows a vein valve transplantation. A 2–3-cm segment of axillary vein is inserted as an interposition graft at the termination of the femoral vein just below the junction of the deep femoral vein with the femoral vein or at the popliteal vein.

Neo Valve

Various techniques for creating a neo valve have been developed. Plagnol[16] constructed a neo bicuspid valve by invagination using the ter-

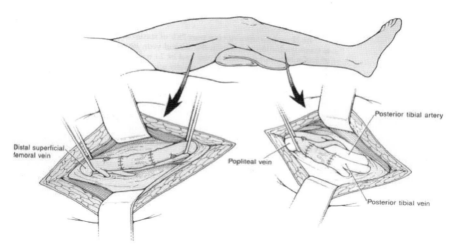

FIGURE 18-4 Valve transplant. Above-knee or below-knee axillary vein-to-popliteal vein transplantation end-to-end anastomosis.

mination of the great saphenous vein, while Maleti[17] created a valvular cusp by dissecting the femoral venous wall to obtain a single or a bicuspid valve (Figure 18-5). Both of these techniques have been used in PTS.

Allograft Cryopreserved Valve

Femoral vein allografts with competent valve(s) removed from qualified donors have been prepared, stored, and implanted in appropriate hosts.[18]

TECHNIQUES WITHOUT PHLEBOTOMY

Wrapping, Banding, Cuffing, and External Stenting

Wrapping, banding, cuffing, and external stenting techniques were initially developed to treat saphenous incompetence and later applied to the treatment of primary deep vein incompetence.[19,20]

External Valvuloplasty

The first step in external valvuloplasty consists in adventitial dissection until the valves' insertion lines are clearly identified as an inverted V. The commissural angle is normally acute but is widened in a refluxing valve.

Transmural Valvuloplasty

Kistner introduced external valvuloplasty in 1990.[21] In the transmural valvuloplasty technique, an external row of sutures is placed along the diverging margins of the valve cusp in the vein wall. Sutures for external repair are begun at each commissure on both sides of the vein. The inter-

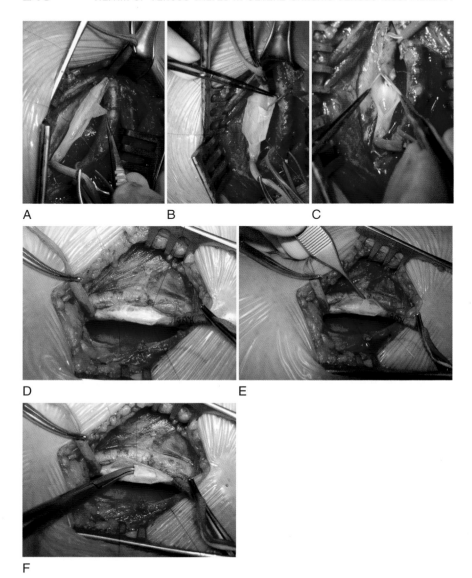

FIGURE 18-5 **A.** Neovalve construction in valve agenesia. Intimal flap is created. **B.** The flap is fixed in the opposite side. **C.** Removing proximal clamp the competence of neovalve is checked. **D.** Postthrombotic wall damage. **E.** After endophlebectomy the flap is created. **F.** The flap is fixed and neovalve is completed.

rupted sutures are carried inferiorly until the valve becomes competent by strip testing.

Transcommissural Valvuloplasty

Transcommissural valvuloplasty, developed by Raju, differs from transmural valvuloplasty by the use of a transluminal suture. "A through–and–through transluminal resuspension suture (7-0 Prolene) is placed obliquely across the inserted commissural V, traversing the valve cusps 'blindly' near their wall attachment to pull them up."[22]

Angioscopy-Assisted External Valve Repair

In angioscopy-assisted external valve repair,[23] an angioscope is introduced through a saphenous side branch and advanced into the proximal femoral vein and positioned directly above the valve. Blood is cleared from the operative field by isolation with vascular clamps and a heparinized solution is infused through the angioscope. Prolene sutures are passed from outside to inside the lumen, directed by video-enhanced, magnified angioscopic imaging, allowing for precise approximation of the valve cusps.

PERCUTANEOUSLY PLACED DEVICES

The percutaneously placed device, the Portland valve,[24] consists of a square stent and porcine small intestinal submucosa covering and is currently in Phase I clinical trial.

RESULTS

DVR surgery results are somewhat difficult to assess because superficial venous surgery and/or perforator surgery have often been performed in combination with DVR surgery.

PRIMARY DVR

In primary DVR, the most frequent procedure used is valvuloplasty. Results are summarized in Table 18-1.[8,14,22,25–29]

On the whole, valvuloplasty is credited with achieving a good result in 70% of cases in terms of clinical outcome, defined as a freedom of ulcer recurrence and the reduction of pain, valve competence, and hemodynamic improvement over a follow-up period of more than 5 years. In all series, good correlation was observed among these three criteria. External transmural valvuloplasty does not seem to be as reliable as internal valvuloplasty in providing long-term valve competence or ulcer-free survival.[26]

TABLE 18-1 Valvuloplasty Results

	Surgical Technique	Number of limbs (Number of Valves Repaired)	Etiology PVI/Total	Follow-up Months (Mean)	Ulcer Recurrence or Nonhealed Ulcer (%)	Hemodynamic Results	
						Competent Valve (%)	☐ AVP ■ VRT
ERIKSSON [28]	I	27	27/27	(49)	not specified	19/27 (70)	☐ ↗ 81% (av) ■ ↗ 50% (av)
MASUDA [25]	I	32	not specified	48–252 (127)	(28)	24/31 (77)	☐ ↗ 81% (av) ■ ↗ 56% (av)
PERRIN [8]	I	85 (94)	65/85	12–96 (58)	10/35 (29)	64/83 (77)	■ Normalized 63% (av)
RAJU [26]	I	68 (71)	not specified	12–144	16/68 (26)	30/71 (42)	not specified
RAJU [26]	TMEV	47 (111)	not specified	12–70	14/47 (30)	72/111	not specified
RAJU [22]	TCEV	141 (179)	98/141	1–42	(37)	(59)	☐ ↗ 15% (av) ↗ Normalized 100%
ROSALES [29]	TMEV	17 (40)	17/17	3–122 (60)	3/7 (43)	(52)	☐ ↗ 50% (av)
SOTTIURAI [27]	I	143	not specified	9–168 (81)	9/42 (21)	107/143 (75)	not specified
TRIPATHI [14]	I	90 (144)	96/118	(24)	(32)	(79.8)	not specified
	TMEV TCEV	12 (19)			(50)	(31.5)	not specified

Abbreviations:
PVI = Primary Venous Insufficiency
AVP = Ambulatory Venous Pressure
VRT = Venous Refill Time
av = Average
I = Internal Valvuloplasty
TMEV = Transmural External Valvuloplasty

TABLE 18-2 Banding, Cuffing, External Stent, and Wrapping Results

Author Material Used	Number of Extremities Treated (Number of Valves Repaired)	Site	Etiology PVI/Total	Follow-up Months (Average)	Ulcer Recurrence or Nonhealed Ulcer (%)	Competent Valve (%)	Hemodynamic Results □ AVP ■ VRT
Akesson (Venocuff I) [19]	20 (27)	F, P	7/20	5–32 (19)	2/10 (20) both PTS	PVI 7/7 (100) PTS 7/10 (4)	**PVI** □ ↗10% (av) ■ ↗10% (av) **PTS** □ ↗10% (av) ■ ↗100% (av)
Camilli (Dacron) [30]	54	F	54/54	4–63	not specified	41/54 (76)	not specified
Lane (Venocuff II) [20]	42 (125)	F, P	36/42	64–141 (93)	(20)	(90)	□ ↗? ■ ↗100 (av)
Raju (Dacron) [26]	28	F, P, T	not specified	12–134	6/22 (27)	60/72 (83)	not specified

Abbreviations:
F = Femoral
P = Popliteal
T = Tibial (Posterior)
PVI = Primary Venous Insufficiency
AVP = Ambulatory Venous Pressure
VRT = Venous Refill Time
av = Average
PTS = Post-Thrombotic Syndrome

TABLE 18-3 Transposition Results

	Number of Extremities Treated	Etiology PTS/Total	Follow-up in Month	Ulcer Recurrence or Nonhealed Ulcer (%)	Hemodynamic Results	
					Competent Valve (%)	☐ AVP ■ VRT
Cardon [31]	18	18/18	24–120	4/9 **(44)**	12/16 **(75)**	not specified
✩**Johnson [32]**	16	16/16	12	4/12 **(33)**	3/12 **(25)**	☐ Unchanged ■ Unchanged
Masuda [25]	14	**not specified**	48–252	7/14 **(50)**	10/13 **(77)**	☐ ↗70% (av) ■ ↗70% (av)
Perrin [8]	18	16/18	12–168	2/8 **(25)**	9/17 **(53)**	not specified
Sottiurai [27]	16	**not specified**	9–149	9/16 **(54)**	8/20 **(40)**	not specified

Abbreviations:
PTS = Post-Thrombotic Syndrome
AVP = Ambulatory Venous Pressure
VRT = Venous Refill Time
av = Average
✩ Transportation as isolated surgical procedure

TABLE 18-4 Transplantation Results

	Number of Extremities Treated	Site	Etiology PTS/Total	Follow-up in Month (Average)	Ulcer Recurrence or Nonhealed Ulcer (%)	Hemodynamic Results Competent Valve (%)	Hemodynamic Results □ AVP ■ VRT
Eriksson [28]	35	F, P	35/35	6–60	not specified	11/35 (31)	■ Unchanged
Mackiewicz [33]	18	F	not specified	43–69	5/14 (36)	not specified	■ Improved?
Nash [34]	25	P	25/25	not specified	3/17 (18)	18/23 (77)	□ ↗18% (av)
Bry [35]	15	P	not specified	15–132	3/14 (21)	7/8 (87)	□ Unchanged ■ Unchanged
Perrin [8]	32	F	31/32	12–124 (66)	9/22 (41)	8/32 (25)	■ ↗19% (av)
Raju [26]	83*	F, P, T	83/83	12–180	(40) 6 yrs	(38) 4 yrs	□ Unchanged
Raju [36]	82	not specified	77/82?	12–180	(25) 6 yrs	not specified	not specified
Sottiurai [27]	18	F, P	not specified	7–144	6/9 (67)	6/18 (33)	not specified
Taheri [37]	71	F, P	not specified	not specified	1/18 (6)	28/31 (90)	□ ↗15% (av)

Abbreviations:
□ AVP = Ambulatory Venous Pressure
■ VRT = Venous Refill Time
av = Average
↗ = Improved
F = Femoral
P = Popliteal
T = Tibial (Posterior)
PTS = Post-Thrombotic Syndrome
* Axillary vein transfer in trabeculated (poorly recanalized) vein

Only two series provide information regarding the outcome of patients who present with severe CVI without ulcer[8,29] and treated by valvuloplasty. No patient in either series developed an ulcer during the long-term follow-up, presumably due to successful prevention of disease progression by the operative intervention.

Other procedures used in primary deep reflux, including angioscopic repair and wrapping, are more difficult to assess because these series have a relatively short follow-up period, with the exception of Lane's series (Table 18-2).[19,20,26,30]

PTS

In PTS, long-term results are available for transposition (Table 18-3)[8,25,27,31,32] and transplantation (Table 18-4).[8,26–28,33–37]

With regard to clinical outcomes and valve competence, a meta-analysis demonstrates that a good result is achieved in 50% of cases over a follow-up period of more than 5 years, with a poor correlation between clinical and hemodynamic outcome.

Plagnol has reported favorable results at the very short follow-up in PTS.[16] Those reported by Maleti are more interesting as the follow-up is longer (mean 22 months). Sixteen out of 18 ulcers healed and none recurred.[2,38] The results with cryopreserved valves are poor.[18]

INDICATIONS

Indications for deep valve repair for reflux are based on clinical severity, hemodynamics, and imaging.

CLINICAL SEVERITY

Most authors recommend surgery in patients classified as C6 and, particularly, with recurrent ulcer. In addition, given the good results associated with valve repair for primary reflux, we feel that valve repair should be considered in young and active patients with severe edema or C4b findings, in order to avoid lifetime use of compression garments. When superficial and perforator reflux occur in association with DVR, they must be treated as well, either as the first step in therapy or in staged fashion. Contraindications to reconstructive surgery include uncorrectable hypercoagulable state or ineffective calf pump of any etiology.

HEMODYNAMICS AND IMAGING

Only reflux graded 3–4 based on Kistner's criteria[39] are usually appropriate for DVR surgery. It is generally recognized that, to be significantly abnormal, the values for venous refill time must be less than 12 seconds, and the difference between pressure at rest and after standardized exercise in the standing position must be less than 40%.

The decision to operate should be based on the clinical status of the patient, not the noninvasive data, since the patient's symptoms and signs may not correlate with the laboratory findings.[40] In addition to meeting the clinical criteria, patients selected for surgery should be highly motivated to participate in their recovery, since their ultimate success is dependent on their compliance with postoperative management.

INDICATIONS ACCORDING TO ETIOLOGY

The indications for surgery can be simplified according to the clinical, hemodynamic, and imaging criteria described previously.

In patients with primary reflux, reconstructive surgery is recommended after failure of conservative treatment and in young and active patients reluctant to wear permanent compression. Valvuloplasty is the most suitable technique, with Kistner, Perrin, Sottiurai, and Tripathi favoring internal valvuloplasty[8,14,25,27]; Raju, transcommissural external valvuloplasty[22]; and Rosales, transmural valuloplasty.[29] Some, but not all, authors recommend repairing several valves.[27,29]

In PTS, obstruction may be associated with reflux; most authors recommend that when significant obstruction is present in the iliac system, this obstruction must be treated first, before addressing the infrainguinal disease. Secondary deep venous reflux (PTS) should be treated only after failure of conservative therapy. Since the results achieved with subfascial endoscopic perforator surgery (SEPS) with or without superficial venous surgery are poor,[41] it is recommended that SEPS be carried out in combination with deep reconstructive surgery.

Given that valvuloplasty is rarely feasible, the order of preference for the currently available techniques is transposition, transplantation, neo valve, and cryopreserved allograft. Patients must be informed that for those with PTS, surgery for reflux has a relatively high failure rate.

CONCLUSIONS

Because large randomized control trials comparing conservative treatment and deep valve repair for DVR may become difficult to conduct, we must rely on the results of existing series of valve repairs. This analysis provides recommendation at grade 2b or 2c.

Better results are obtained in the treatment of primary reflux compared to secondary reflux. Even in the setting of primary reflux, however, surgery is not often indicated, and the procedure, as well as the decision to operate, should be performed in centers with extensive experience in this area.

REFERENCES

1. Eklöf B, Bergan JJ, Carpentier PH, et al. For the American Venous Forum's International Ad Hoc Committee for Revision of the CEAP Classification. Revision of the CEAP classification for chronic venous disorders. A consensus statement. *J Vasc Surg.* 2004;40:1248–1252.
2. Agus GB, Bavera PM, Mondani P, Santuari D. External valve-support for saphenofemoral junction incompetence: Randomized trial at 3 years follow-up. *Acta Phlebologica.* 2002;3:101–106.
3. Belcaro G, Errichi BM. Selective saphenous valve repair: A 5-year follow-up study. *Phlebology.* 1992;7:121–124.
4. Corcos L, De Anna D, Zamboni P, et al. Reparative surgery of valves in the treatment of superficial venous insufficiency. *J Mal Vasc.* 1997;22:128–136.
5. Geier B, Voigt I, Marpe B, et al. External valvuloplasty in the treatment of greater saphenous insufficiency: A five-year follow-up. *Phlebology.* 2003;18:137–142.
6. Lane RJ, Graiche JA, Coroneos JC, Cuzilla ML. Long term comparison of external valvular stenting and stripping of varicose veins. *Anz J Surg.* 2003;73:605–609.
7. Sakatowa H, Hoshino S, Igari T, Takase S, Ogawa T. Angioscopic external valvuloplasty in the treatment of varicose veins. *Phlebology.* 1997;12:136–141.
8. Perrin M. Reconstructive surgery for deep venous reflux. *Cardiovasc Surg.* 2000;8:246–255.
9. Raju S. Operative management of chronic venous insufficiency. In: Rutherford RB, ed. *Vascular Surgery.* Philadelphia: W.B. Saunders, 1995:1851–1862.
10. Kistner RL. Surgical repair of the incompetent femoral vein valve. *Arch Surg.* 1975;110:1336–1342.
11. Kistner RL. Surgical repair of a venous valve. *Straub Clin Proc.* 1968;24:41–43.
12. Raju S. Venous insufficiency of the lower limb and stasis ulceration. *Ann Surg.* 1983;197:688–697.
13. Sottiurai VS. Technique in direct venous valvuloplasty. *J Vasc Surg.* 1988;8:646–648.
14. Tripathi R, Sieunarine K, Abbas M, Durrani N. Deep venous valve reconstruction for non-healing leg ulcers: Techniques and results. *Anz J Surg.* 2004;74:34–39.
15. Kistner RL. Transpositions techniques. In: Bergan JJ, Kistner RL, eds. *Atlas of Venous Surgery.* Philadelphia: W.B. Saunders, 1992:153–156.
16. Plagnol P, Ciostek P, Grimaud JP, Prokopowicz SC. Autogenous valve reconstruction technique for post-thrombotic reflux. *Ann Vasc Surg.* 1999;13:339–342.
17. Maleti O. Venous valvular reconstruction in post-thrombotic syndrome. A new technique. *J Mal Vasc.* 2002;27:218.
18. Neglen P, Raju S. Venous reflux repair with cryopreserved vein valves. *J Vasc Surg.* 2003;37:552–557.
19. Akesson A, Risberg B, Bjôrgell O. External support valvuloplasty in the treatment of chronic deep vein incompetence of the legs. *Intern Angio.* 1999;18:233–238.
20. Lane RJ, Cuzilla ML, McMahon CG. Intermediate to long-term results of repairing incompetent multiple deep venous valves using external stenting. *Anz J Surg.* 2003;73:267–274.
21. Kistner RL. Surgical technique of external venous valve repair. *Straub Found Proc.* 1990;55:15–16.
22. Raju S, Berry MA, Neglen P. Transcommissural valvuloplasty: Technique and results. *J Vasc Surg.* 2000;32:969–976.
23. Gloviczki P, Merrel SW, Bower TC. Femoral vein valve repair under direct vision without venotomy: A modified technique with use of angioscopy. *J Vasc Surg.* 1991;14:645–648.

24. Pavcnik D, Uchida BT, Timmermans HA, et al. Percutaneous bioprosthetic venous valve: A long term study in sheep. *J Vasc Surg.* 2002;35:598–602.
25. Masuda EM, Kistner RL. Long term results of venous valve reconstruction: A 4 to 21 year follow-up. *J Vasc Surg.* 1994;19:391–403.
26. Raju S, Fredericks R, Neglen P, Bass JD. Durability of venous valve reconstruction for primary and postthrombotic reflux. *J Vasc Surg.* 1996;23:357–367.
27. Sottiurai VS. Current surgical approaches to venous hypertension and valvular reflux. *J Int Angiol.* 1996;5:49–54.
28. Eriksson I, Almgren B. Surgical reconstruction of incompetent deep vein valves. *Up J Med Sci.* 1988;93:139–143.
29. Rosales A, Slagsvold CE, Kroese AJ, Stranden E, Risum O, Jorgensen JJ. External venous valve pasty (5EVVP) on patients with primary chronic venous insufficiency (PCVI). *Eur J Vasc Endovasc Surg.* 2006;32:570–576.
30. Camilli S, Guarnera G. External banding valvuloplasty of the superficial femoral vein in the treatment of primary deep valvular incompetence. *Int Angiol.* 1994;13:218–222.
31. Cardon JM, Cardon A, Joyeux A, et al. (in French). La veine saphène interne comme transplant valvulé dans l'insuffisance veineuse post-thrombotique: Résultats à long terme. *Ann Chir Vasc.* 1999;13:284–289.
32. Johnson HD, Queral LA, Finn WR, et al. Late objective assessment of venous valve surgery. *Arch Surg.* 1981;116:1461–1466.
33. Mackiewicz Z, Molski S, Jundzill W, Stankiewicz W. Treatment of postphlebitic syndrome with valve transplantation: five year experience. Eurosurgery 1995. *Bologna Monduzzi.* 1995:305–310.
34. Nash T. Long-term results of vein valve transplants placed in the popliteal vein for intractable post-phlebitic venous ulcers and pre-ulcer skin changes. *J Cardiovasc Surg.* 1988;29:712–716.
35. Bry JD, Muto PA, O'Donnell TF, Isaacson LA. The clinical and hemodynamic results after axillary-to-popliteal valve transplantation. *J Vasc Surg.* 1995;21:110–119.
36. Raju S, Neglen P, Meydrech J, Doolittle EF. Axillary vein transfer in trabeculated postthrombotic veins. *J Vasc Surg.* 1999;29:1050–1064.
37. Taheri SA, Lazar L, Elias S. Status of vein valve transplant after 12 months. *Arch Surg.* 1982;117:1313–1317.
38. Maleti O, Lugli M. Neovalve construction in post-thrombotic syndrome. *J Vasc Surg.* 2006;43:794–799.
39. Kistner RL, Ferris RG, Randhawa G, Kamida CB. A method of performing descending venography. *J Vasc Surg.* 1986;4:464–468.
40. Iafrati MD, Welch H, O'Donnel TF, Belkin M, Umphrey S, MacLaughlin R. Correlation of venous non-invasive tests with the Society for Vascular Surgery clinical classification of chronic venous insufficiency. *J Vasc Surg.* 1994;19:1001–1007.
41. Gloviczki P, Bergan JJ, Menawat SS, et al. Safety, feasibility, and early efficacy of subfascial endoscopic perforator surgery. A preliminary report from the North American registry. *J Vasc Surg.* 1997;25:94–105.

SECTION IV

SPECIAL TOPICS

19

Treatment of Recalcitrant Venous Ulcers with Free Tissue Transfer for Limb Salvage

Stephen J. Kovach and L. Scott Levin

INTRODUCTION

Venous ulcers of the lower extremity are a common problem in clinical practice and are estimated to affect 4% of the population over the age of 65.[1] This represents a significant number of patients who must be dealt with clinically. The majority of patients will be treated conservatively, but patients with large areas of ulceration or those who have ulcers that are resistant to conservative means represent a challenging clinical problem. For many patients, recurrent and recalcitrant venous ulceration may be a limb-threatening condition.

The underlying pathophysiologic state that results in venous ulceration is that of chronic venous hypertension.[2] Initial treatment of the patient with chronic venous ulceration is aimed at correction of this chronic hypertension through nonsurgical and surgical means. This treatment is accomplished through well-recognized techniques, including compression garments, venous stripping, and subfascial endoscopic perforator surgery (SEPS). All patients will have had some attempts at skin graft coverage of their ulcers prior to consideration of free tissue transfer, and some may have had attempted local flap coverage. Patients become candidates for free flap coverage of recalcitrant venous ulcers only after exhausting these more conservative means for healing their ulcers and achieving durable soft tissue coverage.

Chronic venous ulcers typically have poor wound beds for the imbibition and vascular ingrowth needed to support skin grafting. The dermal microcirculation has been shown to have pathologic changes that are not conducive to the healing of skin grafts.[3] Also, simple skin grafting of ulcers is usually doomed to failure because nothing has been done to alter the underlying pathology of venous hypertension and subsequent fibrosis of the surrounding tissues. This surrounding lipodermatosclerosis (LDS) is fibrotic tissue with a tenuous blood supply that must be considered as pathologic and prone to ulceration and poor wound healing.[4] Any surgical intervention to achieve coverage of existing ulcers or prevent the formation of new ones must take the area of LDS into consideration (Figures 19-1A and 19-1B).

Local flaps are limited in the distal third of the leg where the majority of venous ulceration occurs. The local tissue available is typically involved in the disease process and is sclerotic, which limits its mobility and utility. Vascularized tissue is needed to cover the ulcer that is independent of the wound bed or local area for its vascular supply. Free tissue transfer with its own independent arterial supply and venous outflow is a means to achieve this. However, the patient referred for free tissue transfer universally has had multiple prior surgeries and attempted skin grafting without lasting resolution of the chronic wounds and has come to the option of free tissue transfer as a "last resort."

Free tissue transfer has been proven a valuable means of achieving soft tissue coverage for wounds of the lower extremity.[5] Successful free tissue transfer rates are on the order of 96%[6] but may be somewhat less in the lower extremity, particularly after trauma. If one looks at failure rates of free flaps, venous thrombosis is the leading etiology, both clinically and experimentally.[7] Underlying venous pathology in the extremity is therefore obviously concerning, but free tissue transfer to the lower extremity in the setting of chronic venous disease for coverage of recalcitrant ulcerations has been reported to be successful ranging from 77% to 100% of the time[8-16] (see Table 19-1). The literature that comprises free tissue transfer for chronic, nonhealing venous wounds is spartan, with few surgeons having extensive experience. The surgical technique of free tissue transfer for coverage of venous ulcers began as limited case reports[8,13,16] and now has been borne out to be a clinically effective means in smaller case series in properly selected patients.[9-12,15]

PATIENT EVALUATION AND SELECTION

The crucial first step in evaluating the patient with the recalcitrant venous ulcer as a candidate for free tissue transfer is to make sure that venous insufficiency is the true underlying etiology. Venous disease is the

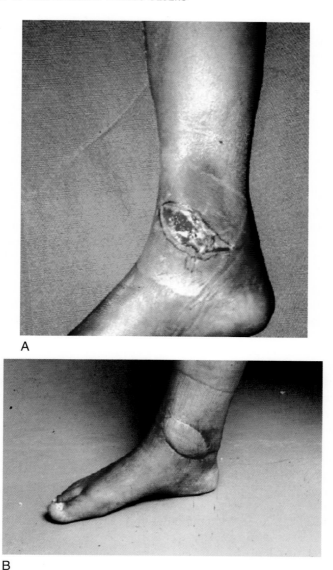

FIGURE 19-1 **A.** Typical-appearing venous ulcer on medial aspect of lower third of extremity. Note the surrounding area of LDS that must be considered as part of the pathologic process of venous ulceration. **B.** Appearance of the venous ulcer following excision of ulcer surrounding area of LDS and coverage with a fasciocutaneous (radial forearm) free flap.

TABLE 19-1 Published Series of Free Tissue Transfer for Recalcitrant Venous Ulcers*

Series	Pts/Flaps	Flap Type	Flap Failure	Recurrence: F/U
Allen (1989)	2/2	1 muscle 1 fasciocutaneous	0%	None: 8 Mos None: 18 Mos
Swartz (1989)	7/7	Muscle	0%	None: 4 Yrs
Ramirez (1992)	13/13	Muscle	23%	None: 1–7 Yrs
Dunn (1994)	6/8	Fasciocutaneous	0%	None: Mean 24 Mos
Weinzweig (1997)	1/1	Muscle	0%	None
Weinzweig (1997)	18/20	Muscle	10%	1 separate site: 32 mos
Steffe (1998)	11/14	Muscle/fascial	14%	100%
Kumins (2000)	22/25	Muscle/omentum	4%	12% separate site
Isenberg (2001)	9/9	Fasciocutaneous	0%	33% separate site

* Separate site equates to ulcer recurrence outside the flap.

most common cause of lower extremity ulcers, but multiple other pathologies must be ruled out prior to undertaking free tissue transfer. One must be sure that the patient does not have arterial insufficiency, infectious ulceration, diabetic ulcer, underlying vasculitis, or clotting diathesis.[17] As outlined in previous chapters in this text, patients should receive a complete evaluation of the patency of their superficial and deep venous systems with noninvasive duplex scanning to confirm the underlying pathophysiology of venous hypertension. The underlying venous hypertension must be addressed, and most patients have undergone a regimen of elevation, unna boots, and local wound care without success.[18,19] These regimens may aid in alleviating the symptomology of their underlying disease, but they do nothing to correct the pathophysiology. Many patients have additionally undergone intervention to correct their venous hypertension through other means such as stripping or SEPS. Once venous insufficiency has been confirmed as the underlying pathology, and the patient has failed conservative and lesser surgical attempts at closure, one can then proceed to discuss the surgical option of free tissue transfer.

Patients who are referred as candidates for free tissue transfer have exhausted the more traditional methods of surgical intervention after prolonged and lengthy periods of conservative management. Essentially, all patients will have undergone multiple prior attempts at closure of their wounds, typically represented as failed excision and skin grafting or as failed regional flaps. Because the majority of venous ulcers occur in the distal one-third of the extremity, local options for rotational muscle or fasciocutaneous flaps are very limited and impractical in the majority of

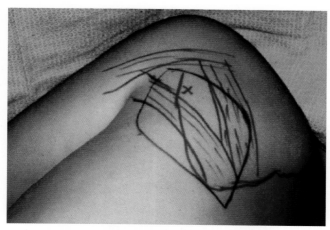

FIGURE 19-2 Outline of a typical scapular flap area prior to harvest. The scapular flap is a fasciocutaneous flap that has been successfully used as a free flap for coverage of recalcitrant venous ulcers.

cases. For this reason, their limited arc of rotation, and the involvement of the surrounding soft tissue with LDS, they are not that useful. Additionally, in ulcers that have exposed tendons or other vital structures, skin grafts or local flaps are not able to achieve adequate soft tissue coverage. The only surgical means for achieving soft tissue coverage after multiple failed attempts is microsurgical free tissue transfer.

Ideal patients who will be candidates for free tissue transfer are those without prohibitive underlying comorbidities that would allow them to undergo a significant surgical undertaking. Patients must have suitable donor vessels for both arterial inflow and venous outflow for the flap. Additionally, the area of ulceration to be covered must be of a size that will allow coverage with a free flap, and they must have a suitable source of donor tissue. The microsurgical flaps that are typically employed are the latissimus dorsi, gracilis, and scapular flap (Figure 19-2). All these flaps have a finite area of coverage, and the largest area that can be covered with a single flap is in the range of 200 cm^2. The area that must be covered is typically underestimated. The visible area of ulceration may only be the "tip of the iceberg" because there is typically a large area of LDS surrounding the ulcer that must be considered part of the underlying pathologic problem that needs to be debrided widely[1] (Figures 19-3 and 19-4). It has been shown that there are multiple derangements within the venous microcirculation, as well as an increased number of leukocytes within the dermis of patients with venous ulcers. The result of these derangements is inflammation with fibrosis in the surrounding tissues and friable tissue prone to

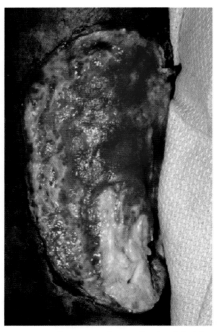

FIGURE 19-3 Appearance of venous ulcer prior to debridement. All areas of LDS must be debrided if at all possible. Debridement should include the underlying fascia, and all ectatic veins that are encountered should be ligated.

bleeding and ulceration with minor trauma.[20-25] For these reasons, the lipodermosclerotic areas must be widely debrided; otherwise, one will face the problem of recurrent ulceration.

Special mention should be made regarding the patient with arterial insufficiency in the setting of concomitant venous ulceration. Free tissue transfer for extremity salvage in the setting of arterial insufficiency has been shown to be successful, and any patient with a combined ulcer should undergo arterial revascularization with comcomitant or subsequent free tissue transfer if needed to reconstruct soft tissue deficits that would otherwise dictate limb loss. Free tissue transfer in the setting of arterial insufficiency has been shown to be safe and effective, albeit with a higher risk of flap loss and complications.[26-29]

FLAP SELECTION

The selection of the appropriate flap for microvascular free tissue transfer is dependent on several variables. One must consider the size of the

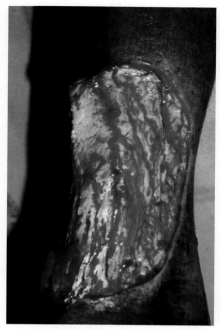

FIGURE 19-4 Appearance of venous ulcer in Figure 19-3 after radical debridement.

recalcitrant ulcer and the surrounding area of LDS in order to appreciate the dimensions of the eventual wound that will have to be covered. It is crucial to take into account the area that will require coverage upon excision of the lipodermosclerotic area. Failure to do this will result in a flap that will not adequately cover the defect. Flaps are classified based on their anatomic components contained within the flap (i.e., muscle, fascial, musculocutaneous, fasciocutaneous).[30] Flaps used in coverage of venous ulcers are typically of two types: fasciocutaneous or muscle. Both of these types of flaps have been shown to be effective in achieving durable coverage of recalcitrant ulcers, and one does not necessarily have an advantage over the other regarding efficacy or recurrence (Figures 19-5A and 19-5B). Both have been shown to possess competent venous valves that may help to restore the appropriate hemodynamic balance.[31,32] Common donor sites for fasciocutaneous flaps include the radial forearm, scapular and parascapular region, and the antero-lateral thigh. Muscle flaps typically will be gracilis or latissimus muscles. The donor site morbidity of the free flap must be considered, as should the dimensions that are obtainable with each flap. Ease of harvesting of the flap must be taken into account and the avoidance, if possible, of having to reposition the patient to harvest the flap while in the operating room.

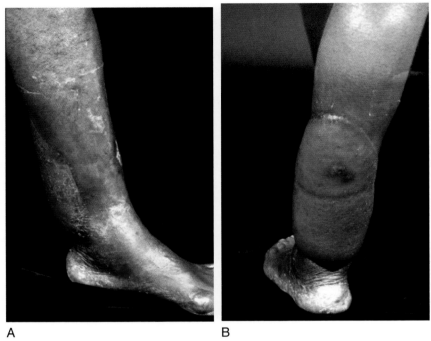

A B

FIGURE 19-5 **A.** Long-term follow-up of patient in Figures 19-3 and 19-4 after undergoing free myocutaneous latissimus dorsi free flap. **B.** The patient has remained ulcer free with stable soft tissue coverage.

SURGICAL TECHNIQUE

The patient is positioned on the table to allow simultaneous excision of the ulcer and surrounding area of LDS and harvesting of the flap. The ulcer and LDS are excised under tourniquet control down to and including the deep fascia. Any veins within the area of excision are ligated and/or divided. If the area of proposed excision exceeds the dimensions of the flap, areas of LDS potentially may be left intact with the understanding that these areas will likely require additional surgical intervention in the future. The donor vessels, typically the posterior tibial or anterior tibial vessels, are dissected free for a suitable length (3 to 5 cm) to allow for microvascular anastamoses. The anterior vessels are chosen if the defect is anteriorly based, and the posterior tibial vessels if medially or posteriorly based. The peroneal artery is used only in the setting of unsuitable anterior or posterior tibial vessels because exposure and the anastamosis are technically more challenging. A heparin bolus is given according to the patient's weight (100 U/kg) prior to division of the pedicle. The microvascular anastamoses are then performed with the use of the operating microscope.

The arterial anastamosis is performed in an end-to-end or end-to-side fashion with interrupted 9-0 or 10-0 nylon, depending on the vascular supply to the distal extremity. The venous anastomosis is performed in an end-to-end fashion to the venous comitantes. If the venous comitantes are deemed inadequate, a suitable vein graft is harvested from the contralateral saphenous vein or cephalic vein for an interposition graft to the popliteal vein. The flap is inset with absorbable sutures to the surrounding skin under a moderate amount of stretch. If a muscle flap is employed, it is skin grafted with a split thickness graft that is meshed at a ratio of 1.5 to 1. The patient is then placed in a loose-fitting splint with a window for monitoring the flap clinically.

POSTOPERATIVE CARE

It is our policy to use the Cook-Swartz implantable Doppler probes for postoperative monitoring of our free flaps (Cook Medical, Bloomington, IN, USA). The probes are left in place for 4 to 5 days and then removed, and the patient is allowed to sit in a chair with elevation of the leg. Elevation of the extremity is continued in the postoperative period, and limited ambulation is allowed 2 weeks postoperatively. Prolonged standing is not allowed for 4 to 6 weeks. Routine skin graft care is employed if a muscle flap and skin graft were used. Compression garments are encouraged once the flap is healed.

DISCUSSION

Allen et al. reported the first successful free flap coverage in two patients with intractable venous ulceration in 1989. Both patients had previously failed skin grafts. One patient underwent an ulnar artery fasciocutaneous flap, and the other a latissimus muscle flap with good results and no recurrence at 18 and 8 months follow-up, respectively.[8] Swartz et al.[13] were the first to publish a larger series of patients. Seven patients with post-traumatic ulcers underwent muscle free flaps with no recurrences at 4 years. Ramirez[16] performed 13 free muscle flaps in 12 patients of which 10 were successful. There were no recurrences in the successful free flaps at a follow-up of 1 to 7 years.

Dunn et al. reported their series of eight flaps in six patients over a 7-year period.[9] Two patients underwent bilateral flaps in separate settings. There was no flap loss and no recurrence of ulcers on the flaps or on the surrounding tissue at a mean follow-up of 24 months, with only minor wound healing issues in two patients. Pathologic examination of the excised areas of LDS revealed collagenization of the entire dermis, capillary proliferation, and deposition of fibrin around the capillaries, confirming the pathologic nature of the surrounding areas of LDS. One of the principles of venous hypertension is that of incompetent venous valves in the lower extremity that allows the hydrostatic pressure from a column of blood to be transmitted to the superficial veins. Interestingly, Aharinejad and Dunn[31] confirmed the presence of competent venous valves in scapular fasciocutaneous flaps in the small veins, with most of the valves in veins less than 500 microns in diameter. This was manifested clinically in Dunn's study by increased filling times in five of the eight patients who underwent photoplethysmography (PPG) with a significant improvement in venous competency in the previously ulcerated areas. This correlates to the introduction of small, competent venous valves contained within the flap that helps to restore venous competency to areas of ulceration and LDS. Biopsy of the flap during the follow-up period revealed no changes consistent with venous insufficiency. Competent venous valves have also been demonstrated in the dominant vascular pedicles of the major muscle groups that may be used as free flaps.[32]

Weinzweig et al. reported their results with 20 free muscle flaps in 18 patients for clinical class 6 ulcers.[15] As is typical in this patient population, the patients in this study had ulceration for average of 3.5 years and had undergone an average of 2.4 skin grafts. The two flap failures were attributed to intractable vasospasm. All patients underwent muscle flaps, in contrast to Dunn et al., with the rectus abdominis being the most common, followed by the latissimus dorsi, gracilis, and serratus. There were no in-flap recurrences at an average follow-up of 32.7 months. Five of the 20 patients had partial skin graft loss.

Steffe et al.[12] performed 14 free tissue transfers to the lower extremities of 11 patients with recalcitrant venous ulcers over a 10-year period. All patients were class 6 chronic venous insufficiency according to the International Consensus Committee on Chronic Venous Disease.[33,34] Wounds were present for an average of 5 years prior to free flap coverage and were post-thrombotic, traumatic, and idiopathic. All patients had normal arteriograms but did not undergo duplex scanning for venous reflux and had undergone multiple surgical procedures. Eleven patients had muscle flaps, and two had fascial flaps. Two flaps were lost secondary to venous thrombosis, and two patients had partial flap loss. What is striking regarding this study is the development of new ulcers in 100% of patients appearing 1 to 72 months from the time of reconstruction at the flap margins. Only two of these recurrences resolved with nonoperative treatment. All these patients required skin grafting, with three patients needing amputations. The reason for the poor outcome in this study in comparison to others is not clear. It may be secondary to incomplete excision of involved lipodermatosclerotic areas.

Kumins et al.[11] reviewed their experience with 25 free flaps (23 muscle, two omentum in the same patient with bilateral ulcers) in 22 patients for nontraumatic, nonosteomyelitic ulcers. All patients failed grafting, and 46% had exposed tendon, bone, or joint and had large areas of ulceration (average 237 cm^2). All patients underwent muscle flaps, with only one flap loss secondary to vasospasm. Seven patients lost areas of flap or graft, three of whom required additional procedures. In the successful free flap patients, no ulceration was noted in areas covered by the flap at a mean follow-up of 58 months. However, three patients had new ulcers arise in the affected extremity after 6 to 77 months in new areas. The unique aspect of this study was the financial analysis performed on these patients for the total cost of their hospitalization. Total charges for patients without complications averaged $30,428. This is quite an investment, but when one compares it to continued wound care punctuated with smaller operative procedures along the way, the costs may be comparable.

Isenberg et al.[10] published his series of nine patients with documented venous insufficiency by duplex scanning. All patients had multiple failed attempts at prior closure of their wounds and underwent coverage with a radial forearm (fasciocutaneous) flap. There was no complete or partial flap loss, but two patients had delayed healing at flap-skin interface for an overall complete wound healing rate of 78%. Additionally, three patients developed new ulceration on the affected extremity outside the area that had been reconstructed with the flap.

The data presented here represent the scope of free tissue transfer for recurrent venous ulceration that has been published. Admittedly, it is a small number of patients in multiple case series. At present, there are still several unanswered questions. First among them is the lack of long-term follow-up data and the longevity of these free flaps. Most of the studies have very short follow-up periods so long-term outcome is impossible to glean from the data. Second, it has been anecdotally stated that fasciocutaneous flaps are preferred to muscle flaps, but there is no good data to make this assumption, and given the paucity of patients with this clinical problem, it is unlikely that a randomized trial to control for confounding variables will be done. Fasciocutaneous flaps may have a slight advantage in that they allow flap harvest without sacrificing any functional muscle.

What can be stated with assurance is that free tissue transfer has become routine in most tertiary centers and that patients with recurrent venous ulceration can achieve coverage of their ulcers. The longevity of the coverage remains uncertain, however.

CONCLUSIONS

Microvascular free tissue transfer has been shown to be a valuable tool for achieving soft tissue coverage of recalcitrant venous ulcers. Patients who have failed to heal their ulcers after trials of conservative therapy, correction of venous hemodynamics, and skin grafting become candidates for free tissue transfer. There have been a total of 89 patients reported in the literature after having 99 free flaps for recalcitrant venous ulceration. The overall success rate is 94%. This number is comparable to accepted rates of success of free tissue transfer in other anatomic sites and for lower extremity reconstruction or limb salvage for other reasons. For some patients with recurrent venous ulceration and exposed vital structures, free tissue transfer and coverage may, in fact, represent limb salvage. The alternative would be amputation. For others, it represents a culmination of failure of all other prior efforts, and free tissue transfer has been deemed a last resort.

The soft tissue of the flap invariably heals well without recurrent ulceration, but the remainder of the leg is prone to recurrent or *de novo* ulceration. This is demonstrated in the published data and our own experience that the transferred flap does not develop new ulceration, but all recurrent ulcerations occur in the surrounding area. This lends credence to the body of evidence that points to the surrounding LDS as a pathologic area that is prone to ulceration. Free tissue transfer is not a small undertaking and certainly small, uncomplicated venous ulcers are still best treated by conservative means. However, for patients with areas of venous ulceration that have not healed despite maximal medical treatment and adjunctive procedures, free tissue transfer is a viable and valuable alternative in achieving soft tissue coverage of these ulcers.

REFERENCES

1. Lawrence PF, Gazak GC. Epidemiology of chronic venous insufficiency. In: Ballard J, Bergan JJ, eds. *Chronic Venous Insufficiency*. London: Springer-Verlag, 2000:3–9.
2. Burnand KG. The physiology and hemodynamics of chronic venous insufficiency of the lower limb. In: Gloviczki P, Yao JST, eds. *Handbook of Venous Disorders*, 2nd ed. London: Arnold, 2001:49–58.
3. Pappas P, Duran WN, Hobson RW. Pathology and cellular physiology of chronic venous insufficiency. In: Gloviczki P, Yao JST, eds., *Handbook of Venous Disorders*, 2nd ed. London: Arnold, 2001:58–67.
4. Ballard J, Bergan JJ. Sparks S. Pathology of chronic venous insufficiency. In: Ballard J, Bergan JJ, eds. *Chronic Venous Insufficiency,* 1st ed. London: Springer-Verlag, 2000:17–24.
5. Heller L, Levin LS. Lower extremity microsurgical reconstruction. *Plast Reconstr Surg.* 2001;108(4):1029–1041; quiz, 1042.
6. Khouri RK, Cooley BC, Kunselman AR, et al. A prospective study of microvascular free-flap surgery and outcome. *Plast Reconstr Surg.* 1998;102(3):711–721.
7. Kroll SS, Schusterman MA, Reece GP, et al. Timing of pedicle thrombosis and flap loss after free-tissue transfer. *Plast Reconstr Surg.* 1996;98(7):1230–1233.
8. Allen RJ, Celentano R, Dupin C, Crais TF. Management of chronic venous insufficiency ulcers with free flaps. *Wounds.* 1989;1(3):193–197.

9. Dunn RM, Fudem GM, Walton RL, et al. Free flap valvular transplantation for refractory venous ulceration. *J Vasc Surg.* 1994;19(3):525–531.
10. Isenberg JS. Additional follow-up with microvascular transfer in the treatment of chronic venous stasis ulcers. *J Reconstr Microsurg.* 2001;17(8):603–605.
11. Kumins NH, Weinzweig N, Schuler JJ. Free tissue transfer provides durable treatment for large nonhealing venous ulcers. *J Vasc Surg.* 2000;32(5):848–854.
12. Steffe TJ, Caffee HH. Long-term results following free tissue transfer for venous stasis ulcers. *Ann Plast Surg.* 1998;41(2):131–137; discussion, 138–139.
13. Swartz WM. Free tissue transfers for intractable chronic venous ulcerations: A long term evaluation. In: Proceedings of the Annual Meeting of the American Association of Plastic Surgeons; 1989; Scottsdale, AZ.
14. Weinzweig N, Schlechter B, Baraniewski H, et al. Lower-limb salvage in a patient with recalcitrant venous ulcerations. *J Reconstr Microsurg.* 1997;13(6):431–437.
15. Weinzweig N, Schuler J. Free tissue transfer in treatment of the recalcitrant chronic venous ulcer. *Ann Plast Surg.* 1997;38(6):611–619.
16. Ramirez OM. The effectiveness of the free muscle flap in the treatment of the recalcitrant venous statis ulceration. *Plast Surg Forum.* 1992;15:77–78.
17. Patel NP, Labropoulos N, Pappas PJ. Current management of venous ulceration. *Plast Reconstr Surg.* 2006;117(7 Suppl):254S–260S.
18. Kitahama A, Elliot LF, Kerstein MD, et al. Leg ulcer. Conservative management or surgical treatment? *JAMA.* 1982;247(2):197–199.
19. Raju S. Venous insufficiency of the lower limb and stasis ulceration. Changing concepts and management. *Ann Surg.* 1983;197(6): 688–697.
20. Junger M, Steins A, Hahn M, et al. Microcirculatory dysfunction in chronic venous insufficiency (CVI). *Microcirculation.* 2000;7(6 Pt 2):S3–12.
21. Leu HJ. Morphology of chronic venous insufficiency—Light and electron microscopic examinations. *Vasa.* 1991;20(4):330–342.
22. Mani R, White JE, Barrett DF, et al. Tissue oxygenation, venous ulcers and fibrin cuffs. *J R Soc Med.* 1989;82(6):345–346.
23. Pappas PJ, Fallek SR, Garcia A, et al. Role of leukocyte activation in patients with venous stasis ulcers. *J Surg Res.* 1995;59(5):553–559.
24. Peus D, Heit JA, Pittelkow MR. Activated protein C resistance caused by factor V gene mutation: Common coagulation defect in chronic venous leg ulcers? *J Am Acad Dermatol.* 1997;36(4):616–620.
25. Thomas PR, Nash GB, Dormandy JA. White cell accumulation in dependent legs of patients with venous hypertension: A possible mechanism for trophic changes in the skin. *Br Med J (Clin Res Ed).* 1988;296(6638):1693–1695.
26. Illig KA, Moran S, Serletti J, et al. Combined free tissue transfer and infrainguinal bypass graft: An alternative to major amputation in selected patients. *J Vasc Surg.* 2001;33(1):17–23.
27. Moran SL, Illig KA, Green RM, et al. Free-tissue transfer in patients with peripheral vascular disease: A 10-year experience. *Plast Reconstr Surg.* 2002;109(3):999–1006.
28. Serletti JM, et al. Atherosclerosis of the lower extremity and free-tissue reconstruction for limb salvage. *Plast Reconstr Surg.* 1995;96(5):1136–1144.
29. Serletti JM, Deuber MA, Guidera PM, et al. Extension of limb salvage by combined vascular reconstruction and adjunctive free-tissue transfer. *J Vasc Surg.* 1993;18(6):972–978; discussion, 978–980.
30. Serafin D. *Atlas of Microsurgical Composite Tissue Transplantation*, 1st ed. Philadelphia: W.B. Saunders Company, 1996.
31. Aharinejad S, Dunn M, Nourani F, et al. Morphological and clinical aspects of scapular fasciocutaneous free flap transfer for treatment of venous insufficiency in the lower extremity. *Clin Anat.* 1998;11(1):38–46.

32. Watterson PA, Taylor GI, Crock JG. The venous territories of muscles: Anatomical study and clinical implications. *Br J Plast Surg.* 1988;41(6):569–585.
33. Beebe HG, Bergan JJ, Bergqvist D, et al. Classification and grading of chronic venous disease in the lower limbs. A consensus statement. *Eur J Vasc Endovasc Surg.* 1996;12(4):487–491; discussion, 491–492.
34. Porter JM, Moneta GL. Reporting standards in venous disease: An update. International Consensus Committee on Chronic Venous Disease. *J Vasc Surg.* 1995;21(4):635–645.

20

THE DIAGNOSIS AND TREATMENT OF MAJOR VENOUS OBSTRUCTION IN CHRONIC VENOUS INSUFFICIENCY

PATRICIA ELLEN THORPE AND
FRANCISCO JOSÉ OSSE

INTRODUCTION

Although the most common cause of chronic venous insufficiency is primary valvular reflux in the superficial venous system, deep venous obstruction is associated with 1% to 6% of patients with ulcers; a combination of obstruction and reflux is reported in approximately 17% of limbs with stasis ulceration.[1] This means that, in the United States, almost 250,000 patients have venous obstruction as a factor in the pathophysiology of ulcer development. The prevalence of venous ulcers in the United States is estimated to be between 0.5% and 1%, and about one-third have a documented history of deep vein thrombosis (DVT).[2] From these data, we can estimate that between 500,000 and 1.5 million venous ulcer patients have had DVT. Venous obstruction, secondary reflux, or both can play a role in development of chronic venous insufficiency. Residual thrombus, with inadequate recanalization and poor collaterals, contributes to development of the edema, pain, and skin changes that comprise the post-thrombotic syndrome (PTS). Whereas obstruction alone can lead to ulceration, a combination of obstruction and reflux is associated with a worse clinical prognosis and greater risk of ulcer formation.[3]

Compression has long been the principal therapy directed at prevention of PTS after deep vein thrombosis (DVT). Pesavento et al. showed that adequate compression can reduce the incidence of post-thrombotic syndrome by 50%.[4] But, as we discover in longitudinal studies, there must be multiple factors that determine whether or not a patient will develop progressive venous hypertension associated with thrombosis. Meissner et al. suggested that PTS determinants include the rate of recanalization, the presence of popliteal obstruction, and the extent of reflux.[5] Clearly, prevention of ipsilateral recurrent DVT has been shown to decrease development of PTS.[4] But many patients have multiple thrombotic episodes. Some DVT episodes are not even diagnosed or treated. The persistence of residual venous obstruction after an episode of DVT has been underestimated as a risk factor in recurrent DVT. Galli et al. found that residual obstruction could be documented in 88% of patients with recurrent DVT compared to 50% of patients with a history of only a single episode.[6] Recurrence often occurs after cessation of anticoagulation, since the existence of residual obstruction, following a course of oral anticoagulation, is not considered a risk factor. But it is. In our culture, venous disease is conveniently covered up with clothing and tolerated by individuals who think poor circulation is inevitable with advancing age. The clinical impact of PTS is becoming more apparent to physicians and the whole medical community as we cope with the growing number of patients with chronic venous disease.[7] Patients with multisegmental thrombosis often fail to improve with standard therapy. A summary of longitudinal studies, with 3- to 10-year follow-up, indicated that 20% to 50% of patients develop PTS within 1 to 2 years of the initial DVT episode.[8] Therefore, the prevalence of chronic venous insufficiency is probably far greater than we have been able to document.[9]

Most physicians are reluctant to treat chronic venous obstruction with anything other than compression.[10] Compression is an important component of therapy but does not address the underlying causes of venous hypertension. Surgical reconstructions for chronic occlusion are rarely performed due to invasiveness and relatively suboptimal long-term clinical results.[11] In 1986, it was recommended that direct venous reconstruction be accompanied by perforator interruption and removal of superficial varicosities.[12] Two decades later, the therapeutic goal remains the same, but the objectives are met with endovascular techniques, rather than traditional surgery. Iliac stenting, percutaneous saphenous ablation, and ultrasound-guided foam sclerotherapy of perforators are available alternatives to bypass surgery, vein stripping, and perforator ligation.[13]

Beginning in the 1990s, catheter-directed thrombolysis, with adjunctive stenting, was reported for treatment of acute iliofemoral thrombosis.[14] Left iliac or May-Thurner compression, often recognized as a chronic lesion underlying iliac thrombosis, was reported to be effectively treated with thrombolysis, mechanical thrombectomy, and self-expanding metallic stents.[15-19] Patient selection has favored acute clinical presentation with thrombus less than 7–10 days old.[20] Complete lysis is possible with acute thrombus, as demonstrated by data in the National Venous Registry.[21] Incomplete lysis may require stenting because balloon angioplasty is not very effective.[15] Thrombolytic agents are thought to soften chronic thrombus and thereby facilitate wire passage through the fenestrations and along the vein wall.[17] In 2006, thrombolysis for DVT remained an off-label use of fibrinolytic agents, despite successful clinical applications.[22] Mechanical thrombectomy devices have proliferated and are, in fact, useful in debulking acute thrombus in iliofemoral segments. Organized thrombus is still not effectively removed by these devices.

Despite advances in endovascular therapy, the standard care for deep vein thrombosis remains heparin (albeit, low-molecular weight) and warfarin. In general, little thought is given to prevention of the post-thrombotic syndrome by removal of acute thrombus and any underlying stenosis. Even though it has been shown that patient quality of life improves after percutaneous thrombolysis and stenting in lower extremity venous occlusion,[23] concerns about complications with thrombolysis and questions about long-term patency of stents prevent greater application of these tools in treating acute or chronic DVT. Admittedly, after nearly 2 decades of clinical experience, Level I data are still lacking regarding the use of thrombolysis and stenting as a primary treatment for iliofemoral thrombosis. Subsequently, although endovascular reconstruction of chronic large vein obstruction is a new therapeutic option for patients in whom the residual thrombosis limits outflow and causes significant venous hypertension, relatively few physicians perform these procedures. Still, patients with chronic venous insufficiency, secondary to obstruction, are likely to seek therapeutic intervention as they discover that compression stockings do not alleviate their post-thrombotic disability and surgical options are not recommended. Fortunately, the technical and clinical results of endovascular venous reconstruction in cases of chronic venous obstruction are encouraging.[24–28]

DIAGNOSIS AND EVALUATION

CLINICAL PRESENTATION

In many individuals, an obstructing lesion at the caval or iliac level may go unnoticed prior to a thrombotic complication (Figure 20-1). The degree of their venous insufficiency can be mild and consist of obvious, but not incapacitating, unilateral or bilateral edema. In the absence of edema, patients may present with early appearance of varicose discomfort in their youth. Others may have venous claudication indicated by unilateral thigh or inguinal discomfort with exercise or walking. The incidence of

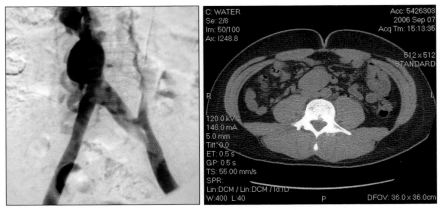

FIGURE 20-1 Example of chronic large vessel obstruction with minimal clinical sequella. Phlebogram of the pelvis and distal inferior vena cava (IVC) demonstrate a large venous aneurysm that appears to arise from the proximal right common iliac or the right side of the bifurcation. The etiology of the lesion remains unclear. It could be congenital or post-traumatic from childhood. The CT image shows the homogenous mass anterior to the bifurcation. The patient has a 5-year history of varicose veins, minimal edema, no pain, and no history of thrombosis.

nonthrombotic venous occlusive disease is surprisingly high, but it is not a well-recognized clinical entity.[29] Clinicians debate whether or not to treat a patient with an asymptomatic left iliac compression. Surely, a mechanical compression of the iliac vein is a risk factor for thrombosis. But, how do we weigh the risk of thrombosis based on identification of an iliac compression that does not cause edema or pain in a patient without prior thrombosis or a known thrombophilia?

One can advocate prevention by eliminating the compression with a metallic stent. The opponents of this therapy argue that placement of a stent in an otherwise asymptomatic patient raises questions about stent longevity and thrombogenicity. Opponents argue that there are too many unknowns regarding long-term viability of metallic stents to warrant placement in asymptomatic patients. Clearly, this is a clinical situation in which a biodegradable stent might have a welcome application.

There are three common clinical presentations among patients with chronic venous obstruction: (1) asymptomatic, (2) acutely symptomatic, and (3) chronically symptomatic.[25] Asymptomatic patients may have transient signs or symptoms that are not disabling. In the Raju series, seven patients had limb, perineal, or abdominal varices indicating venous hypertension. Two presented with focal phlebitis within a varix. Mild edema, intermittent groin pain, and a tiny ulcer were noted in three other patients. None of these conditions were judged to merit intervention at the time.

However, all of these signs and symptoms can increase in intensity and frequency if the venous flow pattern is compromised.

Acutely symptomatic patients present with a new or recurrent episode of acute thrombosis superimposed upon the chronic obstructive lesion. There may be tell-tale signs of chronic venous insufficiency such as hyperpigmentation, dryness, or telangiectasia. Alternatively, the lower limb may be normal in appearance, since the occult obstruction was well compensated with collaterals prior to the acute thrombosis. This clinical scenario is characteristic of left iliac compression. The mechanical risk factor of a narrowed or obstructed common iliac segment combined with immobilization can result in acute left leg DVT. This often occurs after long car or airplane trips, or a relatively minor trauma such as a sprained ankle. Prior radiation therapy, trauma, or surgery as an infant can result in focal caval lesions that cause enough obstruction to increase the risk of DVT when other risk factors are present. In three such patients, we have treated acute bilateral DVT, which developed within 1 week following a diagnostic vascular intervention. Another subset within this group includes young women who get DVT shortly after starting oral contraceptives. A related subset includes intra- or postpartum thrombosis. In both groups, the occurrence of DVT is multifactoral.

The pathophysiology of PTS originates with obstructive thrombus. Although valve injury can occur with persistence of thrombus, the presence of residual obstruction is the main cause of venous hypertension in most patients we treat. The majority of PTS patients present with multisegmental thrombosis involving the iliac and femoral veins. However, there is a specific subset of patients with PTS in which residual obstruction is confined to the popliteal and infrapopliteal veins. Residual obstructive thrombus in the popliteal, soleal, tibial, peroneal, and gastrocnemial veins can be identified with ultrasound and can cause disability in patients with poor recanalization and small caliber saphenous veins or few collaterals.[30] The clinical complaints include calf tightness, cramping, and pain in a pattern that mimics arterial claudication. However, unlike arterial insufficiency, this pain is accompanied by edema and takes longer to resolve with rest. Rest plus elevation diminish the discomfort of venous congestion in the upper calf. Many patients with this condition have had trauma or orthopedic surgery. Their proximal deep veins are patent with normal valves. The strategic obstruction of the popliteal and/or large calf veins feeding the popliteal vein causes calf muscle pump failure, pain, and swelling. Treatment with flow-directed thrombolysis has resulted in clinical improvement in these patients despite the presence of valve damage. Tourniquets (Tiger Surgical Inc., Portland, OR), designed to facilitate thrombolytic therapy, are placed at the malleolar and condylar levels to selectively compress the superficial veins against the bone. When the path of least resistance is eliminated, the thrombolytic agent is "directed" into the higher-resistance deep system. A pedal infusion is preferred, but the technique is effective in optimizing the circulating thrombolytic drug infused through a catheter placed in a femoral or iliac segment. It is important to treat tibial-peroneal and popliteal thrombus, since compromised in-flow can threaten the long-term patency of reopened axial veins and iliac stents. Continuity of deep venous flow is as important in venous reconstruction as run-off is in arterial bypass surgery and stenting.

IMAGING VENOUS DISEASE

The evaluation of obstructive venous disease utilizes multiple imaging modalities. Contrast phlebography provides the clearest understanding of obstructive disease, but it is invasive. Real-time digital phlebography shows rate of contrast clearance and contrast density patterns that can be helpful. Magnetic resonance (MRV) imaging for DVT and obstruction provides sagittal, coronal, and cross-sectional views and is able to detect acute thrombus.[31]

Demonstration of collaterals, pathognomonic for obstruction, is clearly seen in patients with SVC and IVC syndrome. However, for evaluation of acute upper extremity thrombosis, MRV is not favored by most clinicians.[32] MRV can nicely demonstrate left vein iliac compression with multiple oblique views. It can be useful in assessing the patency of the suprarenal cava. However, both gadolinium and CO_2 have been shown to be less effective contrast agents than iodinated contrast in the venous system.[33] Thin-slice CT is widely used for diagnosis of pulmonary embolus, but its application in chronic thrombosis is not always reliable in the absence of thrombus. The chronically thrombosed vein may be isodense with surrounding tissue (Figure 20-2). Particularly in chronic IVC thrombosis, the residual cava may not be distinguished and may be interpreted as congenitally absent. We recommend that cross-sectional imaging be combined with phlebography if intervention is planned. Additional noninvasive imaging techniques can provide useful information before and after intervention. However, such techniques demand skilled technologists and a dedicated lab. Raju noted that the most useful component of air plethysmography (APG) is venous filling index (VFI_{90}/mL/s). He was able to demonstrate a decrease after stenting.[25] However, as a tool for diagnosing occult proximal obstruction, APG is not ideal because large collaterals can alter pressures.[18] In nonthrombotic iliac lesions, intravascular ultrasound (IVUS) is the best tool. Duplex ultrasound is capable of registering high velocities in the compressed left common iliac segment, particularly if there is a focal compression by the right iliac artery. Large collaterals adjacent to the common iliac will often be present. However, they will not have phasic flow. Nonphasic flow is seen in collaterals and axial veins distal to the obstructing lesion.

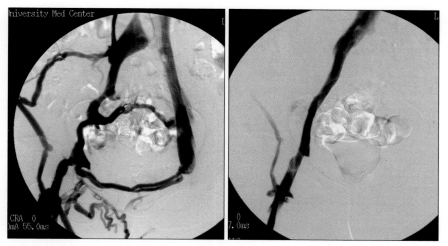

FIGURE 20-2 The patient experienced an acute left leg DVT immediately following minor laparoscopic surgery. It was initially believed that the external iliac vein had been resected, since large vessels had been identified and ligated. The external iliac segment could not be visualized with duplex, CT, or MR. After 1 year, the patient was referred for endovascular reconstruction. The stented left iliac system is seen in the image on the right.

ENDOVENOUS THERAPY

NONTHROMBOTIC OBSTRUCTION

The prevalence of left iliac vein compression may be higher than concluded from autopsy studies in the 1940s. Originally described by Cockett and Thomas,[34] the anatomical lesion was thought to be significant in approximately 30% of the population. The associated clinical syndrome often went unreported until thrombosis occurred in the left leg. This is still true. Recent application of IVUS to investigations of venous causes of limb edema or pain yield has revealed a much higher estimate of the incidence of nonthrombotic iliac compression in patients with symptomatic venous insufficiency.[35] Raju and Neglen reported that a single anatomical lesion was identified in the proximal left common iliac vein in 36% of the 255 patients included in their study. A smaller number, 8%, had a compressive lesion near the internal iliac bifurcation. In 46% of the patients, a combination of proximal and distal lesion was observed. Whereas the proximal lesion is more common on the left, there is an equal incidence of the distal narrowing. The cause of the distal lesions is extrinsic compression from the iliac artery, but relatively little fibrosis occurs at this level compared to the proximal lesions. The arterial compression and proximity of the spine can result in intrinsic vein wall thickening in the proximal left iliac. Bilateral common iliac compression is found in approximately 2% of the

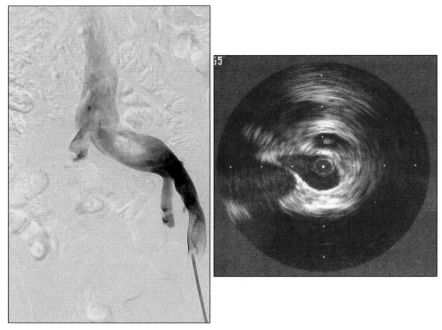

FIGURE 20-3 Example of a nonthrombotic iliac compression. Note multiple channels, stasis of contrast distal to the lesion, and relative widening of the common iliac segment. The image on the right shows how such a lesion looks with intravascular ultrasound.

population. Whereas IVUS measurements and images are definitive, the phlebographic image is often misinterpreted as normal. The wider transverse diameter and lighter contrast density in the common iliac segment are consistent with antero-postero narrowing of the vein (Figure 20-3). But, if collaterals are not seen and the characteristic impression of the right iliac artery is not identified, the diagnosis can be missed with phlebography alone. Collaterals that are visualized with deep valsalva may not be seen with normal respiration in a supine position.

ACUTE THROMBOTIC OBSTRUCTION

Catheter-directed thrombolysis, mechanical thrombectomy, and combination therapy have evolved over the past 2 decades. To further assist clinicians, guidelines for endovenous intervention of acute DVT have recently been published.[36,37] For the most part, patients presenting with phlegmasia cerulean dolens are referred for catheter-directed thrombolysis in centers where interventionalists are available. Patients not responding to standard heparin and warfarin therapy may also be referred for catheter-directed lysis. However, the majority of DVT patients treated in the emergency room are discharged with low-molecular-weight heparin and are no longer routinely hospitalized for bed rest. A multidisciplinary, multicenter clinical trial, under the auspices of the NIH, will be initiated in 2007 to investigate the benefits of early endovascular thrombus removal (with thrombolysis and mechanical thrombectomy and adjunctive use of stents) compared to standard anticoagulation therapy.[38] Prevention of post-thrombotic syndrome is a primary goal of early endovascular intervention. The objective of the study is the acquisition of Level I data, comparing anticoagulation alone to thrombus removal plus anticoagulation.

CHRONIC THROMBOTIC OBSTRUCTION: ILIAC

Prior to 2000, reports detailing the use of thrombolytic therapy and stents for use in venous thrombosis focused on patients with acute thrombosis, in whom about 50% had an underlying iliac lesion or May-Thurner compression.[21] Very often, the acute clinical presentation is deceptive, since the "critical mass" of thrombus that causes the clinical symptoms is composed of newer thrombus superimposed upon a substrate of subacute or even chronic thrombus. Endovascular stenting of chronic iliac occlusions, after removal of acute thrombus, has become familiar to interventionalists treating venous obstruction. Balloon angioplasty proved to be inadequate in treating iliac lesions.[15] The long-term patency of large-caliber (12–16 mm), self-expanding stents placed in the common iliac segment is acceptable and compares favorably to surgical thrombectomy alone and by-pass surgery.[17] Neglen reported early experience with cannulation of totally occluded segments and technical success of stenting as 98% and 97%, respectively, in the treatment of 94 consecutive patients with suspected iliac vein obstruction.[39] In this series, the reported primary, assisted primary, and secondary patency rates at 1 year were 82%, 91%, and 92%, respectively.

The clinical impact of correcting the deep venous obstructive lesions in symptomatic patients can be measured in terms of quality of life[23] because pain and swelling diminish when the venous hypertension is reduced by restoration of axial flow. In 2002, Raju et al. reported on the clinical impact of venous stents in managing chronic venous insufficiency.[40] This was the first mention in the literature of the effect of stenting on the healing of venous stasis ulcers. Stasis dermatitis and ulceration was present in 69/304 (23%) limbs treated with stenting with or without saphenous ablation. They reported that the cumulative recurrence free ulcer healing rate was 63% at 24 months. Furthermore, the rate of healing was not influenced by concomitant treatment of saphenous reflux. The reduction of pain and swelling after intervention was statistically significant ($P < .001$) (Figure 20-4).

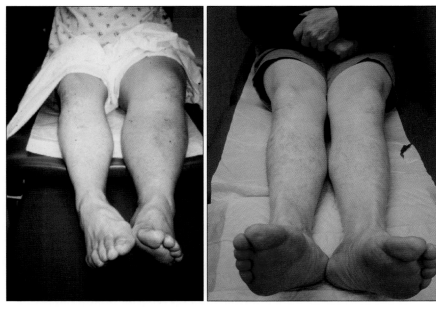

FIGURE 20-4 Pre- and postendovascular therapy in a man with a 5-year history of recurrent DVT. He underwent thrombolysis and stenting of the iliac and femoral segments. He was diagnosed with a thrombophilia that was unknown until referral for thrombolysis. His edema resolved and the image on the right shows his legs at 7-year follow-up. The ulcer recurs if the compression legging is not worn. Deep vein and perforator reflux have been identified.

CHRONIC THROMBOTIC OBSTRUCTION: IVC

Inferior vena caval (IVC) occlusion, in combination with iliac obstruction, can lead to severe venous hypertension when collateral circulation is inadequate. The IVC syndrome can include lower extremity swelling, stasis dermatitis and ulceration, varicosities, and loss of sensation and strength in the legs. Pain can be felt in the lower back, pelvis, or inguinal region in addition to discomfort and heaviness in one or both limbs. It is most common to discover caval obstruction confined to the infrarenal cava. In patients in whom we have suspected an early thrombosis in infancy, the entire inferior vena cava was discovered to be very narrow, as if minimally recanalized following thrombosis. Endovenous reconstruction of chronically occluded long iliocaval segments is feasible and safe, and the long-term results are encouraging.[24-28]

Tables 20-1 through 20-3 summarize data collected from the most recent clinical experience in a series reporting more than 15 patients.

In 2006, Raju reported a series of 120/4,217 (2.8%) patients with obstructive lesions of the IVC, identified among patients with chronic venous

TABLE 20-1 Summary of Current Clinical Experience with Endovascular Treatment for Chronic Iliocaval Thrombosis: Demographics

Year	Author	Ref	#pt	M/F	Age Mean	Age Range	Duration of Symptoms	Malignant	Benign
2000 JVIR	Razavi et al.	26	17	12/5	40.6	15–77	32 mo (1–276)	5	12
2005 JVS	Hartung et al.	28	44	9/35	42	21–80	chronic	0	44
2006 JVS	Raju, Neglen	29	120	1/2	51	14–80	chronic	0	120
2006 AVF	Thorpe, Osse	24	26	20/6	41	17–73	145 mo (12–300)	2	24
2006 JVET	te Riele et al.	27	9	5/4	30	14–58	7 mo (6–142)	3	6

TABLE 20-2 Summary of Current Clinical Experience with Endovascular Treatment for Chronic Iliocaval Thrombosis: Procedural and Clinical Results

	Median F-U months	Range F-U months	Re-Thrombosis	Restenosis at 24 months	Peri-operative Death	Cumulative Primary Patency %	Assisted Primary Patency %	Mean F-U months
Razavi	14	2–60	2	1	0	80	87 (19)	19
Hartung	27	2–103	1	5	0	73	88 (36)	27
Raju	11	3–59	2	15	0	58 @ 2 yr	82 (24)	24
Thorpe	61	8–120	2	9	0	63 @ 1 yr 45 @ 2 yr	89 (48)	48
te Riele	9	4–110	1	1	0	78	78 (9)	21

TABLE 20-3 Summary of Current Clinical Experience with Endovascular Treatment for chronic Iliocaval Thrombosis: CEAP Clinical Classification

	# pts	# limbs	CEAP CLASSIFICATION					
			1	2	3	4	5	6
Razavi et al.	17	—	1	0	15	1	0	0
Hartung et al.	44	48	0	11	31	4	1	1
Raju, Neglen	120	99	0	5	40	27	8	19
Thorpe, Osse	26	50*	1	0	13	16	11	9
te Riele et al.	9	—	—	—	9	—	—	—

* 2 BKA, secondary to stasis ulcer (BKA is Below Knee Amputation).

insufficiency examined between 1997 and 2005.[25] The majority of the lesions were infrarenal (82%), 14% (14/97) were suprarenal but below the diaphragm, and 4 cases extended into the superior aspect of the intrahepatic IVC. In 93% of cases, lesion involved the common iliac segment, but in 7%, the lesion was limited to the IVC. Most of the IVC lesions 85/99 (86%) were stenotic (>60%) and not occlusive. There were 14 occlusions, in which seven of the attempted recanalizations were not successful. The CEAP clinical classification[41] for the stented limbs included C3 (37%), C4 (26%), C5 (7%), and C6 (19%). The seven limbs with unhealed ulcers were found to have residual untreated reflux, which was predominantly in axial veins. There were 19 limbs with active ulcers that underwent stenting. In 12 limbs (63%), the ulcers healed and remained healed at 24 months. Reflux was present in 17, involving both deep and superficial systems. Only two of these limbs underwent concurrent endovascular treatment of saphenous reflux. Although the other limbs had identified reflux, correction of the obstruction alone allowed healing of the ulcer. Stented patients experienced significant clinical improvement. After 3.5 years, the reduction in pain and swelling was 74% and 51%, respectively. At 24 months, cumulative primary and primary assisted patency was 58% and 82%, respectively.

In the series reported by Hartung et al., the ulcerated limb healed completely after stenting and remained healed at 38 months.[28] No reflux intervention was performed. They noted a reduction in the clinical severity score from 8.5 to 2. Assisted primary patency was 88%. In the Thorpe-Osse series of 26 patients (50 limbs, two with prior BKA secondary to venous disease), the CEAP clinical classification included C3 (26%), C4 (32%), C5 (26%), and C6 (17%).[24] All but two patients in the series had iliocaval

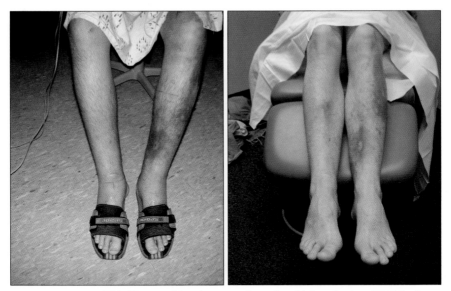

FIGURE 20-5 This 22-year-old patient experienced multiple episodes of DVT prior to referral for evaluation and treatment of a nonhealing ulcer on the left and unresolving DVT on the right. Studies revealed a completely occluded IVC, including the suprarenal segment. The patient has only a left kidney and was born 2 months premature. Central-line placement in infancy or *in utero* thrombosis may have caused the chronic IVC occlusion. The image on the right shows the patient at 6-month follow-up.

occlusion of long duration (Figure 20-5). Two patients presented with acute IVC thrombosis; one had a focal intrahepatic web, and one had no identifiable lesion but had recently begun oral contraceptives. There is a wide range of clinical presentation in patients with IVC occlusion (Table 20-4). We have observed that varicosities in a teenager and ulceration in a young patient can be signs of a proximal venous occlusion in the iliocaval region.

Since most of the patients we treated had chronic occlusion, there was a higher percentage of limbs with CEAP 5–6 classification. Among the seven patients with unhealed ulcers, one had a 25-year history of caval thrombosis and PTS. He had bilateral malleolar ulcers that healed within 40 days following stenting of the IVC and iliac veins (Figures 20-6 and 20-7). Among the other five patients, four underwent successful IVC reconstruction followed by ulcer healing within 30 days. One patient was not successfully recanalized, and although nonelastic compression promotes healing, his ulcer has occasionally reopened. On the other hand, 5/5 (100%) of the successfully stented patients underwent redilatation of the stents within the first 24 months of follow-up. In-stent intimal hyperplasia caused recurrence of the ulcer, which healed following reintervention.

Primary patency in the Thorpe-Osse iliocaval series was 63% at 1 year and 45% at 2 years.[24] Rethrombosis or distal stents occurred in one patient

TABLE 20-4 Clinical Presentations of 26 Patients with Iliocaval Obstruction/IVC Syndrome

	n (%)
Acute bilateral DVT	5 (19)
Acute unilateral DVT	1 (4)
Chronic bilateral edema	3 (12)
Chronic unilateral edema	2 (8)
Nonhealing ulcer in young pt	4 (15)
BKA + ulcer on other limb	2 (8)
PTS <1 year after IVC filter	6 (23)
Hematuria	1 (4)
Abdominal collaterals	6 (23)
Abdominal collateral + ulcer	2 (8)
Varicose veins	13 (50)

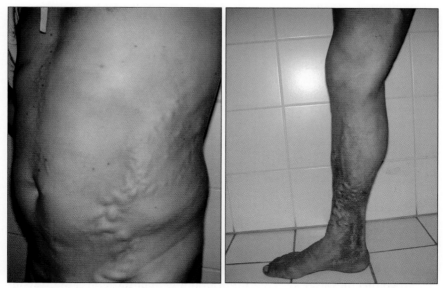

FIGURE 20-6 Abdominal collaterals and distal varices and ulceration in a 52-year-old patient with a 25-year history of IVC syndrome.

with stents extending from the popliteal to the suprarenal cava. The system was reopened with thrombolysis. Despite two additional interventions for restenosis, over the following 3 years, the stents were not patent at 5-year follow-up. This patient did have two successful full-term pregnancies, with C-section deliveries. The manipulation of anticoagulation may have

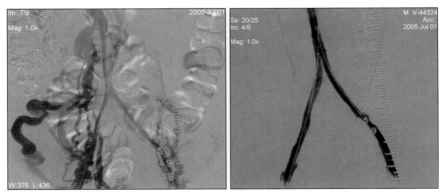

FIGURE 20-7 Pre- and poststenting of the IVC. The phlebogram on the left shows collaterals and a partially recanalized IVC following 36 hours of bilateral catheter-directed thrombolysis with a popliteal approach. The image on the right shows the reconstructed iliocaval system, which reduced the patient's pain and swelling and allowed the bilateral ulcers to heal.

contributed to stent failure along with restenosis. Other patients with equally complex and long stent systems, designed to reconstruct multiple occluded segments, have experienced uninterrupted long-term patency. Patients with multiple stents remain on warfarin. After a mean follow-up of 4 years, the assisted primary and secondary patency rates were 89% and 94%, respectively. The quality of life improvement is most significant as indicated in Figure 20-8 showing the change in disability scores. Acutely thrombosed patients with focal chronic obstructions returned to a relatively normal state following intervention. The patients with long-standing PTS improved measurably.

SPECIAL ISSUES IN ENDOVENOUS THERAPY

IVUS

Intravascular ultrasound (IVUS) is an important tool in endovenous therapy. It has application in two main areas: (1) assessing the nonthrombosed common iliac segment and (2) in the follow-up examination of patients with signs and symptoms of restenosis. In the first instance, IVUS has proven to be the best way to image and evaluate the iliac stenosis in the May-Thurner syndrome.[42] Whereas the phlebogram will generally demonstrate widening of the transverse diameter of the common iliac segment and decrease in the intensity of contrast, IVUS will clearly reveal the severity of the antero-postero narrowing causing high venous resistance. The presence of collaterals was previously thought to be necessary for diagnosis of a hemodynamically significant iliac lesion. Information from IVUS has shown that not all significant narrowings are accompanied

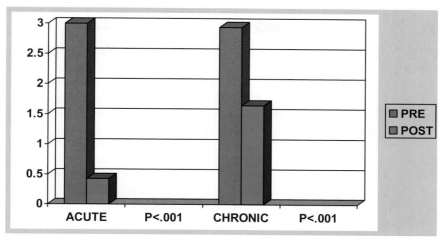

FIGURE 20-8 Significant Difference in Disability Score Before and After Endovascular

by collaterals that can be demonstrated phlebographically in the supine position.

Restenosis

Restenosis is an important issue in stenting. Monitoring the patency of iliac stents has proven less straightforward than the initial placement. Relatively few iliac stents rethrombose, particularly in patients without malignancy, but identification and treatment of restenosis are less obvious. When patients present with recurrent symptoms of limb edema or discomfort, the intensity is generally much less than the initial clinical presentation. The routine duplex exam will generally show patency of the stents. The waveform will remain phasic. Intimal hyperplasia, or in-stent-restenosis, is best imaged with IVUS. Before we had access to IVUS, we would redilate stents in patients with recurrent pain or edema. The symptoms and signs resolved. As little as 30% to 35% luminal decrease by intimal hyperplasia can cause recurrence of symptoms. In 2004, Neglen and Raju reported their observations of restenosis in venous stents and concluded that risk factors included a history of venous thrombosis, a positive test for thrombophilia, and longer stent systems extending below the inguinal ligament.[43] Gender and side had no significance. After 42 months of observation in 324 limbs treated with stenting, 23% had no sign of luminal narrowing on phlebography, 61% showed >20% narrowing, and 15% showed >50% narrowing. The value of identifying restenosis is in maintaining long-term patency of the stents. The study by Neglen and Raju showed that re-intervention optimizes long-term patency. At 3 years, primary, primary-assisted, and secondary patencies were 75%, 90%, and 93%, respectively.

Anticoagulation

Once flow is reestablished with stents, anticoagulation is important in patients with thrombophilia and those with significant residual distal thrombus and vein wall irregularity from previous DVT. Both conditions are risk factors for recurrent DVT despite good iliac flow. However, in patients with normal distal veins, the presence of a single or double iliac stent does not require long-term anticoagulation, per se. The use of anticoagulation and antiplatelet agents varies among clinicians. Low molecular weight heparin is used to bridge subtherapeutic INR levels, particularly in patients who must travel more than 2 hours to return home. Plavix®, 75 mg per day, is started prior to stenting and continued for 30 days. Patients not on warfarin also take 80 mg aspirin. It should be noted that rethrombosis of a metallic stent is a matter of urgency. In our experience, it is much more difficult, and sometimes not possible, to reopen chronic in-stent thrombosis. Unlike organized thrombus in native veins, the old thrombus in occluded stents is very hard to traverse.

Stents

Self-expanding metallic stents are used in venous stenting. The 12–14 mm-Wallstent™ (Boston Scientific, Natick, MA) has been used more than other stents because of the good long-term patency rates, the visibility under fluoro, and the ability to reposition. Stent migration is rare[44,45] but can occur with any stent if it is undersized and the vein expands. Tandem Wallstents will not interlock as the SmartStent™ (Cordis Corporation, New Brunswick, NJ) will. The foreshortening of the Wallstent requires experience, and there are times when a stent that does not foreshorten is ideal. A single, high-grade focal stenosis requires a sufficiently long stent to prevent the stent from slipping (like a watermelon seed) to either side. In our experience, long-term success of these interventions is also related to establishing continuity of flow within the deep system of the entire lower extremity. Poor inflow from the tibio-popliteal segment or the femoral segment may compromise long-term patency as much as poor outflow caused by understenting the iliac compression. Stents must extend into the IVC to completely treat the iliac compression.

IVC Filters

It is not generally thought to be necessary to place a caval filter while performing catheter-directed thrombolysis. The incidence of PE has been remarkably low because the thrombus does not fragment without undue manipulation. However, mechanical thrombectomy for removal of acute

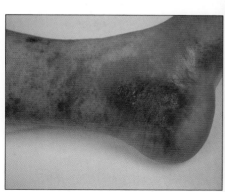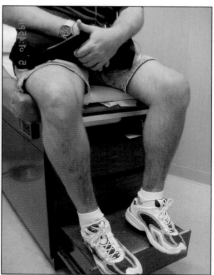

FIGURE 20-9 Ankle ulcer in a 44-year-old man with multiple thrombophilias with a 6-year history of PE and IVC thrombosis complicated by heparin-induced thrombocytopenia. The image on the right was taken 1 month after stenting.

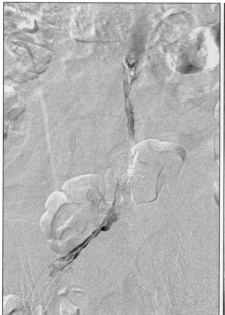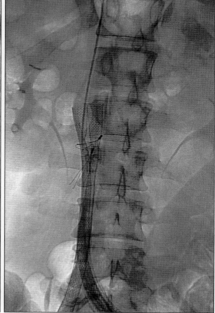

FIGURE 20-10 An occluded Simon Nitinol filter was recanalized with kissing Wallstents 1 year after the initial bilateral iliac reconstruction. The thrombolytic therapy initially cleared the IVC filter, but progressive narrowing occurred at the filter level and this resulted in rethrombosis. The image on the right shows parallel stents passing through the filter.

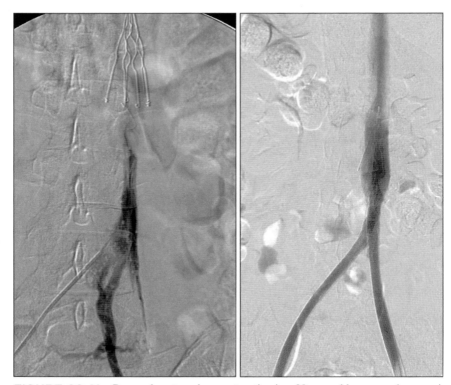

FIGURE 20-11 Pre- and postcaval reconstruction in a 38-year-old woman who experienced postpartum PE/DVT 2 years prior to intervention. The image on the left shows an early contrast image during thrombolysis. Following 48 hours of bilateral catheter-directed thrombolytic therapy, the iliocaval system was reconstructed with stents. Parallel stents pass through the Greenfield filter, which was chronically occluded.

thrombus poses the risk of embolism. In this case, temporary filters have been placed and subsequently removed. Occluded IVC filters can be safely and effectively managed with endovascular methods.

Occluded IVC Filters

We have passed single and kissing Wallstents through and alongside all types of occluded filters without complication. Once the exchange guide wire has traversed this area of occlusion, serial balloon dilation will expand the tissue and allow stent deployment (Figures 20-9 and 20-10). Mechanical thrombectomy and thrombolysis can be effectively and safely used to restore flow in acutely thrombosed IVC-bearing filters.[46] In our iliocaval series, successful recanalization of chronically occluded filters has also been performed with remarkable long-term patency[24] (Figure 20-11).

SUMMARY

Endovenous therapy represents a significant advance in the treatment of chronic venous insufficiency. The clinical options for treatment of residual obstruction, following DVT, now include reconstruction of the native axial veins with stents. The evolution of the tools and techniques is ongoing because the majority of the procedures have been adapted from arterial interventions. But experience and careful observation of clinical outcomes constantly help us refine technique and patient selection. However, there are far too many patients who need treatment given the relatively few physicians who safely and willfully perform endovenous procedures. Chronic venous insufficiency is that "something old" that seems to be in a Renaissance in the midst of new pharmaceutical discoveries and the emerging field of endovenous surgery. We hope that continued innovation, multidisciplinary efforts, and clinical evidence will all support our efforts to better prevent and treat post-thrombotic chronic venous insufficiency.

REFERENCES

1. Tassiopoulos AK, Golts E, Oh DS, Labropoulos N. Current concepts in chronic venous ulceration. *Eur J Vasc Endovasc Surg.* 2000;20(3):227–232.
2. Patel NP, Labropoulos N, Pappas PJ. Current management of venous ulceration. *Plast Reconstr Surg.* 2006;17(Suppl):254S–260S.
3. Johnson BF, Manzo RA, Bergelin RO, Strandness DE Jr. Relationship between changes in the deep venous system and the development of the postthrombotic syndrome after an acute episode of lower limb deep vein thrombosis: A 1- to 6-year follow-up. *J Vasc Surg.* 1995;21(25):307–312.
4. Pesavento R, Bernardi E, Concolato A, Dalla VF, Pagnan A, Prandoni P. Postthrombotic syndrome. *Semin Thromb Hemost.* 2006;32(7):744–751.
5. Meissner MH, Caps MT, Zierler BK, et al. Determinants of chronic venous disease after acute deep venous thrombosis. *J Vasc Surg.* 1998;28(5):826–833.
6. Galli M, Ageno W, Squizzato A, et al. Residual venous obstruction in patients with a single episode of deep vein thrombosis and in patients with recurrent deep vein thrombosis. *Thromb Haemost.* 2005;94(1):93–95.
7. Kahn SR. The post-thrombotic syndrome: The forgotten morbidity of deep venous thrombosis. *J Thromb Thrombolysis.* 2006;21(1):41–48.
8. Heit JA, Kobbervig CE, James AH, Petterson TM, Bailey KR, Melton LJ III. Trends in the incidence of venous thromboembolism during pregnancy or postpartum: A 30-year population-based study. [see comment] [summary for patients in *Ann Intern Med.* 2005;143(10):I12; PMID:16287785]. *Ann Int Med.* 2005;143(10):697–706.
9. Bergan JJ, Schmid-Schonbein GW, Smith PD, Nicolaides AN, Boisseau MR, Eklof B. Chronic venous disease. *N Engl J Med.* 2006;355(5):488–498.
10. Prandoni P, Lensing AW, Prins MH, et al. Below-knee elastic compression stockings to prevent the post-thrombotic syndrome: A randomized, controlled trial. [see comment]. *Ann Int Med.* 2004;41(4):249–256.
11. Jost CJ, Gloviczki P, Cherry KJ Jr., et al. Surgical reconstruction of iliofemoral veins and the inferior vena cava for nonmalignant occlusive disease. *J Vasc Surg.* 2001;33(2):320–327.

12. Bergan JJ, Kumins NH, Owens EL, Sparks SR. Surgical and endovascular treatment of lower extremity venous insufficiency. *J Vasc Intervent Radiol.* 2002;13(6):563–568.
13. Neglen P, Hollis KC, Raju S. Combined saphenous ablation and iliac stent placement for complex severe chronic venous disease. *J Vasc Surg.* 2006;44(4):828–833.
14. Semba CP, Dake MD. Iliofemoral deep venous thrombosis: Aggressive therapy with catheter-directed thrombolysis. *Radiology.* 1994;191(2):487–494.
15. Nazarian GK, Austin WR, Wegryn SA, et al. Venous recanalization by metallic stents after failure of balloon angioplasty or surgery: Four-year experience. *Cardiovasc Intervent Radiol.* 1996;19(4):227–233.
16. Nazarian GK, Bjarnason H, Dietz CA Jr, Bernadas CA, Hunter DW. Iliofemoral venous stenoses: Effectiveness of treatment with metallic endovascular stents. *Radiology.* 1996;200(1):193–199.
17. Bjarnason H, Kruse JR, Asinger DA, et al. Iliofemoral deep venous thrombosis: Safety and efficacy outcome during 5 years of catheter-directed thrombolytic therapy. *J Vasc Intervent Radiol.* 1997;8(3):405–418.
18. Hurst DR, Forauer AR, Bloom JR, Greenfield LJ, Wakefield TW, Williams DM. Diagnosis and endovascular treatment of iliocaval compression syndrome. *J Vasc Surg.* 2001;34(1):106–113.
19. Vedantham S, Vesely TM, Sicard GA, et al. Pharmacomechanical thrombolysis and early stent placement for iliofemoral deep vein thrombosis. *J Vasc Intervent Radiol.* 2004;15(6):565–574.
20. O'Sullivan GJ, Semba CP, Bittner CA, et al. Endovascular management of iliac vein compression (May-Thurner) syndrome. *J Vasc Intervent Radiol.* 2000;11(7):823–836.
21. Mewissen MW, Seabrook GR, Meissner MH, Cynamon J, Labropoulos N, Haughton SH. Catheter-directed thrombolysis for lower extremity deep venous thrombosis: Report of a national multicenter registry. 1999;211(1):39–49; erratum appears in *Radiology.* 1999;213(3):930.
22. Comerota AJ, Throm RC, Mathias SD, Haughton S, Mewissen M. Catheter-directed thrombolysis for iliofemoral deep venous thrombosis improves health-related quality of life. *J Vasc Surg.* 2000;32(1):130–137.
23. Comerota AJ. Quality-of-life improvement using thrombolytic therapy for iliofemoral deep venous thrombosis. *Rev Cardiovasc Med.* 2002;3(Suppl 2):S61–S67.
24. Thorpe PE, Osse FJ. Endovascular reconstruction for chronic IVC occlusion: 1–10 year follow-up. Presented at: 17th Annual Meeting of the American Venous Forum; February 24, 2006; Miami, FL.
25. Raju S, Hollis K, Neglen P. Obstructive lesions of the inferior vena cava: Clinical features and endovenous treatment. *J Vasc Surg.* 2006;44(4):820–827.
26. Razavi MK, Hansch EC, Kee ST, Sze DY, Semba CP, Dake MD. Chronically occluded inferior venae cavae: Endovascular treatment. *Radiology.* 2000;214(1):133–138.
27. te Riele WW, Overtoom TT, Van Den Berg JC, van de Pavoordt ED, de Vries JP. Endovascular recanalization of chronic long-segment occlusions of the inferior vena cava: Midterm results. *J Endovasc Ther.* 2006;13(2):249–253.
28. Hartung O, Otero A, Boufi M, et al. Mid-term results of endovascular treatment for symptomatic chronic nonmalignant iliocaval venous occlusive disease. *J Vasc Surg.* 2005;42(6):1138–1144.
29. Raju S, Neglen P. High prevalence of nonthrombotic iliac vein lesions in chronic venous disease: A permissive role in pathogenicity. *J Vasc Surg.* 2006;44(1):136–143.
30. Meissner MH, Caps MT, Bergelin RO, Manzo RA, Strandness DE Jr. Early outcome after isolated calf vein thrombosis. *J Vasc Surg.* 1997;26(5):749–756.
31. Du J, Thornton FJ, Mistretta CA, Grist TM. Dynamic MR venography: An intrinsic benefit of time-resolved MR angiography. *J Magnetic Resonance Imaging.* 2006;24(4):922–927.

32. Baarslag HJ, Van Beek EJ, Reekers JA. Magnetic resonance venography in consecutive patients with suspected deep vein thrombosis of the upper extremity: Initial experience. *Acta Radiologica.* 2004;45(1):38–43.
33. Brown DB, Pappas JA, Vedantham S, Pilgram TK, Olsen RV, Duncan JR. Gadolinium, carbon dioxide, and iodinated contrast material for planning inferior vena cava filter placement: A prospective trial. *J Vasc Intervent Radiol.* 2003;14(8):1017–1022.
34. Cockett FB, Thomas ML, Negus D. Iliac vein compression—Its relation to iliofemoral thrombosis and the post-thrombotic syndrome. *Br Med J.* 1967;2(5543):14–19.
35. Raju S, Neglen P. High prevalence of nonthrombotic iliac vein lesions in chronic venous disease: A permissive role in pathogenicity. *J Vasc Surg.* 2006;44(1):136–143.
36. Vedantham S, Thorpe PE, Cardella JF, et al. Quality improvement guidelines for the treatment of lower extremity deep vein thrombosis with use of endovascular thrombus removal. *J Vasc Intervent Radiol.* 2006;17(3):435–447.
37. Vedantham S, Grassi CJ, Ferral H, et al. Reporting standards for endovascular treatment of lower extremity deep vein thrombosis. *J Vasc Intervent Radiol.* 2006;17(3): 417–434.
38. Vedantham S, Rundback JH, Khilnani NM, et al. Development of a research agenda for endovenous treatment of lower-extremity venous reflux: Proceedings from a multidisciplinary consensus panel. *J Vasc Intervent Radiol.* 2005; 6(12):1575–1579.
39. Neglen P, Raju S. Balloon dilation and stenting of chronic iliac vein obstruction: Technical aspects and early clinical outcome. [see comment]. *J Endovasc Ther.* 2000;7(2): 79–91.
40. Raju S, Owen S Jr, Neglen P. The clinical impact of iliac venous stents in the management of chronic venous insufficiency. *J Vasc Surg.* 2002;35(1):8–15.
41. Eklof B, Rutherford RB, Bergan JJ, et al. Revision of the CEAP classification for chronic venous disorders: Consensus statement. *J Vasc Surg.* 2004;40(6):1248–1252.
42. Neglen P, Raju S. Intravascular ultrasound scan evaluation of the obstructed vein. *J Vasc Surg.* 2002;35(4):694–700.
43. Neglen P, Raju S. In-stent recurrent stenosis in stents placed in the lower extremity venous outflow tract. *J Vasc Surg.* 2004;39(1):181–187.
44. Slonim SM, Dake MD, Razavi MK, et al. Management of misplaced or migrated endovascular stents. *J Vasc Intervent Radiol.* 1999;10(7):851–859.
45. Mullens W, De KJ, Van DA, et al. Migration of two venous stents into the right ventricle in a patient with May-Thurner syndrome. *Int J Cardiol.* 2006;110(1):114–115.
46. Vedantham S, Vesely TM, Parti N, et al. Endovascular recanalization of the thrombosed filter-bearing inferior vena cava. *J Vasc Intervent Radiol.* 2003;14(7):893–903.

21

THE VENOUS ULCER AND ARTERIAL INSUFFICIENCY

ROBERT B. MCLAFFERTY AND
COLLEEN M. JOHNSON

INTRODUCTION

Venous stasis ulceration (VSU) resulting from chronic venous disease (CVD) affects approximately 1% of the adult population in European and North American nations.[1–3] In contrast, the spectrum of peripheral arterial disease (PAD) affects up to 20% of the same adult population and is often underestimated in the treatment of venous ulcers.[4–7] This incidence increases with age and in the presence of cigarette smoking, diabetes, and renal failure. Despite the number of patients with PAD, the incidence of VSU in combination with PAD is largely unknown. Depending on the population being studied, the incidence varies from 0.8% to 10%.[3,8,9] Nevertheless, the combination of VSU and PAD remains a vexing problem that health care providers will encounter, given the commonality of leg ulcers.

EVALUATION

HISTORY AND PHYSICAL EXAMINATION

When a health care provider evaluates a patient having an ulcer in the gaiter area of the leg that is presumed to be a VSU, several important parts of the history and physical examination may help focus the health care provider to the presence of combined CVD and PAD. The presence of the typical atherosclerotic risk factors (smoking, hypercholesterolemia, hypertension, diabetes mellitus, and renal failure) in association with advancing age should immediately trigger a high suspicion of PAD. Additionally, concomitant coronary artery disease, cerebrovascular disease, or remote history of claudication are important factors in the history. A more serious history of PAD can be gleaned by asking about pain in the forefoot at night while in bed and whether the patient has any concomitant ulcers on the toes. A family history of any of the atherosclerotic occlusive diseases, particularly in patients less than 60 years of age, is equally important in determining the presence of PAD. Lastly, PAD should be confirmed to be normal by some other objective means if a VSU is present for a prolonged period of time (>3 months) and minimally responsive or worsens with compression therapy.

Historical factors important in determining the presence of CVD include those associated with primary or secondary venous insufficiency. Key factors that may clue the health care provider to CVD from primary venous insufficiency include a past history of ankle ulcers or edema in the lower leg that improves rapidly with recumbence and is void in the foot. Important factors indicating secondary venous insufficiency include a past history of deep venous thrombosis (DVT) or a preceding event that may have caused DVT, unbeknownst to the patient. These include a history of major trauma, major abdominal operation, prolonged bed rest, malignancy, or known hypercoagulable state.

The physical examination remains equally important in determining whether or not a VSU is complicated by PAD. Inspection should note the location, size, and depth of the ulcer. Often, in VSU complicated by PAD, the ulcers are multiple, larger, deeper, and concomitantly infected with minimal evidence of granulation tissue. Other stigmata of VSU include the presence of varicose veins, edema, stasis dermatitis, hyperpigmentation, and lipodermatosclerosis. Further scrutiny by inspection in patients with concomitant PAD may reveal chronic nail changes, a paucity of hair growth, toe ulcers often lurking between web spaces, and dependent rubor. Palpation may reveal absent pulses in the femoral, popliteal, dorsal pedal, or posterior tibial arteries. The absence of pulses in the foot should never be solely ascribed to the presence of CVD. Other confounding factors that can worsen edema in the foot such as obesity, chronic lower extremity dependency, chronic renal insufficiency, or congestive heart failure should never lead to the presumption that pulses are probably present but cannot be palpated. The lack of palpable pulses should immediately lead to further objective testing in the vascular laboratory. In contrast, similar objective testing should also be obtained in patients with long-term VSUs that have palpable pulses. A palpable pulse does not guarantee "normal" arterial perfusion and freedom from arterial stenosis. Even in patients with mild to moderate PAD and palpable pulses, standard VSU treatment can be compromised. Auscultation of neck, heart, abdomen, and femoral regions may reveal bruits or murmurs that alert the health care provider to the presence of atherosclerosis.

VASCULAR LABORATORY TESTING

Patients with VSU should undergo vascular laboratory testing to confirm diagnosis of CVD and appropriately ascribe the cause of pathophysiologic dysfunction to reflux, obstruction, or both etiologies. While this topic is covered in detail within other chapters of this textbook, duplex ultrasound should be performed to assess for venous obstruction, including the iliac veins and inferior vena cava. Additionally, vascular laboratory testing should be used to determine presence and scope of venous valvular reflux. Typically, most vascular laboratories perform venous valve closure time testing as described by van Bemmelen or venous refill time testing.[10–12] Duplex can also be used to determine the presence of incompetent perforator veins in the vicinity of VSU.

The presence of PAD can be defined by an ankle-brachial index (ABI). Most vascular laboratories objectively define PAD with an ABI of less than 0.9, with normal ranging from 0.9 to 1.3. Conditions such as morbid obesity, diabetes mellitus, and renal failure may artificially elevate the ratio greater than 1.3. In these cases, a toe-brachial index can be similarly used. Often, because of ulcer location on the lower aspect of the leg, an ABI may not be possible, and a toe-brachial index is preferred. The presence of PAD can also be defined by duplex arterial mapping. A skilled technologist can define a preliminary map from the infrarenal aorta to the tibial arteries. Most often, history, physical examination, and ankle- or toe-brachial indices are sufficient to determine whether other more detailed imaging may be needed in order to guide further CVD or PAD treatment with invasive procedures.

ARTERIOGRAPHY AND VENOGRAPHY

If the ABI is less than 0.9, additional arterial imaging may be warranted to determine whether arterial reconstruction can be performed. Arteriography includes computed tomographic arteriography (CTA), magnetic resonance arteriography (MRA), and contrast arteriography. Discussion as to the strengths and weaknesses of these modalities is beyond the scope of this chapter, but suffice it to say that CTA and MRA may be of use if multilevel and/or bilateral PAD is suspected. Additionally, MRA may play a more vital initial role in those patients with moderate to severe chronic renal insufficiency. Given these two scenarios, contrast arteriography can be better guided for endovascular intervention or avoided altogether because open surgical repair will be indicated.

Evidence is mounting that venous obstruction may be more common than previously assumed and play a significant role in contributing to the recalcitrance of VSU.[13,14] In light of these new findings, venography for evaluation and treatment of this condition may become more commonplace. Venography can also be performed by computed tomography, magnetic resonance, or with contrast. As with CTA and MRA, special protocols and image post-processing specific to venous anatomy are required to obtain reliable information.

TREATMENT

Great strides have been made in the treatment of VSU over the past decade. They include new forms of compression therapy, a wide array of wound care products, skin substitutes, novel medications, percutaneous ablation of superficial venous reflux, and relief of venous obstruction with angioplasty and stenting. Their effectiveness in treating VSU has come with a better understanding of venous physiology and the VSU macro- and micro-environment. In contrast, little is known as to how these modalities fall into the algorithm of treatment when PAD is concomitantly present. Certainly, in a limb with an ankle-brachial index of less than 0.5 or with ipsilateral ischemic rest pain, ulcer, or gangrene of the foot, aggressive arterial revascularization is required in order to further facilitate VSU healing with traditional and newer treatments. In patients with an ankle-brachial index between 0.5 and 0.9, treatment algorithms require more clinical acumen in determining how to correctly proceed. The constellation and severity of patient comorbidities may further guide the health care provider as to the next course of action.

In the presence of severe VSU, a commonsense approach to patients with concomitant mild to moderate PAD might begin with aggressive inpatient care for bed rest, intravenous antibiotics, minor wound debridement, and skin care. Within 3 to 5 days, signs for VSU improvement can be monitored. With improvement, a trial of compression therapy can be instituted. Many of these patients will require continued aggressive global therapy in a skilled nursing facility. If, on the other hand, improvement seems sluggish, arterial intervention is indicated. Given that many of these patients are very elderly and debilitated, endovascular treatment remains the preferred method to bridge the VSU to a more rapid healing rate. Often, such measures may be indicated to simply prevent worsening of the VSU and limb loss. Even in the presence of severe PAD, endovascular therapy such as recanalization of an occluded femoral artery with angioplasty of a tibial artery may be indicated rather than bypass. If open surgery is required and the patient is a surgical candidate, then definitive bypass should be performed. Often, the great saphenous vein cannot be used due to its own diseased state from venous insufficiency. In this case, arm vein or a polytetrafluoroethylene graft with construction of a distal venous hood or patch can be used. It is absolutely imperative that the patient be nutritionally replete and the limb be void of active infection and edema prior to operation. This often requires several days of skilled nursing care in the inpatient setting. Ulcers must be kept out of the operative fields, and surgical approaches may also require nontraditional methods such as partial fibula resection to access the peroneal artery or lateral approach to the popliteal artery.

After arterial revascularization, VSU treatment should proceed accordingly. Depending on the patient's overall function and the care needs, more aggressive VSU treatment other than compression and wound care may not be indicated. In selected patients, percutaneous treatment of great saphenous and perforator ablation can serve to further enhance healing. After arterial reconstruction, failure to facilitate VSU healing by aggressive wound care, compression, and percutaneous ablation of pertinent superficial refluxing and perforating veins should spawn a high suspicion that the patient may have significant venous obstruction. Venography with angioplasty and stenting of any obstructing femoral or iliac vein lesions should be seriously considered. Patients with a history of DVT should be considered sooner for evaluation.

Some patients will have such serious VSU and PAD that primary amputation is the most appropriate treatment recommendation. Although in this often frail population progression to ambulation with a prosthetic is exceedingly rare, the benefit of a healed stump as well as freedom from pain and open wounds often outweighs the risks from a more extensive bypass that lengthens suffering and is doomed for failure. Most often, amputation at or above the knee joint best serves these patients.

REVIEW OF CLINICAL STUDIES

Few clinical studies have addressed the epidemiology, diagnosis, treatment, and outcome of VSU with concomitant PAD. This may be ironic considering the findings of Nelzen et al.[2] In a cross-sectional study encompassing a random population sample of an entire county in Sweden, researchers determined with detailed vascular laboratory examination on 453 ulcerated limbs that 72% had deep venous insufficiency, 38% had purely superficial venous insufficiency, and 40% had ABIs <0.9. One in 10 patients had lower extremity ulceration with combined venous and arterial insufficiency.

In a study by Stacey et al. on leg ulcers, venous refill times measured by photoplethysmography were examined in controls of differing age groups (157), patients with venous insufficiency (133), patients with only PAD (17 limbs), and those with PAD and venous insufficiency (20 limbs).[15] Advancing age in controls was associated with significantly shortening refill times. Limbs with only venous insufficiency had significantly shorter refill times compared to controls. Limbs with only PAD had significantly shorter refill times compared to controls but significantly longer times compared to limbs with only venous insufficiency. Venous refill times in patients with combined VSU and PAD did not differ significantly from those with VSU who had only venous insufficiency.

Treatment with compression therapy remains a challenging question in patients with VSU and PAD, particularly in those patients with moderate PAD (ABI from 0.5 to 0.9). Ghauri et al. examined this question in a study comparing healing rates (wound status at 36 weeks follow-up) using compression in patients with VSU and only venous insufficiency, patients with VSU and moderate PAD (ABI: 0.5 to 0.85), and patients with VSU and severe PAD (ABI < 0.5). Patients in the severe PAD group also underwent arterial revascularization.[9] In the 33 limbs with moderate PAD, 12 (36%) worsened or failed to show any signs of improvement. Nine of these 12 patients agreed to arteriography, of which only five were candidates for revascularization. Four of these five patients healed their ulcers, and one died of a myocardial infarction with an improved VSU at 18 weeks post revascularization. There was no difference in mean ABI between the group of 12 patients with VSU and moderate PAD that did not improve with compression (mean ABI 0.73) and those 21 patients that did improve with compression (mean ABI 0,71). Thirty-six week healing rates for the chronic venous insufficiency group, moderate arterial/venous group, and severe arterial/venous group were 70%, 64%, and 23%, respectively.

In determining what factors may play an important role in ulcer healing in patients with combined arterial and venous insufficiency, Treiman et al. sought to evaluate variables associated with successful ulcer healing and better define criteria for interventional therapy.[16] Fifty-nine patients were identified as having combined disease and were divided into groups depending on arterial reconstruction patency, distribution of venous insufficiency (superficial only or combined), or history of deep venous thrombosis. In the 49 of 59 patients that remained alive in follow-up, 58% healed their ulcers, 31% failed to heal their ulcers, and 12% required amputation. Mean interval from arterial revascularization to healing was 7.9 months. Only two of the 13 patients with a prior history of deep venous thrombosis and arterial revascularization healed their ulcers. The presence of deep venous thrombosis and arterial revascularization patency were the only factors statistically significant in predicting healing of ulcers.

The outcome of VSU when complicated by PAD was further elucidated by Bohannon et al.[8] Using the appropriate discharge diagnosis codes over a 10-year period at two tertiary care hospitals, 14 patients (15 limbs) were identified as having VSU and having undergone an arterial revascularization or amputation. Mean age of this group was 80 years, and mean time of presence of VSU up to arterial revascularization or amputation was 6.4 years. There was a mean number of 2.1 ulcers per affected limb and a mean wound area was 71 cm^2. Excluding the two limbs that required amputation, three limbs healed, of which two recurred. Mean follow-up for the group was 2.8 years, and at last follow-up, 11 of the 12 patients having arterial revascularization had expired. Of the nine patients having an arterial bypass, eight remained patent at the time of death. Patients in this clinical series with combined VSU and PAD were characterized as very elderly, having multiple medical comorbidities, exhibited very large ulcers with minimal healing, and had a high mortality over the short term.

CONCLUSIONS

VSU in combination with PAD represents a morbid condition requiring keen diagnostic and treatment skills of the health care provider. Although somewhat conflicting in its prevelance, the overall ubiquity of leg ulcers requires a methodical approach to identify all factors leading to lack of healing and recalcitrance. Each patient's treatment plan must be tailored to the distribution of arterial and venous insufficiency that exists. Therefore, the health care team should be equipped to offer the full spectrum of treatment techniques in arterial revascularization, venous insufficiency, and venous obstruction. Primary recommendation of amputation should not be considered a failure in selected patients. Freedom from pain, infection, and wounds is often the quickest way to alleviate prolonged suffering.

REFERENCES

1. Baker SR, Stacey MC, Singh G, et al. Aetiology of chronic leg ulcers. *Eur J Vasc Surg.* 1992;6:245–251.
2. Nelzen O, Bergqvis D, Lindhagen A. Leg ulcer etiology—A cross sectional population study. *J Vasc Surg.* 1991;14:557–564.
3. Nelzen O, Bergqvis D, Lindhagen A. Venous and non-venous leg ulcers: Clinical history and appearance in a population study. *Br J Surg.* 1994;81:182–187.
4. Novo S. Classification, epidemiology, risk factors, and natural history of peripheral arterial disease. *Diabetes Obes Metab.* 2002;4:S1–S6.
5. Meijer WT, Hoes AW, Rutgers D, Bots ML, Hofman A, Grobbee DE. Peripheral arterial disease in the elderly: The Rotterdam Study. *Arterioscler Thromb Vasc Biol.* 1998;18:185–192.
6. Stoffers HEJH, Rinkens PELM, Kester ADM, Kaiser V, Knottnerus JA. The prevalence of asymptomatic and unrecognized peripheral arterial occlusive disease. *Int J Epidemiol.* 1996;25:282–290.

7. Cardiovascular Health Study (CHS) Collaborative Research Group. Ankle-arm index as a marker of atherosclerosis in the Cardiovascular Health Study. *Circulation*. 1993;88:837–845.
8. Bohannon WT, McLafferty RB, Chaney ST, et al. Outcome of venous stasis ulceration when complicated by arterial occlusive disease. *Eur J Vasc Endovasc Surg*. 2002;24:249–254.
9. Ghauri ASK, Nyamekye I, Grabs AJ, et al. The diagnosis and management of mixed arterial/venous leg ulcers in community-based clinics. *Eur J Vasc Endovasc Surg*. 1998;16:350–355.
10. van Bemmelen PS, Bedford G, Beach K, Strandness DE. Quantitative segmental evaluation of venous valvular reflux with duplex ultrasound scanning. *J Vasc Surg*. 1989;10:425–431.
11. Barnes RW, Collicot RE, Mozersky DJ, et al. Non-invasive quantitation of venous reflux in the post phlebitic syndrome. *Surg Gynecol Obstet*. 1973;136:769–774.
12. Abramowitz HB, Queral LA, Flinn WR, et al. The use of photoplethysmography in the assessment of venous insufficiency: A comparison to venous pressure measurements. *Surgery*. 1979;Sept:434–441.
13. Neglen P, Thrasher TL, Raju S. Venous outflow obstruction: An underestimated contributor to chronic venous disease. *J Vasc Surg*. 2003;38:879–885.
14. Neglen P, Raju S. Proximal lower extremity chronic venous outflow obstruction: Recognition and treatment. *J Vasc Surg*. 2002;15:57–64.
15. Stacey MC, Rashid P, Hoskin SE, Pearce CA. Influence of arterial disease, age and active ulceration on venous refill time measured by photoplethysmography. *Cardiovasc Surg*. 1996;4:368–371.
16. Treiman GS, Copland S, McNamara RM, et al. Factors influencing ulcer healing in patients with arterial and venous insufficiency. *J Vasc Surg*. 2001;33:1158–1164.

22

Topical Negative Pressure Techniques in Chronic Leg Ulcers

Jeroen D. D. Vuerstaek, Tryfon Vainas, and Martino H. A. Neumann

Introduction

Chronic leg ulcers (CLUs) affect approximately 1% of the adult population in developed countries.[1] The prevalence increases with age and is estimated to be 4% to 5% in octagenerians.[1,2] The course and prognosis of patients with leg ulcers differs according to the underlying pathogenesis, which is venous disease in up to 80%.[2]

Several treatment modalities and protocols have been reported to date, all of them mainly focusing on ambulatory treatment of venous ulcers.[2–4] Compression therapy and restoration of the etiology are the cornerstones of these regimens.[2–5] It is broadly accepted now that ulcers should be debrided of necrotic and fibrous tissue to allow formation of granulation tissue, adequate epithelialization, and to decrease the chances of infection.[2–4,6] Apart from surgical or chemical debridement, there is not much evidence for the use of special dressings underneath the compression bandages.[2–4,7–10] When the currently available protocols are followed, about 50% of ulcers will heal within 4 months, around 20% do not heal within 2 years, and about 8% do not heal even after 5 years.[11–12] Owing to these poor healing results, many patients will be admitted to the hospital for inpatient treatment. Furthermore, various studies reported a 1-year recurrence rate after wound healing with nonoperative techniques of up to 57%.[13]

Although the use of suction on wounds has been known since the late 1950s, it was not until 1993 that Fleischmann introduced vacuum sealing for soft-tissue damage in open fractures.[14,15] In 1995, Kinetic Concepts Inc., San Antonio, Texas, introduced into the U.S. market a commercial system, VAC®, for promoting vacuum-assisted closure. The VAC system exerts a controlled, local subatmospheric pressure in the wound.[16]

HOW TO USE VAC THERAPY

VAC therapy provides subatmospheric pressure to a wound through a polyurethane ether (PU) or polyvinyl alcohol (PVA) foam dressing that is custom fitted to the wound size at the bedside. The choice of foam depends on the aspect of the wound bed, the size, and the type of the wound. The foam is covered with an adhesive drape to create an airtight seal. Application of subatmospheric pressure creates a moist wound environment. An evacuation tube embedded in the foam is connected to a fluid collection canister contained within a portable computer-controlled VAC pump. In order to provide optimal tissue tension, moisture level, and enhanced capillary flow to the wound, the VAC pump can provide either continuous or intermittent negative pressures between −75 and −125 mmHg. Initially, the negative pressure may be applied in a continuous mode for 48–72 hours to remove larger amounts of fluids and debris and to create a vital wound bed that is likely to accept skin grafts. Subsequently, in order to provide intimate contact between the granulation tissue and the transplants, negative pressure is again applied continuously through a PVA foam until the skin grafts show adequate adhesion to the wound bed.

VAC THERAPY IN CHRONIC LEG ULCERS: A RANDOMIZED CLINICAL TRIAL

Chronic leg ulcers constitute a major health care problem for patients, physicians, and heath care systems alike. The patients suffer from markedly reduced quality of life, slow healing tendencies, and high recurrence rates, while the treatments are costly. Thus, this clinical problem requires renewed research into its pathophysiology and novel therapeutic approaches. Owing to our own promising retrospective preliminary data and the relative paucity of prospective randomized controlled trials, we started a randomized controlled trial to study the efficacy of VAC in wound healing compared to standard wound dressings in hospitalized patients with recalcitrant CLUs as defined in the inclusion criteria.[38] We also evaluated the effect of VAC therapy on recurrence rate, quality of life (QOL), pain, comfort, and costs of treatment.[17–19]

PATIENTS AND METHODS

The study was conducted in the departments of Dermatology at the University Hospital Maastricht and the Atrium Medical Centre, Heerlen, the Netherlands. All patients hospitalized with chronic venous, combined venous and arterial, or microangiopathic (arteriolosclerotic) leg ulcers of >6 months' duration were eligible for entry in the study, after surgical treatment options had been exhausted and extensive ambulatory treatment (>6 months) in an outpatient clinic according to the Scottish Intercollegiate Guideline Network (SIGN) had failed.[20]

Duplex ultrasound scans of the deep and superficial venous system, an arterial work-up consisting of Doppler ultrasound scans, ankle-brachial pressure index (ABI), and transcutaneous oxygen pressure, as well as bacterial cultures, were performed in all patients. In those with no apparent venous or macroangiopathic arterial insufficiency but with decreased transcutaneous oxygen pressure, biopsy specimens were taken to demonstrate microangiopathic (i.e., arteriolosclerotic) changes. On the basis of these findings, patients were divided into groups: (1) those with chronic venous insufficiency of the deep or superficial system without an arterial incompetence, (2) those with combined venous and arterial insufficiency of the deep or superficial system (ABI, 0.60–0.85), or (3) arteriolosclerotic (Martorell's ulcer, biopsy proven) leg ulcers.

Before inclusion in the study, underlying venous and arterial insufficiency was dealt with, and patients underwent ambulatory conservative local treatment for at least 6 months. This treatment consisted of ambulatory debridement (whenever necessary), daily or weekly (whenever necessary) cleansing with tap water, and nonadherent wound dressings, creating a controlled moist wound environment (SIGN guidelines). Before no wound infections were observed, no topical or systemic antibiotics were used.

Patients with venous or combined venous/arterial leg ulcers were treated with multilayer, short, stretch bandages. If the ulcer did not reduce in size after 6 months of ambulatory treatment, patients were hospitalized to add bed rest and skin grafting to their treatment and became eligible for entry in the study.

Patients meeting one of the exclusion criteria like ulcer chronicity shorter than 6 months, age >85 years old, the use of immune suppression, allergy to wound products, malignant or vasculitis origin, or ABI smaller than 0.60 were excluded from the study. In patients presenting with multiple ulcerations, the clinically most severe CLU, according to the staging system described by Falanga,[21] was included in the study.

Every patient was provided with trial information sheets and written informed consent was obtained. Full ethics approval from the respective local medical ethical committees was obtained, and the study was performed in accordance with guidelines set forth by the Declaration of Helsinki.[22]

Procedures

Hospitalized CLU patients were randomly assigned to the VAC group or to the control group (standard wound care) by a computer program using random permuted blocks of eight. Randomization was carried out within three strata, one for each ulcer type: venous, combined venous/arterial, and arteriolosclerotic ulcers. Treatment allocation occurred through telephone calls to the coordinating center. In both study groups and both study centers, an initial necrosectomy was performed by sharp debridement under local anesthesia until pinpoint bleedings appeared.[6]

In patients assigned to treatment with VAC, a PU foam was applied to the wound during the preparation stage. This foam was appropriately trimmed to fit each individual wound. A noncollapsible drainage tube embedded in the foam was connected to the VAC pump; thereafter, an airtight adhesive drape was applied on top of the foam, and a permanent negative pressure of 125 mmHg was exerted. The tube drained the wound secretion into a collection canister. In this way, a previously open wound was temporarily converted into a controlled, closed, and moist wound. A wound was considered to be prepared when granulation tissue covered 100% of the surface and wound secretion was minimal.[21] Transplantation of full-thickness punch skin grafts was then performed.[23] This autologous grafting, in which 4-mm superficial pieces of skin are normally taken from the thigh, was performed under local anesthetic (Xilocaine 1%; AstraZeneca). In conjunction with a punch graft, the skin is picked up using a rounded biopsy knife (Biopsy punch, Kai Medical Europe, Solingen, Germany) and cut off. The pieces of skin are placed on the ulcer, spaced 5 mm apart, and covered with a nonadhesive dressing, such as a PVA foam. After 4 days of continuous subatmospheric pressure, once all skin grafts attached well, standard wound care was continued using a nonadhesive dressing (Atrauman; Hartmann, Nijmegen, the Netherlands) and a multiple-layer compression bandage (Rosidal K; Lohmann & Rauscher, Rengsdorf, Germany), when possible, until complete healing.

Patients assigned to standard wound care (SWC) received daily local wound care according to the SIGN guideline and compression therapy (especially patients with venous or combined venous/arterial ulcers) until complete healing.[20] Compression therapy consisted of double-layered, short, stretch bandages (Rosidal K), applied from the toe to below the knee, creating a submalleolar pressure of 25 to 35 mmHg. Two basic types of commercially available wound dressings were used in this study, including hydrogels (Nu-gel; Johnson & Johnson, Amersfoort, the Netherlands) and alginates (Sorbalgon; Hartmann, Paul Hartmann B.V., Nijmegen, the

Netherlands). The choice of dressing mostly depended on the wound bed and the amount of exudate.[2-4,7] Once 100% granulation was achieved and minimal wound secretion was seen, these patients also received punch skin-graft transplantation covered with a nonadhesive dressing (Atrauman) and compression therapy. The inner dressing was not changed for 4 days.[7,21] Once all skin grafts attached well, standard wound care was continued using a nonadhesive dressing (Atrauman) and a multiple-layer compression bandage (Rosidal K), when possible, until complete healing.

In both treatment groups, only toilet and basic hygiene mobility was permitted during the wound bed preparation and transplantation stage. After complete wound healing, community-grade elastic support stockings class 2 (Medi 550, Medi Nederland B V, Breda, the Netherlands; or Eurostar, Varodem, Horn, the Netherlands) or 3 (Euroform, Varodem, Horn, the Netherlands) were prescribed, depending on the cause of the ulcer.

Evaluation Criteria

Because masking the interventions was not possible, patients were reviewed clinically by the same independent research physician and consultant dermatologist twice a week until wound closure. Thereafter, the same independent research physician prospectively monitored the patients at 3, 6, and 12 months after discharge. All participating clinicians completed standardized case record forms during their control visits, treatments, and follow-ups.

Two time spans in wound healing were distinguished: (1) the wound bed preparation stage and (2) the time to complete healing. The stage for wound bed preparation was defined as the time between surgical debridement and application of the punch skin grafts. The time to complete healing, being the primary endpoint, was defined as the period between surgical debridement and 100% epithelialization (wound closure). Patients were discharged only after complete healing; therefore, length of hospital stay equaled total healing time. Secondary endpoints included (1) duration of the wound bed preparation; (2) percentage of ulcer recurrences within 1 year after discharged, defined as an epithelial breakdown anywhere along or within the index leg ulcer region; and (3) skin-graft survival, defined as the percentage of successfully adhered skin grafts after 4 days of complete bed rest and local therapy. Furthermore, we compared between both experimental groups the quality of life (QOL), pain scores, the total time needed for wound care until complete wound closure, and the costs per ulcer.

Quality of Life and Pain Scores

Quality of life was measured at baseline and once a week during hospitalization using the EuroQuol Derived Single Index (EQ-DSI). This index is a generic questionnaire consisting of five dimensions: mobility, self-care, usual activities, pain and discomfort, and anxiety and depression.[24] Scoring of these parameters can be transformed into a single index of health status, with the highest score reflecting the best possible health status. During dressing change, pain scores were derived using the Short Form-McGill Pain Questionnaire (SF-MPQ) and the Present Pain Intensity (PPI) score as an adjectival pain scale, as previously reported.[25] The SF-MPQ is designed to provide quantitative measures of clinical pain. It consists of 15 descriptors of 11 sensory and four affective items with a maximum response of three for each item (maximum possible sensory and affective scores are 33 and 12, respectively). Pain intensity using the PPI score was recorded on a scale from 1 to 5, with 1, mild; 2, discomforting; 3, distressing; 4, horrible; and 5, excruciating.

Time Consumption

The total time needed for wound care until total wound closure was based on empirical time registration by the attending dermatologist and the nurse who performed wound care. Both time spans were calculated in minutes.

Cost Analysis

The total costs related to wound care for each ulcer included the salary for personnel and the material costs for wound care procedures until the moment of complete healing. Personnel costs for each procedure were based on empirical time registrations by the dermatologist and nurse who performed the wound care. Because we had chosen to keep patients hospitalized until complete healing was achieved, in contrast to normal clinical practice, we did not incorporate admission costs in the budget. Costs related to materials were based on actual cost prices as calculated by the financial department of the University Hospital Maastricht. The costs of both treatment modalities were calculated in dollars ($) for each ulcer.

Statistical Analysis

Data from a retrospective study showed that the mean ± SD duration of the >90% wound closure period was 50 days ± SD 12 days in standard wound care versus 31 days ± SD 7 days in VAC.[17] In order to detect a minimal difference of 7 days in preparation time with a power of 95% ($\alpha = 5\%$) the number of patients required in each treatment group was 30, as derived from sample-size calculations.

Results were analyzed on an intention-to-treat basis.[26] Time to complete healing (Figure 22-2), duration of wound bed preparation (Figure 22-3), and recurrence rates (Figure 22-4) were compared using the Kaplan-Meier survival analysis.[27] The log-rank test was used to test for statistically significant differences between both groups.[27] To adjust for small imbalances in the baseline distribution of relevant prognostic factors (Table 22-1) to wound healing, multivariate analysis was performed using Cox's propor-

TABLE 22-1 Demographic and Clinical Characteristics

	Treatment:		
	Standard wound care (n = 30)	VAC (n = 30)	P
Male	7 (23%)	7 (23%)	ns
Female	23 (77%)	23 (77%)	
Age	72 (45–83)	74 (53–81)	ns#
Median ulcer chronicity at inclusion (months)	7 (6–12)	8 (6–24)	ns#
Median ulcer surface (length × width = cm^2)	43 (3–250)	33 (2–150)	ns#
Smoking	9 (30%)	6 (21%)	ns
Diabetes mellitus type II	5 (17%)	5 (17%)	ns
Immobility	13 (43%)	12 (41%)	ns
Hypertension	12 (40%)	13 (45%)	ns
No ulcer history	14 (47%)	12 (40%)	ns
Ulcer environment (0–4)*	2 (0–4)	2 (0–4)	ns
Infection signs (pseudomonas aeruginosa)	6 (20%)	8 (28%)	ns
Median ankle/brachial index (%)	100 (59–130)	100 (59–100)	ns#
Medication use:			ns
— Antibiotics	1 (3%)	1 (4%)	
— Nonselective β-blockers	5 (17%)	4 (15%)	
— ACE inhibitors	9 (30%)	6 (23%)	
— Selective β1-blockers	5 (17%)	4 (15%)	
— Anticlotting therapy	7 (23%)	10 (39%)	
— Ca^{++}-antagonists	4 (13%)	1 (4%)	
Ulcer type:			
— Venous origin	13 (43%)	13 (43%)	ns
ceAsP	7 (54%)	9 (69%)	
ceApp	6 (46%)	3 (23%)	
ceAdp	0	1 (8%)	
— Combined venous/arterial origin	4 (13%)	4 (13%)	ns
— Arteriolosclerotic origin	13 (43%)	13 (43%)	ns

Data are median (minimum-maximum) or number (%) unless otherwise specified.
* Ulcer environment (surrounding skin symptoms like petechiae, eczema, erythema, and/or maceration).
# Mann-Whitney U test.
Chi-square test or Fisher exact test as appropriate.

tional hazards model (Tables 22-2A and 22-2B).[13,28,29] The regression coefficient expresses the independent contribution of potential determinants to duration of cleaning and wound healing. Hazard ratios and their 95% confidence intervals (CI) are presented. $P < 0.05$ was considered to be statistically significant.

Percentages were compared by the chi-square test, and continuous variables were compared using the independent samples t-test for normally

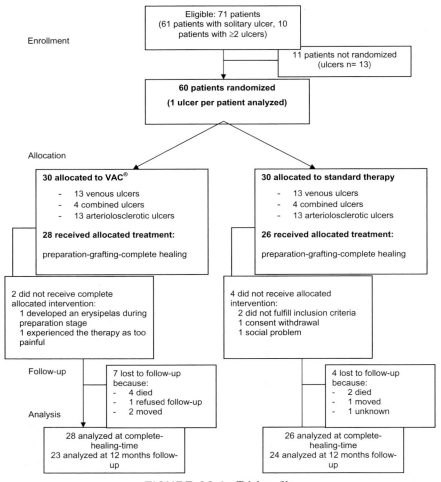

FIGURE 22-1   Trial profile.

distributed variables and the two-independent sample test for non-normally distributed variables. The paired t-test was used to compare QOL scores before treatment and at the end of follow-up.

All data were analyzed using the SPSS 11.0 software package (SPSS Inc., Chicago, IL).

RESULTS

Recruitment

From May 2001 to May 2003, 71 patients with 85 CLUs were hospitalized and screened for inclusion (Figure 22-1). Eleven patients with

13 ulcers were not included because they were not interested in participating.

Protocol Violations

Four randomized patients (n = 4 ulcers), two in the VAC group and 2 in the control group, did not complete the protocol due to reasons mentioned in Figure 22-1. In addition, one patient switched from conventional to VAC. therapy after 8 weeks due to an unsatisfactory therapeutic outcome. Finally, during analysis, it became apparent that two patients were falsely included because they had peripheral arterial disease. Still, the therapeutic outcome in these patients was analyzed according to the intention-to-treat principle.[26]

During follow-up, seven VAC patients and four controls were lost due to reasons outlined in Figure 22-1.

Patient and Ulcer Characteristics

Among the 60 patients, 51 had one ulcer, six had two ulcers, and three had three ulcers (n = 72 ulcers). All patients gave informed consent for inclusion in the study and were randomly allocated between the two treatment arms (Figure 22-1).

Patient and ulcer characteristics of both groups are shown in Table 22-1. Although the median ulcer surface area was larger in the control group, no statistical differences were in common risk factors known to be associated with delayed ulcer healing or in demographic and clinical characteristics.[13,28,29]

Effectiveness of VAC Versus Conventional Wound Care

Kaplan-Meier survival analysis showed that the time to complete healing was reduced in the VAC group (P = 0.0001). The median total healing time was 29 days (95% CI, 25.5 to 32.5) in the VAC group versus 45 days (95% CI, 36.2 to 53.8) in the control group. Ninety percent of the ulcers treated with VAC healed within 43 days. At that point, only 48% of all ulcers in the control group had healed (Figure 22-2). Because patients were discharged only after complete healing, length of hospital stay equalled total healing time. Only one ulcer failed to heal in both the control and VAC-treated patient groups, respectively.

VAC therapy resulted in a significantly shorter wound bed preparation time (P = 0.005). The median preparation time was 7 days (95% CI, 5.7 to 8.3) in the VAC group versus 17 days in the control group (95% CI, 10 to 24). Moreover, 90% of the ulcers treated with VAC could be cleaned within 14 days. By contrast, only 37% of the ulcers in the control group could be cleaned within this time span (Figure 22-3).

Despite the Cox multivariate regression analysis (Tables 22-2A and 22-2B) the treatment by VAC was still associated with significantly faster time

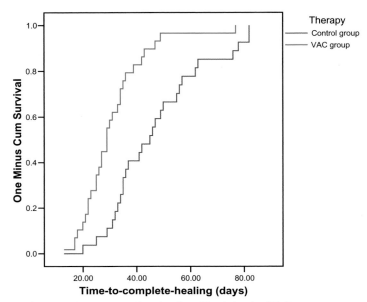

FIGURE 22-2 Time to complete healing (days) in the VAC group as compared to standard wound care.

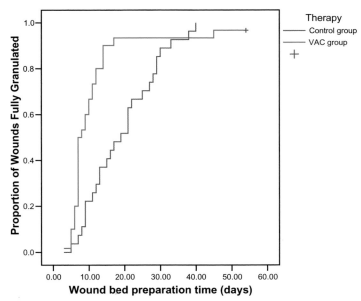

FIGURE 22-3 Wound bed preparation time (days) in the VAC group as compared to standard wound care.

TABLE 22-2A Hazard Ratios with 95% Confidence Intervals for Determinants of Duration of Wound Bed Preparation

Variable	Hazard Ratio	95% CI	P Value
Ulcer area (cm^2)	0.56	0.29–1.09	0.09
Smoking	0.5	0.20–1.21	0.30
Infection signs	0.45	0.21–1.21	0.12
Ulcer history	1.08	0.61–1.91	0.79
Therapy	2.4	1.19–4.71	<0.01*
ACE inhibitors	1.9	0.57–2.07	0.80
Anticlotting therapy	1.03	0.53–2.01	0.92

TABLE 22-2B Hazard Ratios with 95% Confidence Intervals for Determinants of Duration of Wound Healing

Variable	Hazard Ratio	95% CI	P Value
Ulcer area (cm^2)	0·50	0.25–1.01	0.058
Smoking	0.40	0.16–0.98	0.056
Infection signs	0.99	0.46–2.16	0.98
Ulcer history	0.93	0.50–1.71	0.80
Therapy	3.22	1.66–6.21	<0.000*
ACE inhibitors	0.95	0.49–1.82	0.88
Anticlotting therapy	0.69	0.35–1.38	0.30

to complete healing (HR = 3.2; 95% CI, 1.7 to 6.2), and preparation time (HR = 2.4; 95% CI, 1.2 to 4.7) independently small imbalances in prognostic factors (e.g., the ulcer surface area) to wound healing between both study groups.

Secondary Outcomes

Kaplan-Meier survival analysis (Figure 22-4, Table 22-6) showed a median recurrence rate at month 4 (95% CI, 0.7 to 7.4) after the VAC therapy versus month 2 (95% CI, 0.5 to 3.6) in the control group (P = 0.47). After 1-year follow-up (Table 22-6), 52% (n = 12) of all healed VAC ulcers relapsed compared to 42% (n = 10) in the control group (P = 0.47). All differences were not statistically significant.

The median percentage of successful skin grafts (Table 22-6) differed significantly (P = 0.011) between the VAC and control groups, with 83% ± SD 14% versus 70% ± SD 31%.

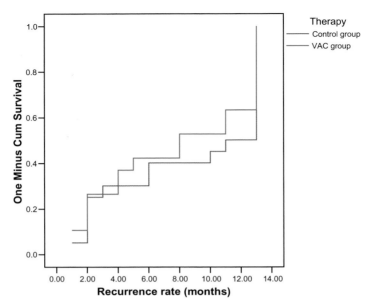

FIGURE 22-4 Recurrence rate (months) in the VAC. group as compared to standard wound care.

The total nursing time consumption (Table 22-6) was significantly longer during standard wound care (386 ± 178 minutes) than during VAC therapy (232 ± 267 minutes). There were no significant differences (P = 0.937) between VAC and control groups with respect to time consumption for medical attention through the physician (177 ± 76 minutes versus 181 ± 91 minutes).

Quality of Life

During the hospitalization period, 56 ulcers healed. Changes in QOL and in pain scores are presented in Table 22-3. The study groups showed significant increases in QOL at the end of therapy. During the first week, the QOL score was significantly lower in the VAC group (P = 0.031); however, this difference had already disappeared in the second week and, during follow-up, life quality was similar in both groups.

With respect to pain scores, both groups showed a significant decrease at the end of follow-up, too. Comparison of pain scores revealed that the scores were initially similar during the first weeks of treatment. From week 5 onwards, however, PPI scores were significantly lower in the VAC group.

Cost Analysis

We listed wound care costs in Table 22-4. Costs for personnel were calculated on the basis of average fees for medical and nursing personnel

TABLE 22-3 Mean Quality of Life Scores (SD) in the VAC Versus Control Group

Week	0	1	2	3	4	5	6	7	8
EQ-5D									
VAC	40 (13)	39 (16)*	47 (15)	62 (13)	70 (17)	74 (16)	73 (17)	76 (17)	77 (14)
Control group	45 (19)	49 (19)*	51 (18)	60 (16)	62 (18)	68 (17)	74 (15)	76 (12)	76 (17)
SF-MPQ									
VAC	9 (4)	10 (5)	6 (3)	5 (3)	4 (2)	2 (2)	2 (2)	1 (1)	1 (1)
Control group	10 (3)	10 (4)	7 (4)	6 (3)	4 (2)	3 (2)	2 (2)	2 (2)	1 (1)
PPI									
VAC	2.5 (1)*	3 (1)	1.9 (1)	1.4 (1)	1 (1)	0.5 (1)*	0.3 (0.5)*	0.2 (0.4)*	0.2 (0.7)*
Control group	3.1 (1)	3.2 (1)	2.6 (1)	1.8 (1)	1.3 (1)	1.4 (1)*	1 (1)	0.5 (1)	0.4 (0.6)

*Mann-Whitney U test, $P < 0.05$.

TABLE 22-4 Average Costs of Treatment

	Standard Wound Care (n = 30)	VAC (n = 30)	P
VAC related products ($)	0	847	
Bandages and dressings ($)	4,770	2,391	
Personal costs ($)	508	583	
Nurse costs ($)	175	124	
Total ($):	**5,452**	**3,881**	0.001*

* Mann-Whitney U test, P < 0.05.

TABLE 22-5 Adverse Events

	Standard Wound Care (n = 30)	VAC (n = 30)
Erysipelas	0	1
Pain	1	3
Cutaneous damage secondary to therapy	2	7
Wound infection	1	0
Postoperative bleeding at donor site	2	0
Nonhealing ulcer	1	1

* Mann-Whitney U test, P < 0.05.

adjusted for the clocked time spent by medical doctors and nurses with the patients. The total wound care costs for a hospitalized CLU were 25% to 30% lower for VAC than for standard wound care (Mann-Whitney U test, P = 0.001). Because we chose to keep patients hospitalized until complete healing was achieved, in contrast to normal clinical practice, we did not incorporate admission costs in the cost analysis. Considering that length of hospital stay was significantly longer for patients with standard wound care, the actual cost efficiency of VAC treatment would be even greater.

Complications

The complication rate was higher in the VAC group (40%) compared with the control group (23%), but the difference was not statistically significant (P = 0.17). Table 22-5 summarizes all adverse events.

VAC THERAPY IN CHRONIC LEG ULCERS: MODES OF ACTION

The pathophysiology of wound healing is complex and involves at least four major factors: local tissue hypoxia, bacterial colonization, repetitive ischemia-reperfusion injury, and altered cellular and systemic stress responses, mainly, in the aged patient. The molecular biology of wound healing is increasingly complex and involves a wide range of inflammatory mediators and a diverse array of processes such as matrix synthesis, angiogenesis, and wound contraction.

In the past, two main factors have been documented to exert an important influence on wound healing. First, it has been shown that chronic wound exudate may disturb normal wound healing because these fluids contain lower amounts of growth-promoting cytokines, persistent elevated inflammatory cytokines, and proteolytic enzymes.[3,14] Second, the considerable effects of mechanical forces on wound shape and tissue growth have been reported.[32] VAC therapy has been repeatedly utilized successfully to positively influence both factors by removing excess interstitial fluid and transmitting mechanical forces to the surrounding tissue.[15,16,33]

In support of this notion, a recent report indicated that reduced levels of matrix metalloproteinases 1, 2, and 13 during VAC presumably result in reduced collagen breakdown.[34] Further, active reduction of excessive wound fluid also results in decompression of small blood vessels, restores microcirculation, and increases oxygen and nutrient delivery to the wound. All these factors notably improve the rate of granulation tissue formation.[16,33] The positive influences of mechanical forces on the growth of tissues, especially in stimulating cell migration and mitosis during VAC, are also documented.[16,33,35] Further, an occlusive environment, as created by adhesive transparent dressings during VAC, markedly supports maintenance of a controlled, moist wound environment and reduces the infection chances.[16,33,35] All these factors have been documented to result in improved wound healing.

LIMITATIONS OF VAC THERAPY

Although VAC has shown to be of great importance in the treatment of chronic leg ulcers, one must not lose sight of certain limitations inherent to the procedure. Success of the treatment depends on thorough knowledge of the material and its indications and extensive clinical experience. It cannot be stressed enough that any necrosis or heavy slough needs to be removed prior to treatment. Occasionally, the connection tube and adhesive tape can cause pressure erosions and irritation eczema on the wound edges. Lifting the tube and protecting the edges by hydrocolloid dressings and zinc oil may overcome this. The large size of the equipment may keep the patient bedridden. Some patients experience discomfort and sharp pain sensations when a topical negative pressure is applied (especially patients with arteriolosclerotic ulcers), but reducing pressure temporarily and increasing slowly can alleviate these problems. VAC is generally contraindicated in cancerous wounds and untreated osteomyelitis.

CONCLUSIONS

Despite the development of modern diagnostic tools and remarkable therapeutic improvement, many CLUs do not heal satisfactorily in an outpatient clinic setting within a limited time period.[2-5,6-10,30] Furthermore, current treatment modalities are time consuming and expensive.[2] In this study, we used a prospective and comparative model to evaluate the efficacy of VAC treatment in recalcitrant CLUs treated in an inpatient facility compared with current standard therapeutic regimens.

Our study shows that VAC therapy results in a significant reduction of both the wound preparation time and the time to complete healing when compared to common treatment modalities. Because ambulatory VAC pump units are available, this treatment might be offered on an outpatient basis. Providing that patients adhered strictly to the treatment protocol and trained wound care nurses are employed, the same efficacy may be reached in an outpatient setting.

Mouës at al.[31] randomized patients with acute traumatic or infected wounds, pressure ulcers, and other chronic leg ulcers to VAC versus standard wound care. They did not find any difference in wound bed preparation time; however, they did observe a significant reduction in wound area surface by VAC, implying a positive effect of VAC on wound healing. Their failure to observe decreased wound bed preparation time may be related to inclusion of patients with acute infected or traumatic wounds.

For practical reasons, we subdivided the time to complete healing and distinguished the preparation phase and the overall complete healing time. Apparently, in our study the greatest benefit of VAC is reflected by a reduced median preparation time of 58% during the first time period (Figure 22-3). The overall complete healing time was reduced by 35% (Figure 22-2). These data demonstrate that VAC is an extremely valuable tool in wound bed preparation.

With this randomized clinical trial, we sought to assess the effects of VAC therapy on wound healing in patients with recalcitrant CLU irrespective of underlying etiology. This approach may have increased the external validity of our findings, but it imposes some limitations on the interpretation of the results, especially since the number of patients is insufficient to perform subgroup analyses for each type of CLU individually. Neverthe-

TABLE 22-6 Secondary Outcomes

	Standard Wound Care	VAC	P
Recurrence percentage (n)	42% (10)	52% (12)	Chi-square test P = 0.405
Median recurrence moment month	2nd	4th	Log rank test P = 0.47
Median percentage skin graft survival (SD) *	70% (31)	83% (14)	Mann-Whitney U test P = 0.011
Median wound care time (SD)			Mann-Whitney U test
Nurse*	330 (178)	140 (267)	P = 0.001
MD	174 (91)	155 (76)	P = 0.887

* Significant difference P < 0.05.

less, with this limitation in mind, VAC therapy resulted in a decreased wound preparation time and total wound healing time within each CLU group (i.e., venous and arteriolosclerotic, or combined venous/arterial). For the venous and arteriolosclerotic ulcers, but not for the combined venous/arterial ulcers, this difference reached statistical significance (data not shown).

A critical aspect in the survival of a skin graft is an adequate contact to the granulated wound bed. With conventional treatment using compression bandages, the survival rates of skin grafts are usually much lower owing to excessively exudative wound bed surfaces and surfaces subjected to motion.[23] These difficulties can be overcome, though, by application of VAC as demonstrated by Scherer et al.[23] and confirmed in our study (Table 22-6) by the higher percentage of adhered skin grafts (80% versus 70%, P < .011).[23]

One of the major problems in leg ulcer treatment is the high frequency of relapses.[13] Interestingly, recurrence rates in CLUs after treatment with VAC therapy compared with standard wound care modalities have not been reported to date, to the best of our knowledge. Without any doubt, it would be of great advantage in daily practice if VAC revealed lower recurrence rates than standard wound care regimens. We observed 12 recurrences (52%) in the VAC group and 10 relapses (42%) after SWC at 12 months follow-up (Table 22-6). Although this difference was not significant, larger study cohorts are certainly needed to determine a putative positive effect of VAC with respect to ulcer recurrence rates.

Recently, socioeconomic and QOL studies in patients with CLUs revealed the broad impact of this long-neglected health care problem on everyday routine.[36] Persoon et al.[36] convincingly demonstrated a strong correlation between the amount of time spent on ulcer care, feelings of anger, and the QOL of patients. In our study, the overall subjective QOL measured by the EQ-DSI and pain scores improved significantly in both groups within 8 weeks but did not differ significantly. Still, the improvement in QOL was achieved more rapidly in the VAC group despite an initial decrease during the first week of treatment (Table 22-3). This may be explained by an accelerated wound preparation phase through VAC. A possible explanation for the initial decrease during VAC treatment might be that patients experienced the necessity of strict bed rest as a factor negatively influencing QOL.

Wound care costs for the treatment of CLUs following standard protocols were significantly higher than those in the VAC group. The biggest part of this cost difference was caused by higher personnel costs and longer hospitalization time due to slower wound healing in the control group. We are aware of the fact, though, that this wound care time might not represent the real situation because, in daily practice, most patients with leg ulcers will be discharged once the ulcers are almost closed.

In conclusion, the data of the randomized clinical trial presented here strongly advocate the use of VAC therapy in a mixed population of ulcers. The preparation of the wound bed is faster, time to complete healing with VAC therapy is a mean of 16 days shorter, and, not surprisingly, the costs are less. Rapid improvement in quality of life and pain are noted with VAC therapy, and this is critical for this patient population, who are generally demoralized by their inability to heal their wounds. This is good news for clinicians, administrators, and, most importantly, patients.[37]

REFERENCES

1. Bergan JJ, Schmid-Schonbein GW, Smith PD, Nicolaides AN, Boisseau MR, Eklof B. Chronic venous disease. *N Engl J Med.* 2006;355:488–498.
2. Mekkes JR, Loots MA, Van Der Wal AC, Bos JD. Causes, investigation and treatment of leg ulceration. *Br J Dermatol.* 2003;148:388–401.
3. Schultz GS, Sibbald RG, Falanga V, et al. Wound bed preparation: A systematic approach to wound management. *Wound Repair Regen.* 2003;11(1):S1–S28.
4. Harold B, Robert K, Falanga V. Protocol for the successful treatment of venous ulcers. *Am J Surg.* 2004;188:1S-8S.
5. Barwell JR, Davies CE, Deacon J, et al. Comparison of surgery and compression with compression alone in chronic venous ulceration (ESCHAR study): Randomised controlled trial. *Lancet.* 2004;363:1854–1859.
6. Bradley M, Cullem N, Sheddon T. The debridement of chronic wounds: A systematic review. *Health Tech Assess.* 1999;3(17 Pt 1):1–78.
7. Bradley M, Cullum N, Nelson EA, Petticrew M, Sheldon T, Togerson D. Systemic reviews of wound care management: Dressings and topical agents used in the healing of chronic wounds. *Health Tech Assess.* 1999;3(17 Pt 2):1–78.
8. Howell-Jones RS, Wilson MJ, Hill KE, Howard AJ, Price PE, Thomas DW. A review of the microbiology, antibiotic usage and resistance in chronic skin wounds. *J Antimicrob Chemother.* 2005;55(2):143–149.

9. Jones JE, Nelson EA. Skin grafting for venous leg ulcers. *Cochrane Database Syst Rev.* 2005;25(1):CD001737.
10. Falanga V, Margolis D, Alvarerez O, et al. Rapid healing of venous ulcers and lack of clinical rejection with an allogenic cultured human skin equivalent. *Arch Dermatol.* 1998;134:293–300.
11. Nicolaides AN. Investigation of chronic venous insufficiency: A consensus statement. *Circulation.* 2000;102:123–163.
12. The Venous Forum of the Royal Society of Medicine and Societas Phlebologica Scandinavica. The management of chronic venous disorders of the leg: An evidence-based report of an international task force. *Phlebology.* 1999;14:23–25.
13. Erickson CA, Lanza DJ, Karp DL, et al. Healing of venous ulcers in an ambulatory care program: The roles of chronic venous insufficiency and patient compliance. *J Vasc Surg.* 1995;22:629–636.
14. Renaud J. Postoperative drainage of wounds by means of negative pressure. *Ned Milit Geneeskd Tijdschr.* 1959;103:1845–1847.
15. Fleischmann W, Strecker W, Bombelli M, Kinzl L. Vacuum sealing as treatment of soft tissue damage in open fractures. *Unfallchirurg.* 1993;96:488–492.
16. Argenta LC, Morykwas MJ, Vacuum-assisted closure: A new method for wound control and treatment: Clinical experience. *Ann Plast Surg.* 1997;38:562–576.
17. Vuerstaek JDD, Wuite J, Veraart JCJM, Neumann HAM. Un nouveau concept de cicatrisation active des ulcères rebelles veineux ou mixtes: Le vacuum assisted closure. *Phlébologie.* 2001;55(1):15–19.
18. McCallon SK, Knight CA, Valiulus JP, Cunningham MW, McCulloch JM, Farinas LP. The effectiveness of vacuum assisted closure vs. saline moistened gauze in the healing of post-operative diabetic foot wounds. *Ostomy Wound Manage.* 2000;46:28–32.
19. Evans D, Land L. Topical negative pressure for treating chronic wounds. *Cochrane Database Syst Rev.* 2001;1:CD001898.
20. Finnie A. The Scottish Leg Ulcer Project. *J Tissue Viability.* 1999;9:85–87.
21. Falanga V. Classifications for wound bed preparation and stimulation of chronic wounds. *Wound Repair Regen.* 2000;8(5):347–352.
22. World medical association declaration of Helsinki. *Cardiovasc Res.* 1997;35:2–3.
23. Scherer LA, Shiver S, Chang M, Meredith JW, Owings JT. The vacuum assisted closure device: A method of securing skin grafts and improving graft survival. *Arch Surg.* 2002;137(8):930–933.
24. Euroquol Group. Euroquol—A new facility for the measurement of health-related quality of life. *Health Pol.* 1990;16:199–208.
25. Nemeth KA, Harrison MB, Graham ID, Burke S. Understanding venous leg ulcer pain: Results of a longitudinal study. *Ostomy Wound Manage.* 2004;50:34–46.
26. Hollis S, Campbell F. What is meant by intention to treat analysis? Survey of published randomized controlled trials. *Br Med J.* 1999;319:670–674.
27. Hop WCJ, Hermans J. Statische analyse van overlevingsduren. *T soc. Geneesk.* 1981;59:279–287.
28. Phillips TJ, Machado F, Trout R, Porter J, Olin J, Falanga V, and the Venous Ulcer Study Group. Prognostic indicators in venous ulcers. *J Am Acad Dermatol.* 2000;43:627–630.
29. Gohel MS, Taylor M, Earnshaw JJ, Heather BP, Poskitt KR, Whyman MR. Risk factors for delayed healing and recurrence of chronic venous leg ulcers: An analysis of 1,324 legs. *Eur J Vasc Endovasc Surg.* 2005;29:74–77.
30. Mantoni M, Larsen L, Lund JO, et al. Evaluation of chronic venous disease in the lower limbs: Comparison of five diagnostic methods. *Br J Radiol.* 2002;75(895):578–583.
31. Mouës CM, Vos MC, van de Bemd GJ, Stijnen T, Hovius SE. Bacterial load in relation to vacuum-assisted closure wound therapy: A prospective randomized trial. *Wound Repair Regen.* 2004;12(1):11–17.

32. Albuquerque ML, Waters CM, Savla U, Schnaper HW, Flozak AS. Shear stress enhances human endothelial cell wound closure in vitro. *Am J Physiol Heart Circ Physiol.* 2000;279:H293–302.
33. Morykwas MJ, Argenta LC, Shelton-Brown EI, McGuirt W. Vacuum-assisted closure: A new method for wound control and treatment: Animal studies and basic foundation. *Ann Plast Surg.* 1997;38:553–562.
34. Shi B. Effects of vacuum assisted closure (VAC) on the expressions of MMP-1, 2, 13 in human granulation wounds. *Zhongua Zheng Wai Ke Ka Zhi.* 2003;19:279–281.
35. Morykwas MJ. External application of sub-atmospheric pressure and healing: Mechanisms of action. *Wound Healing Soc Newsletter.* 1998;8:4–5.
36. Persoon A, Heinen MM, van der Vleuten CJ, de Rooij MJ, van de Kerkhof PC, van Achterberg T. Leg ulcers: A review of their impact on daily life. *J Clin Nurs.* 2004; 13:341–354.
37. Thomas F. Lindsay. Invited commentary. *JVS.* 2006;44:1038
38. Vuerstaek JDD, Vainas T, Wuite J, Nelemans P, Neuman HAM, Veraart J. State-of-the-art treatment of chronic leg ulcers: A randomized controlled trial comparing vacuum-assisted closure (V.A.C.) with modern wound dressings. *JVS.* 2006;44: 1020–1037.

Epilogue

John J. Bergan, MD, FACS, FRCS, Hon. (Eng)

It is generally agreed that venous leg ulcers are important. This has been true since the early 19th century when, as Loudon reports, between 1760 and 1800, leg ulcers accounted for 16% to 23% of all inpatient admissions at the Devon and Exeter Hospitals.[1] It takes a medically trained person to differentiate between the various underlying causes, but some 80% are venous in origin.

David Negus, in his book on leg ulcers, has said, "it is sad but true that there are only four facts about leg ulcers which can be stated without fear of contradiction: they are common, their treatment is time consuming and tedious, they are not life threatening and most surgeons would prefer someone else to be looking after them."[2]

Although the basic knowledge of tissue repair and regeneration has expanded exponentially, it is well recognized that only correction of the underlying pathophysiology will heal the ulcer and keep it from recurring. Now, it can be said that investigations into the pathophysiology have become scientific with the advent of Doppler ultrasound, that research into fundamental causation is at the molecular level and that correction of venous pathophysiology has become increasingly less invasive.

It was intended that this book would provide a stepping-stone into the future, and a glance back through the chapter titles reveals that it has done so. Exciting times are here for the investigation and treatment of venous leg ulcers. Perhaps it is still true that we surgeons would prefer that someone else care for them, but there are new and increasingly powerful tools at our disposal. These are detailed in this book, and the future will disclose others that are even more effective. If this book stimulates the generation of new ideas, new techniques, and more effective methods, we who have gathered the present thoughts will be gratified indeed.

REFERENCES

1. Loudon ISL. Leg ulcers in the eighteenth and early nineteenth centuries. *J Roy Coll Gen Pract*. 1981;31: 263–273. Cited in: Ryan TJ, ed. in *Leg Ulcers: Diagnosis and Treatment*. Esterhoff D. Amsterdam: Elsevier, 1993.
2. Negus D. Introduction, p vii, in *Leg Ulcers, A Practical Approach to Management*. Oxford: Butterworth/Heinemann, 1991.

Index

A

ABI. *See* Ankle-brachial index
Acquired thrombophilia conditions, 56t, 60–62
   antiphospholipid syndrome, 58t, 60–61
   hyperhomocysteinemia, 58t, 61
   recurrent DVT, 58t, 61–62
Adjuvant therapy, 93–95
   allografts, 93–94
   collagen dressings and growth factor therapy, 95
   HSEs, 94
   VAC, 95
   xenografts, 93
Age, venous ulcers distribution by, 35–36, 35f
Air plethysmography (APG)
   overview of, 152
   saphenous reflux correction with, 162
   for VLUs, 152–154, 153–154f
Allergic reaction, with foam sclerotherapy, 208
Allograft cryopreserved valve, for venous valve repair, 247
Allografts, for venous ulcers, 93–94
Angioscopy-assisted external valve repair, for venous valve repair, 249
Ankle-brachial index (ABI), for PAD, 301
Anticoagulation, for DVT, 62–63, 276, 292
Antioxidants, in venous ulcers, 21
Antiphospholipid syndrome, as thrombophilia condition, 58t, 60–61
Antithrombin
   deficiency as thrombophilia condition, 57–59, 58t
   DVT and, 56–58
APG. *See* Air plethysmography
Apligraf, for venous ulcers, 94
Arterial insufficiency, and VLUs, 299–305
   evaluation of, 299–302
Arteriovenous shunting, venous ulcers from, 16
Atherosclerosis, from homocystein, 61

B

Bacterial burden, with negative pressure dressings, 107–108
Balloon angioplasty, for iliac lesions, 283
Blood flow, with negative pressure dressings, 107
Blood pressure, in venous disease, 10
B-lymphocytes
   in venous disease, 5
   in venous ulcers, 19
Bradykinin, in leukocyte activation, 5

C

Capillary occlusion, by leukocytes, 18–19
Capillary permeability, edema and, 71–72
CEAP classification, 173, 174f
CHIVA. *See* Hemodynamic correction of varicose veins

Chronic leg ulcers (CLUs). *See also* Venous leg ulcers
  randomized clinical trial for VAC therapy and, 308–320
    complications of, 320, 320t
    cost analysis of, 312, 318–320, 320t
    evaluation criteria for, 311
    patient and ulcer characteristics in, 314f, 315
    procedures for, 310–311
    protocol violations in, 314f, 315
    quality of life and pain scores in, 311–312, 318, 318f, 319t, 323t
    recruitment for, 314–315, 314f
    secondary outcomes of, 317–318, 318f, 323t
    setup, 309–310
    statistical analysis of, 312–314, 313t, 316f, 317t, 318f
    time consumption in, 312
    VAC v. conventional wound care in, 315–317, 316f, 317t
  VAC therapy for, 307–324
    concluding remarks on, 316f, 319t, 322–324, 323t
    limitations of, 321–322
    modes of action, 321
Chronic venous disease (CVD)
  clinical study review for, 303–305
  concluding remarks on, 305
  duplex ultrasound for, 43–52
    deep veins in, 50–51
    findings and treatment implications of, 51–52
    mapping for guiding treatment, 52
    perforating veins in, 49–50
    requirements and settings for, 43–45
    superficial veins in, 46–49
    ulcer bed in, 51
  evaluation of, 299–302
    arteriography and venography, 301–302
    history and physical examination, 299–301
    vascular laboratory testing, 301
  pathophysiology of, 3–11
    inflammation as repair mechanism, 8–9
    inflammatory mediators in, 5–6
    leukocyte activation in, 4–5
    proteolytic degradation, 6–8
    tissue restructuring, 6
    trigger mechanism in, 9–10
  quality of life with, 162
  treatment of, 302–303
  VSU with, 299
Chronic venous insufficiency (CVI)
  acquired thrombophilia conditions in, 56t, 60–62
    antiphospholipid syndrome, 58t, 60–61
    hyperhomocysteinemia, 58t, 61
    recurrent DVT, 58t, 61–62
  etiology of, 243–244
  foam sclerotherapy for, 185–196
    clinical applications of, 192–195
    foam for, 187–192
    philosophical conclusions for, 195–196
  following DVT, 55–56, 56t, 62
  hypercoagulable states with, 55–64
  inherited thrombophilia conditions in, 56–60, 56t
    antithrombin deficiency, 57–59, 58t
    factor V Leiden mutation, 57, 58t
    other conditions, 59–60
    protein C and S abnormalities, 58t, 59
  perforator reflux in, 166
  PPG for, 152
  presentation of, 149–150
  venous obstruction in, 275–295
    diagnosis and evaluation of, 277–280, 278f, 281f
    endovenous therapy for, 281–294
  venous valve repair in, 243–255
    concluding remarks on, 255
    indications for, 254–255
    methods for, 244–249
    rationale for, 244
    results of, 249, 250–253t, 254
  VLUs from, 55, 149–150
Chronically swollen leg
  clinical presentation of, 72–74, 73f, 73t
  edema mechanisms, 69–72, 70f
  finding cause of, 67–74
  fluid exchange in microcirculation, 68–69, 68f
Circ-Aid, for venous ulcers, 84
CLUs. *See* Chronic leg ulcers
Coban two-layer bandage, for venous ulcers, 84

Collagen
  decreased production of, 22
  MMP-2 degradation of, 21
  in pericapillary bed, 17
  restructuring of, 6
Collagen dressings, for venous ulcers, 95
Color flow imaging, with duplex ultrasound, 43–44
Compression bandages
  after EVLA, 180
  after foam sclerotherapy, 218
  after RFA, 179
  for venous ulcers, 79–80, 80t, 126
Compression bandaging systems, for venous ulcers, 83–85, 83f
Compression therapy
  devices for, 77–86, 78t
    compression bandages, 79–80, 80t
    compression bandaging systems, 83–85, 83f
    elastic and inelastic bandages, 80–82, 81t, 82f
    intermittent pneumatic compression, 85
    medical compression stockings, 77–79, 79t
    pelottes and pads, 85–86, 86f
    single-component and multiple-component bandages, 83
    single-layer and multilayer bandages, 82–83
  after DVT, 275–276
  minimally invasive techniques v., 239
  PTS prevention with, 275–276
  recurrence prevention with, 86
  with surgery
    alone v., 135–146
    concluding remarks on, 145–146
    ESCHAR study, 136–145
    other studies, 145
  techniques for, 86–89, 87f
  for venous ulcers, 77–89
Computed tomographic arteriography (CTA), for PAD, 301–302
Concomitant varicosity ablation, need for, 162
Contrast arteriography, for PAD, 301–302
Contrast phlebography, for venous obstruction, 280
CTA. See Computed tomographic arteriography
CVD. See Chronic venous disease

CVI. See Chronic venous insufficiency
Cytokines
  in leukocyte activation, 5
  in venous disease, 3
  in venous ulcers, 20

D

Deep vein thrombosis (DVT)
  anticoagulation for, 62–63, 276, 292
  antithrombin for, 56–58
  compression therapy for, 275–276
  CVI following, 55–56, 56t, 62
  with foam sclerotherapy, 193, 206–207
  heparin for, 277
  history of, 37
  PTS and, 113–127
  recurrent, 58t, 61–62
  from RFA, 182
  risk of, 114–116, 115t
  thrombophilias and, 56, 56t
  treatment for, 62–63
  VLUs and, 275
Deep veins
  reflux of, 156, 157t, 244
  in venous ulcers, 50–51
Deep venous incompetence (DVI)
  SVI v., 38
  venous ulcer from, 36–37
DU. See Duplex ultrasound
Duplex ultrasound (DU)
  for chronic venous disease, 43–52
  for deep veins, 50–51
  description of, 43
  for endovenous treatment, 227–239
    equipment for, 228
    monitoring by, 231–234, 232–234f
    preoperative assessment for, 227–228
    venous mapping for, 228–231, 229f
  findings and treatment implications of, 51–52
  for foam sclerotherapy, 202, 210
    monitoring with, 234–239, 235–238f
  mapping for guiding treatment, 52
  for perforator veins, 49–50
  requirements and settings for, 43–45, 45–48f
  for superficial veins, 46–49
  ulcer bed in, 51
  for VLUs, 150–151
DVI. See Deep venous incompetence
DVT. See Deep vein thrombosis

E

Edema
  with chronic venous wounds, 106–107
  clinical presentation of, 72–74, 73f, 73t
  mechanisms of, 69–72, 70f
  with severe CVI, 243
  swollen leg and, 67–68
EF. *See* Ejection fraction
Ejection fraction (EF), calculation of, 152–153, 153f
Elastic bandages, for venous ulcers, 80–82, 81t, 82f
Elastic compression stockings, for post-thrombotic syndrome, 63
Elastic fibers, MMP-2 degradation of, 21
Elastin fiber, restructuring of, 6
Endothelium, leukocyte attachment to, 4
Endovascular techniques, for superficial vein ablation, 173–183
  complications of, 181–182
  concluding remarks on, 182–183
  history of, 173–174
  mechanisms of action, 175–177
  obliterative techniques, 178–180
  options for, 175
  outcomes of, 180–181
  patient selection for, 177–178
Endovenous laser ablation (EVLA)
  comments on, 239
  complications of, 182
  DU for, 227–239
    equipment for, 228
    monitoring by, 231–234, 232–234f
    preoperative assessment for, 227–228
    venous mapping for, 228–231, 229f
  methods for, 180
  outcomes of, 181
  overview of, 176–177, 177f
  RFA v., 182
  for saphenous reflux, 164
Endovenous therapy
  anticoagulation in, 292
  concluding remarks on, 295
  IVC filters in, 292–294
  IVUS for, 290–291
  occluded IVC filters in, 293–294f, 294
  restenosis in, 291–292
  stents in, 292
  for venous obstruction
    acute thrombotic obstruction, 282–283
    chronic thrombotic obstruction: iliac, 283, 284f
    chronic thrombotic obstruction: IVC, 284–290, 285–287t, 288–291f, 289t
    nonthrombotic obstruction, 281–282, 282f
    special issues in, 290–294
Enoxaparin, as thromboprophylaxis, 121t, 123–124
Epidemiology
  description of, 27–28
  incidence v. prevalence in, 28–30, 29t
ESCHAR study, 136–145
  clinical outcomes in, 137–140, 139f, 141f
    anatomic changes, 140–143, 142f
    hemodynamic changes, 143–144, 144f
    surgery impact on healing, 137–140, 139f
    surgery impact on recurrence, 140, 141f
  critique of, 144–145
  patient flow in, 137
  study design for, 136–137, 138f
EVLA. *See* Endovenous laser ablation

F

Factor V Leiden mutation, as thrombophilia condition, 57, 58t
Femoral vein, SSV termination at, 161
Fibrin, in pericapillary bed, 17–18, 17f
Fibrinogen, in venous ulcers, 17–18
Fibroblasts, in venous ulcers, 22
Fibronectin
  degradation of, 21
  in pericapillary bed, 17
Fluid shear stress
  in inflammatory and thrombotic phenotype, 9–10
  in pseudopod formation, 4
  in venous disease, 4, 10
Foam sclerotherapy
  from classic to, 201
  comments on, 209–211, 221–223, 222f, 222t, 239
  complications of, 181–182, 194–195
  contraindications for, 209
  for CVI, 185–196
  DU monitoring for, 234–239, 235–238f
  foam for, 187–192
    clinical applications of, 192–195
    philosophical conclusions for, 195–196
  methods for, 178, 186–187, 202–203, 216–218, 217f

outcomes of, 180, 203–207, 204–205f, 206t, 207f, 218–221, 219–220f, 219t
overview of, 175, 186
safety of, 208–209
statistical analysis of, 218
for VLUs, 146, 199–211, 215–223
Focal paresthesias, from RFA, 182
Fondaparinux, as thromboprophylaxis, 121t, 124–125
Four-layer bandage, for venous ulcers, 85
Free tissue transfer
  free flap selection for, 265–267, 265f, 268f
  postoperative care for, 269
  surgical technique for, 268–269
  tissue debridement for, 265–266, 266–267f
  for VLUs treatment, 261–272
    concluding remarks on, 271–272
    discussion of, 269–271
    overview of, 261–262, 264t
    patient evaluation and selection for, 262–266
Frullini's method, for foam sclerotherapy, 193

G

G-protein, pseudopods from, 4
Granulocytes, activating factor for, 5
Great saphenous vein (GSV)
  imaging of, 44, 45–48f, 46–49
  location of, 48–49
  mapping of, 230
  reflux of, varicose veins from, 173
  removal of, 157–158
    to ankle or knee, 160–166
    history of, 174
    with saphenofemoral ligation, 158–160
  sclerotherapy in, 188
Growth factor therapy, for venous ulcers, 95
GSV. *See* Great saphenous vein

H

Hemodynamic changes
  with saphenous surgery, 162
  with surgery, 143–144, 144f
  venous valve repair and, 254–255
  with VFI reduction, 162
Hemodynamic correction of varicose veins (CHIVA), for saphenous reflux, 163

Heparin
  antithrombin and, 58
  for DVT, 277, 292
Homocysteine, function of, 61
HSEs. *See* Human skin equivalents
Human skin equivalents (HSEs), for venous ulcers, 94
Hydraulic pressure gradient, edema and, 70–71
Hydrogen peroxide, plasma production of, 5
Hyperhomocysteinemia, as thrombophilia condition, 58t, 61
Hypoxia, negative pressure dressings for, 109

I

ICAM-1. *See* Intracellular adhesion molecules
IL-6. *See* Interleukin-6
IL-1α. *See* Interleukin-1α
Iliac, chronic thrombotic obstruction of, 283, 284f
Incidence
  prevalence v., 28–30, 29t
  of venous ulcer, 34
Incompetent perforator veins (IPVs), surgical ligation of, 166–169, 167f
  concluding remarks on, 169, 170f
  indications for, 168–169
  procedure for, 166–168, 167f
Inelastic bandages, for venous ulcers, 80–82, 81t, 82f
Infection, from RFA, 182
Inferior vena cava (IVC), chronic thrombotic obstruction of, 284–290, 285–287t, 288–291f, 289t
Inflammation, venous disease pathophysiology and, 3–11
  inflammatory mediators in, 5–6
  leukocyte activation in, 4–5
  proteolytic degradation, 6–8
  as repair mechanism, 8–9
  tissue restructuring, 6
  trigger mechanism for, 9–10
Inflammatory cascade
  interference of, 8–9
  triggering of, 9
Inflammatory mediators
  abdominal obesity and, 11
  digestive enzymes and, 11
  in venous disease, 3, 5–6

Inherited thrombophilia conditions, 56–60, 56t
  antithrombin deficiency, 57–59, 58t
  factor V Leiden mutation, 57, 58t
  other conditions, 59–60
  protein C and S abnormalities, 58t, 59
Integrins
  in leukocyte and endothelium attachment, 4
  in tissue restructuring, 6
Interleukin-6 (IL-6), in venous ulcers, 20
Interleukin-1α (IL-1α), in venous ulcers, 20
Interleukins, in venous ulcers, 20
Intermittent pneumatic compression (IPC)
  for DVT prevention, 125
  for venous ulcers, 85
Internal valvuloplasty, for venous valve repair, 244, 245f
Intracellular adhesion molecules (ICAM-1), in leukocyte migration and activation, 19
Intravascular ultrasound (IVUS)
  for endovenous therapy, 290–291
  for intimal hyperplasia, 291
IPC. *See* Intermittent pneumatic compression
IPVs. *See* Incompetent perforator veins
Iron, in venous ulcers, 21
IVC. *See* Inferior vena cava
IVUS. *See* Intravascular ultrasound

K
Kikuhime tester, for compression bandages, 80
Klippel-Trenaunay syndrome, SSV and, 161

L
Lactoferrin, in venous hypertension, 7
Laminin, in pericapillary bed, 17
LDS. *See* Lipodermatosclerosis
Leg Club, for venous ulcer care, 99–100
Leg ulcer. *See also* Venous leg ulcers
  description of, 29–30
  etiologic spectrum of, 30–31
  prevalence of, 30
Leukocytes
  activation of
    inflammatory molecules in, 5
    microvascular infiltration and, 4–5
  in capillary occlusion, 18–19
  circulating, in venous disease, 4
  neutrophilic, entrapment in microcirculation, 4–5
  in venous ulcers, 19–20
Leukolysin, from MMPs, 7
Leukotrienes, in leukocyte activation, 5
Ligation
  with GSV stripping, 159
  of IPVs, 166–169, 167f
  SEPS v. open, 174
  at SFJ, 157–160
Linton procedure
  for perforator reflux, 166–167, 174
  SEPS with, 167–168
Lipedema, edema v., 74
Lipodermatosclerosis (LDS)
  fibroblasts in, 22
  in free tissue transfer, 262, 263f
  MMP-2 in, 21
LMWH. *See* Low molecular weight heparin
Low molecular weight heparin (LMWH), as thromboprophylaxis, 121t, 122–124
Lymphatic flow, edema and, 72
Lymphedema, edema v., 72

M
α-Macroglobulin, in pericapillary cuff, 18
Macrophages
  abdominal obesity and, 11
  MMP production by, 8
  in repair mechanism, 8
  in venous ulcers, 19
Magnetic resonance arteriography (MRA), for PAD, 301–302
Magnetic resonance imaging, for venous obstruction, 280
Mast cells
  in venous disease, 5
  in venous ulcers, 19
Matrix metalloproteinases-2 (MMP-2), degradation by, 21
Matrix metalloproteinases (MMPs)
  activation of, 7–8
  activity of, 22
  control of, 6–7
  in inflammation and skin changes, 6–8
  sources of, 6–7
  tissue inhibitors of, 7
  in venous ulcers, 21–22

May-Thurner syndrome, IVUS for, 290
Medical compression stockings, for venous ulcers, 77–79, 79t
Medication, edema caused by, 72, 73t
Membrane type metalloproteinases (MT-MMPs), in venous ulcers, 21
Microcirculation
  in chronic venous insufficiency, 15–23
  fluid exchange in, 68–69, 68f
Microvascular infiltration, leukocyte activation and, 4–5
Migraines, SF and, 194, 208–209
MMP-2. *See* Matrix metalloproteinases-2
MMPs. *See* Matrix metalloproteinases
Monfreux method, for foam sclerotherapy, 178, 208
Monocytes, in venous disease, 5
MRA. *See* Magnetic resonance arteriography
MT-MMPs. *See* Membrane type metalloproteinases
Multilayer bandages, for venous ulcers, 82–83, 135
Multiple-component bandages, for venous ulcers, 83

N
Negative pressure dressings
  overview of, 105–106
  in venous ulcers, 105–111
    adjunct to assisted closure, 109–110
    bacterial burden, 107–108
    chronic wound fluid, 106
    concluding remarks on, 110–111
    direct wound closure, 108
    indications and contraindications, 110
    local blood flow, 107
    local edema, 106–107
    wound bed deformation, 108
    wound healing environment, 108–109
    wound hypoxia, 109
Neo valve
  results of, 249
  for venous valve repair, 246–247, 248f
Nerve damage, with foam sclerotherapy, 208
Neutrophil collagenase, from MMPs, 7
Neutrophil elastase, in venous hypertension, 7
Neutrophils, in venous ulcers, 19
Nonsaphenous veins, with CVD, 49
Nonthrombotic obstruction, 281–282

O
Obesity
  abdominal, venous disease and, 10–11
  venous ulcers and, 37
Oncotic pressure gradient, edema and, 71
Open cell sponge, as negative pressure dressing, 106
Orthostatic collapse, with foam sclerotherapy, 208
Osteoporosis, from homocystein, 61
Overall prevalence, 29
Oxygen free radicals. *See* Reactive oxygen species

P
PAD. *See* Peripheral arterial disease
Pads, for venous ulcers, 85–86, 86f
Patent foramen ovale (PFO), SF with, 194, 209
PDGFR-α. *See* Platelet derived growth factor receptor α
PDGFR-β. *See* Platelet derived growth factor receptor β
Pelottes, for venous ulcers, 85–86, 86f
Percutaneous ultrasound-guided coil embolization, for VLUs, 201
Perforator interruption, for VLUs, 149–170
Perforator veins
  ligation of, 166–169
    concluding remarks on, 169, 170f
    indications for, 168–169
    procedure for, 166–168, 167f
  mapping of, 230–231
  reflux of, 156, 157t
    in CVI, 166
  linton procedure for, 166–167, 174
  in venous ulcers, 49–50
Perforator venous incompetence (PVI), venous ulcer from, 36
Period prevalence, 29
Peripheral arterial disease (PAD)
  clinical study review for, 303–305
  concluding remarks on, 305
  evaluation of, 299–302
    arteriography and venography, 301–302
    history and physical examination, 299–301
    vascular laboratory testing, 301
  prevalence of, 299
  treatment of, 302–303

PFO. *See* Patent foramen ovale
Phlebectomies, for varicose veins, 195
Phlebitis, from RFA, 182
Phlebography
  IVUS v., 290
  for venous obstruction, 280
Phlebotomy
  venous valve repair with, 244–247, 245–248f
    allograft cryopreserved valve, 247
    internal valvuloplasty, 244, 245f
    neo valve, 246–247, 248f
    vein valve transplantation, 246, 247f
    venous segment transfer, 244–246, 246f
  venous valve repair without, 247–249
    external valvuloplasty, 247–249
    wrapping, banding, cuffing, and external stenting, 247
Photoplethysmography (PPG)
  overview of, 151
  for VLUs, 151–152
Pigmentation, with foam sclerotherapy, 206–208
Plasma, granulocytes in, 5
Plasmin, MMP activation by, 7–8
Platelet derived growth factor receptor α (PDGFR-α), in venous ulcers, 20
Platelet derived growth factor receptor β (PDGFR-β), in venous ulcers, 20
Plethysmography
  air, for VLUs, 152–154, 153–154f
  overview of, 151
  PPG, for VLUs, 151–152
Point prevalence, 29
POL. *See* Polidocanol
Polidocanol (POL)
  for foam sclerotherapy, 193, 202, 217
  for liquid sclerotherapy, 201
  for varicose veins, 186
Popliteal vein, SSV termination at, 161
Post-thrombotic syndrome (PTS)
  compression therapy for, 275–276
  DVT and, 113–127
  elastic compression stockings for, 63
  obstruction in, 254–255, 279
  overview of, 113–114, 114t
  prevention of, 126–127
  results
    for transplantation, 253t, 254
    for transposition, 252t, 254

  risk of, 114–115, 116t
  VLUs and, 275
PPG. *See* Photoplethysmography
Pregnancy, telangiectases in, 11
Prevalence
  incidence v., 28–30, 29t
  of leg ulcer, 30
  types of, 29
  of venous ulcer, 31–34, 32–34t
Prostaglandin, in leukocyte activation, 5
Protein C abnormality, as thrombophilia condition, 58t, 59
Protein S abnormality, as thrombophilia condition, 58t, 59
Pseudopods, with circulating leukocytes, 4
PTS. *See* Post-thrombotic syndrome
Pulmonary embolism (PE)
  DVT and, 113
  from RFA, 182
Pütter bandage, for venous ulcers, 84
PVI. *See* Perforator venous incompetence

R

Radiofrequency ablation (RFA)
  comments on, 239
  complications of, 182
  EVLA v., 182
  for foam sclerotherapy, 179
  methods for, 179
  outcomes of, 180–181
  overview of, 175–176, 176f
  for saphenous reflux, 164
RAM. *See* Risk assessment model
Reactive oxygen species
  in leukocyte activation, 5
  from neutrophilic leukocytes, 4
  production of, 9
  in venous disease, 3
  in venous ulcers, 20–21
Regranix, for venous ulcers, 95
Residual volume fraction (RVF), calculation of, 152–153, 153f
Restenosis, in endovenous therapy, 291–292
RFA. *See* Radiofrequency ablation
Risk assessment model (RAM), for VTE thromboprophylaxis, 117–118, 119–120f
Rosidal sys, for venous ulcers, 84
RVF. *See* Residual volume fraction

S

Saphenofemoral junction (SFJ)
  imaging of, 46
  ligation at, 157–162
  mapping of, 230
  in surgery for GSV removal, 157–158
  varicose veins and, 173
Saphenofemoral ligation, saphenous stripping with, 158–160
Saphenopoliteal junction (SPJ), varicose veins and, 173–174
Saphenous banding, for saphenous reflux, 163
Saphenous stripping
  to ankle or knee, 160–166
    concomitant varicosity ablation in, 162
    conventional saphenous surgery for, 162
    deep and superficial venous insufficiency, 165–166
    nonconventional treatment for, 163–164
    perforator reflux in CVI, 166
    small saphenous reflux in, 161–162
    saphenofemoral ligation with, 158–160
Scintillating scotomas, with foam sclerotherapy, 208
Sclerosing foam (SF). *See also* Foam sclerotherapy
  clinical applications of, 192–195
  overview of, 186–187, 235–236
  philosophical conclusions of, 195–196
  properties of, 190–192, 207–208
  safety of, 208–209
  sclerotherapy improvements with, 189–191f
Sclerotherapy. *See also* Foam sclerotherapy
  from classic to foam, 201
  complications of, 208–209
  overview of, 175, 185–186
  for saphenous reflux, 163–164
  telangiectases and, 188
Selectins, in leukocyte and endothelium attachment, 4
SEPS. *See* Subfacial perforator ligation
Serine proteases, MMP activation by, 7–8
Sex, venous ulcers distribution by, 35–36, 35f
SF. *See* Sclerosing foam
SFJ. *See* Saphenofemoral junction
Sigat-tester, for compression bandages, 80
Single-component bandages, for venous ulcers, 83
Single-layer bandages, for venous ulcers, 82–83
Skin burns, from RFA, 182
Skin necrosis, with foam sclerotherapy, 208
Small saphenous vein (SSV)
  imaging of, 44, 45–48f, 46–49
  location of, 49
  mapping of, 230
  reflux of, 156, 157t
    deep and superficial venous insufficiency, 165–166
    nonconventional management of, 163–164
    perforator reflux in CVI, 166
    in saphenous stripping, 161–162
    surgery for, 162
    varicose veins and, 173–174
  removal of, history of, 174
SmartStent, 292
Sodium tetradecylsulfate (STS), for foam sclerotherapy, 193
Sotradecol, for liquid sclerotherapy, 201
SSV. *See* Small saphenous vein
Stem cells, in repair mechanism, 8
Stent
  restenosis in, 291
  for venous obstruction, 278, 283, 288–290, 292
STS. *See* Sodium tetradecylsulfate
Subfacial perforator ligation (SEPS)
  foam sclerotherapy v., 210
  with Linton procedure, 167–168
  open ligation v., 174
  for VLUs, 200–201, 210
Superficial vein ablation, endovascular techniques for, 173–183
  complications of, 181–182
  concluding remarks on, 182–183
  history of, 173–174
  mechanisms of action, 175–177
  obliterative techniques, 178–180
  options for, 175
  outcomes of, 180–181
  patient selection for, 177–178
Superficial veins
  reflux of, 156–158, 157t
    incidence of, 175
    overview of, 173
  in venous ulcers, 46–49

Superficial venous incompetence (SVI)
  DVI v., 38
  venous ulcer from, 36
Surgery
  anatomic changes with, 140–143, 142f
  with compression therapy
    v. compression therapy alone, 135–146
    concluding remarks on, 145–146
    ESCHAR study, 136–145
    other studies, 145
  for GSV removal, 157–158
  healing impact of, 137–140, 139f
  hemodynamic changes with, 143–144, 144f
  rationale for, 136
  recurrence impact of, 140, 141f
  superficial, for VLUs, 149–170
    conventional ligation of IPVs, 166–169, 167f
    GSV removal, 157–158
    saphenofemoral ligation and saphenous stripping, 158–160
    saphenous stripping to ankle or knee, 160–166
SVI. See Superficial venous incompetence

T

Telangiectases
  in pregnancy, 11
  sclerotherapy and, 188
  SF and, 193
  varicose veins from, 8
Tessari method, for foam sclerotherapy, 178, 193, 208
Thermal techniques, for varicose veins, 195
Thrombin, inhibition by antithrombin, 57–58
Thromboembolism, with foam sclerotherapy, 208
Thrombophilia conditions. See also
  Acquired thrombophilia conditions;
  Inherited thrombophilia conditions
  summary of, 63–64
Thrombophlebitis
  from EVLA, 182
  with foam sclerotherapy, 208
Thromboprophylaxis
  methods of, 118–126
    fondaparinux, 121t, 124–125
    LMWH, 121t, 122–124
    mechanical, 121t, 125–126
    UFH, 121t, 122
    warfarin, 118–120, 121t
  for VTE, 116–126
    RAM for, 117–118, 119–120f
Thrombotic obstruction
  acute, 282–283
  chronic
    iliac, 283, 284f
    IVC, 284–290, 285–287t, 288–291f, 289t
  non, 281–282, 282f
Thrombus formation, with leukocytes, 4
TIMP-2. See Tissue inhibitor metalloproteinase-2
Tissue inhibitor metalloproteinase-2 (TIMP-2), in venous ulcers, 21
T-lymphocytes
  in venous disease, 5
  in venous ulcers, 19
TNF-α. See Tumor necrosis factor-α
Transcommissural valvuloplasty, for venous valve repair, 249
Transforming growth factor beta (TGF-β)
  in venous ulcer bed, 18
  in venous ulcers, 6
Transient visual disturbance, with foam sclerotherapy, 208–209
Transmural valvuloplasty, for venous valve repair, 247–248
Treatment
  DU for endovenous, 227–239
    equipment for, 228
    preoperative assessment for, 227–228
    venous mapping for, 228–231, 229f
  DU mapping for, 52, 231–234, 232–234f
  superficial surgery and perforator interruption in, 149–170
    conventional surgical procedures, 157–169
    CVI presentation, 149–150
    diagnostic testing, 150–157
  for VLUs, 200–201
Treatment adherence, in wound healing, 96–99
Trypsin, MMP activation by, 7–8
Tumor necrosis factor-α (TNF-α), in venous ulcers, 5, 20

U

UFH. See Unfractionated heparin
Ulcer bed

evaluation of, 51
negative pressure dressings for, 108
Ultrasound. *See also* Duplex ultrasound
   for foam sclerotherapy, 178, 202, 216–218
   SSV mapping with, 161
Unfractionated heparin (UFH), as thromboprophylaxis, 121t, 122
Unna boot bandage. *See* Zinc paste bandages

V

VAC. *See* Vacuum-assisted closure
Vacuum-assisted closure (VAC)
  for CLUs, 307–324
    concluding remarks on, 316f, 319t, 322–324, 323t
    limitations of, 321–322
    modes of action, 321
    randomized clinical trial with CLUs and, 308–320
      complications of, 320, 320t
      cost analysis of, 312, 318–320, 320t
      evaluation criteria for, 311
      patient and ulcer characteristics in, 314f, 315
      procedures for, 310–311
      protocol violations in, 314f, 315
      quality of life and pain scores in, 311–312, 318, 319t
      recruitment for, 314–315, 314f
      secondary outcomes of, 317–318, 318f, 323t
      setup, 309–310
      statistical analysis of, 312–314, 313t, 316f, 317t, 318f
      time consumption in, 312
      VAC v. conventional wound care in, 315–317, 316f, 317t
  use of, 308
  for venous ulcers, 95
Valsalva maneuver, for duplex ultrasound, 44
Valvuloplasty, external
  results of, 249, 250–251t, 254
  for venous valve repair, 247–249
    angioscopy-assisted, 249
    transcommissural, 249
    transmural, 247–249
Varicose veins
  foam sclerotherapy for, 216
  POL for, 186

risk factors for, 173–174
SF treatment of, 195–196
surgery for SVI/DVI patients, 36
from telangiectases, 8
tissue restructuring in, 6
venous ulcers from, 27, 31
Vascular adhesion molecules (VCAM-1), in leukocyte migration and activation, 19
Vascular endothelial growth factor (VEGF)
  in venous inflammation, 5–6
  in venous ulcers, 20
VCAM-1. *See* Vascular adhesion molecules
VCSS. *See* Venous clinical severity score
VEGF. *See* Vascular endothelial growth factor
Vein of Giacomini, SSV termination at, 161
Vein valve transplantation, for venous valve repair, 246, 247f
Velcro-band gaiters, for venous ulcers, 84
Venous clinical severity score (VCSS), for VLCs, 218
Venous disease. *See* Chronic venous disease
Venous filling index (VFI)
  calculation of, 152–153, 153f
  outcome prediction with, 154, 154f
Venous hypertension (VH)
  reversal of, 200–201
  VLUs caused by, 199–200, 209–210, 261
Venous insufficiency, chronic, microcirculation in, 15–23
Venous leg ulcers (VLUs)
  adjuvant therapy for, 93–95
    allografts, 93–94
    collagen dressings and growth factor therapy, 95
    HSEs, 94
    VAC, 95
    xenografts, 93
  arterial insufficiency and, 299–305
    clinical study review for, 303–305
    concluding remarks on, 305
    evaluation of, 299–302
    treatment of, 302–303
  cause of, 15–23, 149
    arteriovenous shunting, 16
    cytokines and growth factors, 20
    diffusion barrier, 16–18
    historical theories for, 15–16

Venous leg ulcers (VLUs) *(continued)*
  leukocytes and endothelial cells, 19–20
  MMP, 21–22
  oxidative stress, 20–21
  summary of, 23
  varicose veins, 27, 31
  white cell trapping, 18–19
 collaborative models for care of, 99–100
 compression therapy for, 77–89
  devices for, 77–86, 78t
 from CVI, 55, 149–150
 diagnostic testing for, 150–157
  APG, 152–154, 153–154f
  duplex ultrasound, 150–151
  intervention indications with, 154–157, 155–156f, 157t
  photoplethysmography, 151–152
 epidemiology of, 27–39
  age and sex distribution of, 35–36, 35f
  description of, 27–28
  DVT history in, 37
  etiologic spectrum of leg ulcers, 30–31
  future trends in, 38–39
  healing and survival, 38, 38f
  history for, 36–37, 36f
  incidence of, 34
  incidence v. prevalence, 28–30, 29t
  obesity in, 37
  prevalence of, 31–34, 32–34t
 EVLA v. RFA, 164
 fibroblasts in, 22
 fluids of, 21
 foam sclerotherapy for, 146, 199–211, 215–223
  comments on, 209–211, 221–223, 222f, 222t
  contraindications for, 209
  methods for, 202–203, 216–218, 217f
  outcomes of, 203–207, 204–205f, 206t, 207f, 218–221, 219–220f, 219t
  safety of, 208–209
  statistical analysis of, 218
 free tissue transfer, 261–272
  concluding remarks on, 271–272
  discussion of, 269–271
  flap selection for, 266–267, 268f
  overview of, 261–262, 264t
  patient evaluation and selection for, 262–266
  postoperative care for, 269
  surgical technique for, 268–269
  indications for intervention of, 154–157, 155–156f, 157t
 management of, 91–93
 MMP expression of, 7
 negative pressure dressings in, 105–111
  adjunct to assisted closure, 109–110
  bacterial burden, 107–108
  chronic wound fluid, 106
  concluding remarks on, 110–111
  direct wound closure, 108
  indications and contraindications, 110
  local blood flow, 107
  local edema, 106–107
  wound bed deformation, 108
  wound healing environment, 108–109
  wound hypoxia, 109
 overview of, 199
 perforator veins in, 49–50
 recurrence prevention of, 86
 superficial surgery and perforator interruption for, 149–170
  diagnostic testing for, 150–151
 superficial veins in, 46–49
 surgery for, 136–145, 157–169
  healing impact of, 137–140, 139f
  recurrence impact of, 140, 141f
 TGF-β in, 6
 TNF-α in, 5
 treatment options for, 200–201
Venous obstruction
 in CVI, 275–295
 diagnosis and evaluation of, 278f, 281f
  clinical presentation of, 277–280, 278f
  imaging of, 280–281, 281f
 endovenous therapy for
  acute thrombotic obstruction, 282–283
  chronic thrombotic obstruction: iliac, 283, 284f
  chronic thrombotic obstruction: IVC, 284–290, 285–287t, 288–291f, 289t
  concluding remarks on, 295
  nonthrombotic obstruction, 281–282, 282f
  special issues in, 290–294
 endovenous treatment for, 281–294
Venous refill time (VRT), after surgery, 143–144, 144f
Venous segment transfer, for venous valve repair, 244–246, 246f
Venous stasis ulceration (VSU)
 clinical study review for, 303–305
 concluding remarks on, 305

evaluation of, 299–302
  arteriography and venography, 301–302
  history and physical examination, 299–301
  vascular laboratory testing, 301
 prevalence of, 299
 treatment of, 302–303
Venous thromboembolism (VTE)
 risk of, 114–116
 thromboprophylaxis for, 116–126
  methods of, 118–126
  RAM for, 117–118, 119–120f
Venous thrombosis, antiphospholipid syndrome and, 58t, 60–61
Venous valve repair, in CVI, 243–255
 concluding remarks on, 255
 indications for, 254–255
  according to etiology, 255
  clinical severity, 254
  hemodynamics and imaging, 254–255
 methods for, 244–249
  with phlebotomy, 244–247, 245–248f
  without phlebotomy, 247–249
 rationale for, 244
 results of, 249, 250–253t, 254
  primary deep vein reflux, 249, 250–251t, 254
  PTS, 252–253t, 254
Venous volume (VV), measurement of, 152, 153f

VFI. *See* Venous filling index
VH. *See* Venous hypertension
Vitronectin, degradation of, 21
VLUs. *See* Venous leg ulcers
VRT. *See* Venous refill time
VSU. *See* Venous stasis ulceration
VTE. *See* Venous thromboembolism
VV. *See* Venous volume

W

Wallstent, 292
Warfarin, as thromboprophylaxis, 118–120, 121t
Wound healing
 adjuvant therapy in, 93–95
 collaborative models for, 99–100
 management of venous leg ulcers in, 91–93
 with negative pressure dressings, 108–109
 treatment adherence in, 96–99

X

Xenografts, for venous ulcers, 93

Z

Zinc paste bandages, for venous ulcers, 83–84, 83f